The Logic of Knowledge Bases

The Logic of Knowledge Bases

Hector J. Levesque and Gerhard Lakemeyer

The MIT Press
Cambridge, Massachusetts
London, England

© 2000 Massachusetts Institute of Technology

All rights reserved. No part of this book may be reproduced in any form by any electronic or mechanical means (including photocopying, recording, or information storage and retrieval) without permission in writing from the publisher.

This book was set in Times Roman by the authors using the LaTeX document preparation system.
Printed on recycled paper and bound in the United States of America.

Library of Congress Cataloging-in-Publication Data

Levesque, Hector J., 1951–
   The logic of knowledge bases / Hector J. Levesque, Gerhard Lakemeyer
     p. cm.
   Includes bibliographical references and index.
   ISBN 0-262-12232-4 (alk. paper)
   1. Knowledge representation (Information Technology) 2. Expert systems (Computer Science) 3. Logic, symbolic and mathematical.
I. Lakemeyer, Gerhard II. Title.
Q387 L48   2000
006.3′32—dc21   00-025413
                                                               CIP

*To Marc and Michelle,
Jan and Leilani*

Contents

Preface			xv
Acknowledgements			xvii

PART I

1 Introduction — 3
- 1.1 Knowledge — 3
  - 1.1.1 Propositions — 3
  - 1.1.2 Belief — 4
  - 1.1.3 Representation — 5
  - 1.1.4 Reasoning — 6
- 1.2 Why knowledge representation and reasoning? — 6
  - 1.2.1 Knowledge-based systems — 7
  - 1.2.2 Why knowledge representation? — 8
  - 1.2.3 Why reasoning? — 10
- 1.3 Knowledge representation systems — 11
  - 1.3.1 The knowledge and symbol levels — 12
  - 1.3.2 A functional view: TELL and ASK — 13
  - 1.3.3 The interaction language — 13
- 1.4 The rest of the book — 14
- 1.5 Bibliographic notes — 16
- 1.6 Exercises — 17

2 A First-Order Logical Language — 19
- 2.1 Why first-order logic? — 19
- 2.2 Why standard names? — 21
- 2.3 The syntax of the language $\mathcal{L}$ — 23
- 2.4 Domain of quantification — 25
- 2.5 World state — 26
- 2.6 Term and formula semantic evaluation — 27

	2.7	Satisfiability, implication and validity	28
	2.8	Properties of logic $\mathcal{L}$	29
	2.9	Why a proof theory for $\mathcal{L}$?	34
	2.10	Universal generalization	35
	2.11	The proof theory	35
	2.12	Example derivation	37
	2.13	Bibliographic notes	39
	2.14	Exercises	40
3	**An Epistemic Logical Language**		**43**
	3.1	Why not just use $\mathcal{L}$?	43
	3.2	Known vs. potential instances	44
	3.3	Three approaches to incomplete knowledge	44
	3.4	The language $\mathcal{KL}$	46
	3.5	Possible worlds	48
	3.6	Objective knowledge in possible worlds	49
	3.7	Meta-knowledge and some simplifications	51
	3.8	The semantics of $\mathcal{KL}$	54
	3.9	Bibliographic notes	55
	3.10	Exercises	56
4	**Logical Properties of Knowledge**		**57**
	4.1	Knowledge and truth	57
	4.2	Knowledge and validity	59
	4.3	Known individuals	61
	4.4	An axiom system for $\mathcal{KL}$	64
	4.5	A Completeness proof	67
		4.5.1 Part 1	69
		4.5.2 Part 2	70

		4.5.3	Variant systems	72
	4.6	Reducibility		73
	4.7	Bibliographic notes		76
	4.8	Exercises		77
5	**The TELL and ASK Operations**			**79**
	5.1	Overview		79
	5.2	The ASK operation		80
	5.3	The initial epistemic state: e_0		81
	5.4	The monotonicity of knowledge		82
	5.5	The TELL operation		83
	5.6	Closed world assertions		85
	5.7	A detailed example		88
		5.7.1	Examples of ASK	88
		5.7.2	Examples of TELL	90
	5.8	Other operations		92
		5.8.1	Definitions	92
		5.8.2	Wh-questions	93
	5.9	Bibliographic notes		94
	5.10	Exercises		95
6	**Knowledge Bases as Representations of Epistemic States**			**97**
	6.1	Equivalent epistemic states		97
	6.2	Representing knowledge symbolically		99
	6.3	Some epistemic states are not representable		101
	6.4	Representable states are sufficient		102
	6.5	Finite representations are not sufficient		105
	6.6	Representability and TELL		107
	6.7	Bibliographic notes		108

	6.8	Exercises	109
7		**The Representation Theorem**	111
	7.1	The method	112
	7.2	Representing the known instances of a formula	113
	7.3	Reducing arbitrary sentences to objective terms	117
	7.4	TELL and ASK at the symbol level	118
	7.5	The example KB reconsidered	119
	7.6	Wh-questions at the symbol level	124
	7.7	Bibliographic notes	125
	7.8	Exercises	125
8		**Only-Knowing**	127
	8.1	The logic of answers	127
	8.2	The language $\mathcal{OL}$	129
	8.3	Some properties of $\mathcal{OL}$	130
	8.4	Characterizing ASK and TELL	133
	8.5	Determinate sentences	134
	8.6	Bibliographic notes	138
	8.7	Exercises	140

PART II

9		**Only-Knowing and Autoepistemic Logic**	143
	9.1	Examples of autoepistemic reasoning in $\mathcal{OL}$	143
	9.2	Stable sets and stable expansions	148
	9.3	Relation to stable sets	149
	9.4	Relation to stable expansions	151
	9.5	Computing stable expansions	153
	9.6	Non-reducibility of $\mathcal{OL}$	155
	9.7	Generalized stability	158

9.8	Bibliographic notes	159
9.9	Where do we go from here?	160
9.10	Exercises	161

10 On the Proof Theory of $\mathcal{OL}$ — 163

10.1	Knowing at least and at most	163
10.2	Some example derivations	165
10.3	Propositional completeness	168
10.4	Incompleteness	173
10.5	Bibliographic notes	176
10.6	Where do we go from here?	176
10.7	Exercises	176

11 Only-Knowing-About — 179

11.1	The logic of only-knowing-about	179
	11.1.1 A formal semantics	180
	11.1.2 Some properties of only-knowing-about	181
	11.1.3 Prime implicates	185
11.2	ASK and TELL	186
11.3	Relevance	187
11.4	Bibliographic notes	192
11.5	Where do we go from here?	192
11.6	Exercises	193

12 Avoiding Logical Omniscience — 195

12.1	The propositional case	197
	12.1.1 A proof theory	199
	12.1.2 Computing explicit belief	203
12.2	The first-order case	204
	12.2.1 Some properties	208

	12.2.2 Deciding belief implication	211
12.3	Bibliographic notes	220
12.4	Where do we go from here?	220
12.5	Exercises	221

13 The logic $\mathcal{EOL}$ — 223

13.1	Semantics	223
	13.1.1 Equality	223
	13.1.2 Nested explicit belief	224
	13.1.3 Explicitly only-knowing	226
13.2	Some properties of $\mathcal{EOL}$	227
13.3	Representing the explicitly believed instances of a formula	229
13.4	Reducing arbitrary sentences to objective terms	231
13.5	Decidability results	234
13.6	ASK and TELL	237
	13.6.1 ASK	237
	13.6.2 TELL	238
	13.6.3 Decidability	239
	13.6.4 Examples of ASK	239
13.7	Bibliographic notes	241
13.8	Where do we go from here?	241
13.9	Exercises	242

14 Knowledge and Action — 245

14.1	A theory of action	245
	14.1.1 Action preconditions	246
	14.1.2 Action sensing	247
	14.1.3 Action effects	248
	14.1.4 The frame problem and a solution	248

14.1.5 Basic action theories	250
14.2 The logic $\mathcal{AOL}$	251
14.3 Knowledge after action	255
14.3.1 An informal characterization	256
14.3.2 Knowing in $\mathcal{AOL}$	257
14.3.3 Only-knowing in $\mathcal{AOL}$	259
14.4 Using $\mathcal{AOL}$	259
14.5 An axiomatization of $\mathcal{AOL}$	262
14.6 Bibliographic notes	269
14.7 Where do we go from here?	269
14.8 Exercises	270
Epilogue	273
References	275
Index	281

Preface

The idea that defines the very heart of "traditional" Artificial Intelligence (AI) is due to John McCarthy: his imagined ADVICE-TAKER was a system that would decide how to act (in part) by running formal reasoning procedures over a body of explicitly represented knowledge, a *knowledge base*. The system would not so much be programmed for specific tasks as told what it needed to know, and expected to infer the rest somehow. Knowledge and advice would be given declaratively, allowing the system to operate in an undirected manner, choosing what pieces of knowledge to apply when they appeared situationally appropriate. This vision contrasts sharply with that of the traditional programmed computer system, where what information is needed and when is anticipated in advance, and embedded directly into the control structure of the program.

This is a book about the *logic* of such knowledge bases, in two distinct but related senses. On the one hand, a knowledge base is a collection of sentences in a representation language that entails a certain picture of the world represented. On the other hand, *having* a knowledge base entails being in a certain state of knowledge where a number of other epistemic properties hold. One of the principal aims of this book is to develop a detailed account of the relationship between symbolic representations of knowledge and abstract states of knowledge.

This book is intended for graduate students and researchers in AI, database management, logic, or philosophy interested in exploring in depth the foundations of knowledge, knowledge bases, knowledge-based systems, and knowledge representation and reasoning. The exploration here is a mathematical one, and we assume some familiarity with first-order predicate logic (and for motivation at least, some experience in AI).

The book presents a new mathematical model of knowledge that is not only quite general and expressive (including but going well beyond full first-order logic), but that is much more workable in practice than other models that have been proposed in the past. A reader can expect to learn from this book a style of semantic argument and formal analysis that would have been quite cumbersome, or even outside the practical reach of other approaches.

From a computer science point of view, the book also develops a new way of specifying what a knowledge representation system is supposed to do in a way that does not make assumptions about how it should do it. The reader will learn how to treat a knowledge base like an *abstract data type*, completely specified in an abstract way by the knowledge-level operations defined over it.

The book is divided into two sections: Part I, consisting of Chapters 1 to 8, covers the basics; Part II, consisting of Chapters 9 to 14, considers a number of more-or-less independent research topics and directions. (The contents of these chapters are described at the end of Chapter 1.) The material in the book has been used in graduate level courses at the authors' institutions in Canada and Germany. In one semester, it should be possible to cover all of Part I and at least some of the advanced chapters of Part II. Exercises and bibliographic notes are included at the end of each chapter. Suggestions for further research are made at the end of the chapters of Part II. An index of the important technical terms, whose first use is underlined in the text, appears at the end of the book. Comments and corrections are most welcome and can be sent to the authors at

> `hector@cs.toronto.edu`
> `gerhard@cs.rwth-aachen.de.`

Although every effort has been made to keep the number of errors small, this book is offered as is, with no warranty expressed or implied.

Acknowledgements

This book has been in the works for about twenty years. It began in the late seventies, when John Mylopoulos suggested that we consider a small extension to the PSN representation system then under development at the University of Toronto to allow for knowledge bases with incomplete knowledge. He realized that merely extending a classical true/false semantics to include a third value for "unknown" did not work properly. Among other things, tautologies could come out unknown. What sort of semantic account would assign unknown to just those formulas whose truth values really were unknown?

In a sense this book is an attempt to answer this question in a clean and general way. It incorporates the doctoral theses of both authors at the University of Toronto, as well as a number of related conference and journal papers.

Along the way, many people contributed directly and indirectly to the research reported here. We wish to thank John Mylopoulos, of course, and other thesis committee members, Faith Fich, Graeme Hirst, Alberto Mendelzon, Ray Perrault, Charles Rackoff, Ray Reiter, John Tsotsos, Alasdair Urquhart, as well as Gerhard's external examiner, Joe Halpern, and Hector's external examiner, longtime friend, and co-conspirator, Ron Brachman. A special thanks to Alex Borgida and Jim des Rivières for the many discussions at the early stages of this work.

As the work began to progress and further develop in the eighties, we began a long-term dialogue with Joe Halpern which greatly influenced the work in very many ways, first at the Knowledge Seminar held at IBM Almaden, and then at the TARK Conferences at Asilomar. We are very grateful for the help and profound insight of Joe and his colleagues Yoram Moses, Ron Fagin, and Moshe Vardi. We also acknowledge the many illuminating discussions at these meetings with Yoav Shoham, Kurt Konolige, and Bob Moore.

Three further developments shifted the work in fruitful directions. First, in the mid eighties, we began to consider the problem of logical omniscience and a solution in terms of a computationally limited notion of belief. We again thank Joe for his contributions here, as well as Peter Patel-Schneider and Greg McArthur, both at the University of Toronto. Second, at the start of the nineties, we began to see how Bob Moore's syntactic notion of autoepistemic logic could be given a semantic characterization in terms of only-knowing. We are grateful to Vladimir Lifschitz, Victor Marek, and Grigori Schvarts, for ideas there. Third, in the late nineties, we began to explore the connection between knowing, only-knowing and action. This was greatly influenced by the ongoing work on the situation calculus and Golog at Toronto's Cognitive Robotics Group. We are indebted to Ray Reiter, Yves Lespérance, Fangzhen Lin, Richard Scherl, and the other members of the group for their insights and encouragement.

Earlier versions of this book were presented as part of a graduate course at the University of Toronto, the University of Bonn, and Aachen University of Technology. We are grateful to Bruno Errico, Koen Hindricks, Gero Iwan, Eric Joanis, Daniel Marcu, Richard

Scherl, Steven Shapiro, and Mikhail Soutchansky for comments on drafts of chapters, and to all the other students who helped debug many of the ideas.

Over the years, many other friends and colleagues contributed, in one way or another, to this project. Gerhard would like to extend a special thanks to Diane Horton and Tom Fairgrieve for their sustained friendship and hospitality during his many visits to Toronto, and to Armin B. Cremers for providing a stimulating research environment while Gerhard was at the University of Bonn. Hector would like to acknowledge Jim Delgrande, Jim des Rivières, Pat Dymond, and Patrick Feehan for their continued friendship and support.

We would also like to thank Bob Prior, Katherine Innis, and the other staff members at MIT Press, who helped in the production of this book.

Financial support for this research was gratefully received from the Natural Sciences and Engineering Research Council of Canada, the Canadian Institute for Advanced Research, the World University Service of Canada, and the German Science Foundation.

Last but nowhere near least we would like to thank our families, Pat, Margot, Marc, Michelle, Jan, and Leilani, who stood by us over all these years while this book was in the making.

Hector Levesque and Gerhard Lakemeyer
Toronto and Aachen, September 2000

Part I

1 Introduction

In a book about the logical foundations of knowledge bases, it is probably a good idea to review if only briefly how concepts like knowledge, representation, reasoning, knowledge bases, and so on are understood informally within Artificial Intelligence (AI), and why so many researchers feel that these notions are important to the AI enterprise.

1.1 Knowledge

Much of AI does indeed seem to be concerned with *knowledge*. There is knowledge representation, knowledge acquisition, knowledge engineering, knowledge bases and knowledge-based systems of various sorts. In the early eighties, during the heyday of commercial AI, there was even a slogan "Knowledge is power" used in advertisements. So what exactly is knowledge that people in AI should care so much about it? This is surely a philosophical issue, and the purpose of this chapter is not to cover in any detail what philosophers, logicians, and computer scientists have said about knowledge over the years, but only to glance at some of the issues involved and especially their bearings on AI.

1.1.1 Propositions

To get a rough sense of what knowledge is supposed to be, at least outside of AI, it is useful to look at how we talk about it informally. First, observe that when we say something like "John knows that ...," we fill in the blank with a simple *declarative sentence*. So we might say that "John knows that Mary will come to the party" or that "John knows that dinosaurs were warm blooded." This suggests that, among other things, knowledge is a relation between a knower (like John) and a *proposition*, that is, the idea expressed by a simple declarative sentence (like "Mary will come to the party").

Part of the mystery surrounding knowledge is due to the abstract nature of propositions. What can we say about them? As far as we are concerned, what matters about propositions is that they are abstract entities that can be *true* or *false*, right or wrong.[1] When we say that "John knows that p," we can just as well say that "John knows that it is true that p." Either way, to say that somebody knows something is to say that somebody has formed a judgement of some sort, and has come to realize that the world is one way and not another. In talking about this judgement, we use propositions to classify the two cases.

---

1 Strictly speaking, we might want to say that the *sentences* expressing the proposition are true or false, and that the propositions themselves are either factual or non-factual. Further, because of linguistic features such as indexicals (that is, words like "me" and "yesterday"), we more accurately say that it is actual tokens of sentences or their uses in contexts that are true or false, not the sentences themselves.

A similar story can be told about a sentence like "John hopes that Mary will come to the party." The same proposition is involved, but the relationship John has to it is different. Verbs like "knows," "hopes," "regrets," "fears," and "doubts" all denote *propositional attitudes*, relationships between agents and propositions. In all cases, what matters about the proposition is its truth: if John hopes that Mary will come to the party, then John is hoping that the world is one way and not another, as classified by the proposition.

Of course, there are sentences involving knowledge that do not mention a proposition. When we say "John knows who Mary is taking to the party," or "John knows how to get there," we can at least imagine the implicit propositions: "John knows that Mary is taking so-and-so to the party," or "John knows that to get to the party, you go two blocks past Main Street, turn left," and so on. On the other hand, when we say that John has a skill as in "John knows how to play piano," or a deep understanding of someone or something as in "John knows Bill well," it is not so clear that any useful proposition is involved. We will have nothing further to say about this latter form of knowledge in the book.

1.1.2 Belief

A related notion that we are concerned about, however, is the concept of *belief*. The sentence "John believes that p" is clearly related to "John knows that p." We use the former when we do not wish to claim that John's judgement about the world is necessarily accurate or held for appropriate reasons. We sometimes use it when we feel that John might not be completely convinced. In fact, we have a full range of propositional attitudes, expressed by sentences like "John is absolutely certain that p," "John is confident that p," "John is of the opinion that p," "John suspects that p," and so on, that differ only in the level of conviction they attribute. For now, we will not distinguish among *any* of them.[2] What matters is that they all share with knowledge a very basic idea: John takes the world to be one way and not another.

So when we talk about knowledge or any other propositional attitude, we are implicitly imagining a number of different ways the world could be. In some of these, Mary comes to the party; in others, she does not. When we say that John knows or believes or suspects that Mary will come to the party, we are saying that John takes it (with varying degrees of conviction) that those where Mary does not come to the party are fantasy only; they do not correspond to reality.

In this very abstract and informal picture, we can already see emerging two very different but related views of knowledge or belief. First, we can think of knowledge (or belief) as a collection of propositions held by an agent to be true. Second, we can think in terms

---

2 One way to understand (subjective) probability theory is as an attempt to deal in a principled way with these levels of conviction as numeric degrees of belief. This is the last we will say on this subject in the book.

Introduction 5

different possible ways the world could be, and knowledge (or belief) as a classification of these into two groups, those that are considered incorrect, and those that are candidates for the way the world really is.

1.1.3 Representation

The interest of AI in knowledge is obviously that we want to design and build systems that know a lot about their world, enough, in fact, that they do not act unintelligently.[3] But there is more to it. Any system, AI-based or not, can be said to have knowledge about its world. Any Java compiler, for example, knows a lot about the details of the Java language. There's even the joke about a thermos "knowing" whether the liquid it contains is hot or cold, and making sure it preserves the correct one. This idea of attributing knowledge to a more-or-less complex system (or person) is what the philosopher Dennett calls "taking the intentional stance." But when people in AI talk about knowledge bases, knowledge engineering and so on, they mean more than this. They have in mind a system that not only knows a lot in the above sense, but also a system that does what it does using a representation of that knowledge.

The concept of representation is no doubt as philosophically problematic as that of knowledge. Very roughly speaking, *representation* is a relationship between two domains where the first is meant to "stand for" or take the place of the second. Usually, the first domain, the representer, is more concrete, immediate, or accessible in some way than the second. The type of representer that we will be most concerned with here is that of a formal *symbol*, that is, a character or group of them taken from some predetermined alphabet. The digit "7," for example, stands for the number 7, as does the group of letters "VII," and in other contexts, the words "sept," "sieben," and "shichi." As with all representation, it is assumed to be easier to deal with symbols (recognize them, distinguish them from each other, display them *etc.*) than with what the symbols represent. In some cases, a word like "John" might stand for something quite concrete; but many words, like "love" or "truth," stand for abstractions.

Of special concern to us is when a group of formal symbols stands for a proposition: "John loves Mary" stands for the proposition that John loves Mary. Again, the symbolic English sentence is concrete: it has distinguishable parts involving the 3 words, for example, and a recognizable syntax. The proposition, on the other hand, is abstract: it is something like a classification of the ways the world can be into two groups: those where John loves Mary, and those where he does not.

Knowledge Representation, then, is this: it is the field of study within AI concerned

---

[3] What we call "commonsense" clearly involves considerable knowledge of a variety of sorts, at least in the sense of being able to form a judgement about different ways the world could be.

with using formal symbols to represent a collection of propositions believed by some putative agent. As we will see however, we would not want to insist that there be symbols to represent *each* of the propositions believed by the agent. There may very well be an infinite number of propositions believed, only a finite number of which are ever represented. It will be the role of reasoning to bridge the gap between what is represented and the full set of propositions believed.

1.1.4 Reasoning

So what is *reasoning*? In general, it is the formal manipulation of the symbols representing a collection of believed propositions to produce representations of new ones. It is here that we use the fact that symbols are more accessible than the propositions they represent: they must be concrete enough that we can manipulate them (move them around, take them apart, copy them, string them together) in such a way as to construct representations of new propositions.

The analogy here is with arithmetic. We can think of binary addition as being a certain formal manipulation: we start with symbols like "1011" and "10," for instance, and end up with "1101." The manipulation here is addition since the final symbol represents the sum of the numbers represented by the initial ones. Reasoning is similar: we might start with the sentences "John loves Mary" and "Mary is coming to the party," and after a certain amount of manipulation produce the sentence "Someone John loves is coming to the party." We would call this form of reasoning logical inference because the final sentence represents a logical entailment of the propositions represented by the initial ones. According to this view (first put forward, incidentally, by the philosopher Leibniz in the 17th century), reasoning is a form of calculation, not unlike arithmetic, but over symbols standing for propositions rather than numbers.

1.2 Why knowledge representation and reasoning?

Let's talk motivation: why do people in AI who want their systems to know a lot, also want their systems to represent that knowledge symbolically? The intentional stance above says nothing about what is or is not represented within a system. We can say that a system knows that p without claiming that there is anything represented within the system corresponding to that proposition. The hypothesis underlying much (but not all) of the work in AI, however, is that we want to construct systems that do contain symbolic representations with two important properties. First is that we (from the outside) can understand them as standing for propositions. Second is that the system is designed to behave the way that it does *because* of these symbolic representations. This is what Brian Smith has called the

Knowledge Representation Hypothesis:

> Any mechanically embodied intelligent process will be comprised of structural ingredients that a) we as external observers naturally take to represent a propositional account of the knowledge that the overall process exhibits, and b) independent of such external semantic attribution, play a formal but causal and essential role in engendering the behaviour that manifests that knowledge.

In other words, the Knowledge Representation Hypothesis is that we will want to construct systems for which the intentional stance is grounded by design in symbolic representations. A system of this sort is called a *knowledge-based system* and the symbolic representation involved its *knowledge base* (or KB).

1.2.1 Knowledge-based systems

To see what a knowledge-based system amounts to, it is helpful to look at two very simple Prolog programs with identical behaviour.[4] The first is:

```
printColour(snow)  :- !, write("It's white.").
printColour(grass) :- !, write("It's green.").
printColour(sky)   :- !, write("It's yellow.").
printColour(X)     :- write("Beats me.").
```

The second is:

```
printColour(X) :- colour(X,Y), !,
       write("It's "), write(Y), write(".").
printColour(X) :- write("Beats me.").
colour(snow,white).
colour(sky,yellow).
colour(X,Y) :- madeof(X,Z), colour(Z,Y).
madeof(grass,vegetation).
colour(vegetation,green).
```

Observe that both programs are able to print out the colour of various items (getting the sky wrong, as it turns out). Taking an intentional stance, both might be said to "know" that the colour of snow is white. The crucial point, however, is that only the second program is designed according to the Knowledge Representation Hypothesis.

Consider the clause `colour(snow,white)`, for example. This is a symbolic structure that we can understand as representing the proposition that snow is white, and moreover,

---

4 No further knowledge of Prolog is assumed beyond this motivating example.

we know, by virtue of knowing how the Prolog interpreter works, that the system prints out the appropriate colour of snow precisely *because* it bumps into this clause at just the right time. Remove the clause and the system would no longer do so.

There is no such clause in the first program. The one that comes closest is the first clause of the program which says what to print when asked about snow. But we would be hard-pressed to say that this clause literally represents a belief, except perhaps a belief about what ought to be written.

So what makes a system knowledge-based, as far as we are concerned, is not the use of a logical formalism (like Prolog), or the fact that it is complex enough to merit an intentional description involving knowledge, or the fact that what it believes is true; rather it is the presence of a KB, a collection of symbolic structures representing what it believes and reasons with during the operation of the system.

1.2.2 Why knowledge representation?

So an obvious question arises when we start thinking about the two Prolog programs of the previous section: what advantage, if any, does the knowledge-based one have? Would it not be better to "compile out" the KB and distribute this knowledge to the procedures that need it, as we did in the first program? The performance of the system would certainly be better. It can only slow a system down to have to look up facts in a KB and reason with them at runtime in order to decide what actions to take. Indeed advocates within AI of so-called "procedural knowledge" take pretty much this point of view.

When we think about the various skills we have, such as riding a bicycle or playing a piano, it certainly *feels* like we do not reason about the various actions to take (shifting our weight or moving our fingers); it seems much more like we just know what to do, and do it. In fact, if we try to think about what we are doing, we end up making a mess of it. Perhaps (the argument goes) this applies to most of our activities, making a meal, getting a job, staying alive, and so on.

Of course, when we first learn these skills, the case is not so clear: it seems like we need to think deliberately about what we are doing, even riding a bicycle. The philosopher Hubert Dreyfus first observed this paradox of "expert systems." These systems are claimed to be superior precisely because they are knowledge-based, that is, they reason over explicitly represented knowledge. But novices are the ones who think and reason, claims Dreyfus. Experts do not; they learn to recognize and to react. The difference between a chess master and a chess novice is that the novice needs to figure out what is happening and what to do, but the master just "sees" it. For this reason (among others), Dreyfus believes that the development of knowledge-based systems is completely wrong-headed, if it is attempting to duplicate human-level expertise.

So why even consider knowledge-based systems? Unfortunately, no definitive answer can yet be given. We suspect, however, that the answer will emerge in our desire to build systems that deal with a set of tasks that is *open-ended*. For any fixed set of tasks, it might work to "compile out" what the system needs to know; but if the set of tasks is not determined in advance, the strategy will not work. The ability to make behaviour depend on explicitly represented knowledge seems to pay off when we cannot specify in advance how that knowledge will ever be used.

The best example of this, perhaps, is what happens when we read a book. Suppose we are reading about South American geography. When we find out for the first time that approximately half of the population of Peru lives in the Andes, we are in no position to distribute this piece of knowledge to the various routines that might eventually require it. Instead, it seems pretty clear that we are able to assimilate the fact in declarative form for a very wide variety of potential uses. This is the prototypical case of a knowledge-based system.

From a system design point of view, the knowledge-based approach seems to have a number of desirable features:

- We can add new tasks and easily make them depend on previous knowledge. In our Prolog program example, we can add the task of enumerating all objects of a given colour, or even of painting a picture, by making use of the KB to determine the colours.
- We can extend the existing behaviour by adding new beliefs. For example, by adding a clause saying that canaries are yellow, we automatically propagate this information to any routine that needs it.
- We can debug faulty behaviour by locating the erroneous belief of the system. In the Prolog example, by changing the clause for the colour of the sky, we automatically correct any routine that uses colour information.
- We can concisely explain and justify the behaviour of the system. Why did the program say that grass was green? It was because it believed that grass is a form of vegetation and that vegetation is green. Moreover, we are justified in saying "because" here since if we removed either of the two relevant clauses, the behaviour would indeed change.

Overall, then, the hallmark of a knowledge-based system is that by design it has the ability to be *told* facts about its world and adjust its behaviour correspondingly. We will take this up again below.

This ability to have some of our actions depend on what we believe is what the cognitive scientist Zenon Pylyshyn has called *cognitive penetrability*. Consider, for example, responding to a fire alarm. The normal response is to get up and leave the building. But we would not do so if we happened to believe that the alarm was being tested, say. There are any number of ways we might come to this belief, but they all lead to the same effect. So our response to a fire alarm is cognitively penetrable since it is conditioned on what we can be made to believe. On the other hand, something like a blinking reflex as an object approaches your eye does not appear to be cognitively penetrable: even if you strongly believe the object will not touch you, you still blink.

1.2.3 Why reasoning?

To see the motivation behind reasoning in a knowledge-based system, it suffices to observe that we would like action to depend on what the system believes about the world, as opposed to *just* what the system has explicitly represented. In the Prolog example, there was no clause representing the belief that the colour of grass was green, but we still wanted the system to know this. In general, much of what we expect to put in a KB will involve quite general facts, which will then need to be applied to particular situations.

For example, we might represent the following two facts explicitly:

1. Patient x is allergic to medication m.
2. Anyone allergic to medication m is also allergic to medication m'.

In trying to decide if it is appropriate to prescribe medication m' for patient x, neither represented fact answers the question. Together, however, they paint a picture of a world where x is allergic to m', and this, together with other represented facts about allergies, might be sufficient to rule out the medication. So we do not want to condition behaviour only on the represented facts that we are able to *retrieve*, like in a database system. The beliefs of the system must go beyond these.

But beyond them to where? There is, as it turns out, a simple answer to this question, but one that we will argue in later chapters is too simplistic. The simple answer: the system should believe p if, according to the beliefs it has represented, the world it is imagining is one where p is true. In the above example, facts (1) and (2) are both represented. If we now imagine what the world would be like if (1) and (2) were both true, then this is a world where

3. Patient x is allergic to medication m'

is also true, even though this fact is only implicitly represented.

This is the concept of <u>entailment</u>: we say that the propositions represented by a set of sentences S entail the proposition represented by a sentence p when the truth of p is implicit in the truth of the sentences in S. In other words, if the world is such that every element of S comes out true, then p does as well. All that we require to get some notion of entailment is a language with an account of what it means for a sentence to be true or false. As we argued, if our representation language is to represent knowledge at all, it must come with such an account (again: to know p is to take p to be true). So any knowledge representation language, whatever other features it may have, whatever syntactic form it may take, whatever reasoning procedures we may define over it, ought to have a well-defined notion of entailment.

The simple answer to what beliefs a knowledge-based system should exhibit, then, is that it should believe all and only the entailments of what it has explicitly represented. The job of reasoning, then, according to this account, is to compute the entailments of the KB.

What makes this account simplistic is that there are often quite good reasons not to calculate entailments. For one thing, it is too *difficult* computationally to decide if a sentence is entailed by the kind of KB we will want to use. Any procedure that gives us answers in a reasonable amount of time will occasionally either miss some entailments or return too many. In the former case, the reasoning process is said to be *incomplete*; in the latter case, the reasoning is said to be *unsound*.

But there are also conceptual reasons why we might consider unsound or incomplete reasoning. For example, suppose p is not entailed by a KB, but is a reasonable guess, given what is represented. We might still want to believe that p is true. To use a classic example, suppose all I know about an individual Tweety is that she is a bird. I might have a number of facts about birds in the KB, but likely they would not *entail* that Tweety flies. After all, Tweety might turn out to be an ostrich. Nonetheless, it is a reasonable assumption that Tweety flies. This is unsound reasoning since we can imagine a world where everything in the KB is true but where Tweety does not fly.

As another example, a knowledge-based system might come to believe a collection facts from various sources which, taken together, cannot all be true. In this case, it would be inappropriate to do logically complete reasoning, since *every* sentence would then be believed. This is because for any sentence p, any world where all the sentences in the set are true is one where p is also true, since there are no such worlds. An incomplete form of reasoning would clearly be more useful here until the contradictions are dealt with, if ever.

But despite all this, it remains the case that the simplistic answer is by far the best starting point for thinking about reasoning, even if we intend to diverge from it. So while it would be a mistake to *identify* reasoning in a knowledge-based system with logically sound and complete inference, it is the right place to begin.

1.3 Knowledge representation systems

The picture of a knowledge-based system that emerges from the above discussion is one where a system performs some problem-solving activity such as deciding what medicine to prescribe, and does so intelligently by appealing at various points to what it knows: is patient x allergic to medication m? what else is x allergic to? The mechanism used by the system to answer such questions involves reasoning from a stored KB of facts about the world. It makes sense in this scenario to separate the management of the KB from the rest of the system. The data structures within a KB and the reasoning algorithms used are not really of concern to the problem-solving system. Ultimately, what a medical system needs to find out is whether or not x is allergic to m (and perhaps how certain we are of that fact), not whether or not a certain symbolic structure occurs somewhere or can be processed in a certain way.

We take the view that it is the role of a *knowledge representation system* to manage the KB within a larger knowledge-based system. Its job is to make various sorts of information about the world available to the rest of the system based on what information it has obtained perhaps from other parts of the system and whatever reasoning it can perform.[5] So its job is smaller than that of a full knowledge-based problem solver, but larger than that of a database management system which would merely retrieve the contents of the KB. According to this view, the contents of the KB and the reasoning algorithms used by the knowledge representation system are its own business; what the rest of knowledge-based problem solver gets to find out is just what is and is not known about the world.

1.3.1 The knowledge and symbol levels

Allen Newell suggested that we can look at the knowledge in a knowledge-based system in at least two ways. At the *knowledge level*, we imagine a knowledge-based system as being in some sort of abstract epistemic state. It acquires knowledge over time, moving from state to state, and uses what it knows to carry out its activities and achieve its goals. At the *symbol level*, we also imagine that within the system somewhere there is a symbolic KB representing what the system knows, as well as reasoning procedures that make what is known available to the rest of the system. In our terms, the symbol level looks at knowledge from *within* a knowledge representation system where we deal with symbolic representations explicitly; the knowledge level looks at knowledge from *outside* the knowledge representation system, and is only concerned with what is true in the world according to this knowledge. So at the knowledge level, we are concerned with the logic of *what* a system knows; at the symbol level, within a knowledge representation system, we are concerned with *how* a system does it.

There are clearly issues of adequacy at each level. At the knowledge level, we deal with the expressive adequacy of a representation language, the characteristics of its entailment relation, including its computational complexity; at the symbol level, we ask questions about the computational architecture, the properties of the data structures and algorithms, including their algorithmic complexity.

This is similar in many ways to the specification/implementation distinction within

---

5 In this most general picture, we include the possibility of a knowledge representation system *learning* from what it has observed, as well as it having various levels of confidence in what it believes.

traditional computer science. The symbol level provides an implementation for the more abstract knowledge level specification. But what exactly does a knowledge level specify? What would a symbol level need to implement? In a sense, being precise about these is the topic of this book.

1.3.2 A functional view: TELL and ASK

We said that the role of a knowledge representation system was to make information available to the rest of the system based on what it had acquired and what reasoning it could perform. In other words, we imagine that there are two main classes of operations that a knowledge representation system needs to implement for the rest of the system: operations that absorb new information as it becomes available, and operations that provide information to the rest of the system as needed. In its simplest form, a knowledge representation system is passive: it is up to the rest of the system to *tell* it when there is something that should be remembered, and to *ask* it when it needs to know something. It is up to the knowledge representation system to decide what to do with what it is told, and in particular, how and when to reason so as to provide answers to questions as requested.

For a large section of the book, we will be concerned with a very simple instance of each of these operations: a **TELL** operation and an **ASK** operation each of which take as argument a sentence about the world. The idea is that the **TELL** operation informs the knowledge representation system that the sentence in question is true; the **ASK** operation asks the system whether the sentence in question is true. We can see immediately that any realistic knowledge representation system would need to do much more. At the very least, it should be possible to ask *who* or *what* satisfies a certain property (according to what is known). We will examine operations like these later; for now, we stick to the simple version.

So the idea at the knowledge level is that starting in some state of knowledge, the system can be told certain sentences and move through a sequence of states; at any point, the system believes the world is in a certain state, and can be asked if a certain sentence is true. In subsequent chapters, we will show how these two operations can be defined precisely but in a way that leaves open how they might be implemented. We will also discuss simple implementation techniques at the symbol level based on automated theorem-proving, and be able to prove that such implementations are correct with respect to the specification.

1.3.3 The interaction language

With this functional view of knowledge representation, we can see immediately that there is a difference at least conceptually between the *interaction language*, that is, the language used to tell the system or to ask it questions about the world, and the *representation lan-*

guage, the collection of symbolic structures used at the symbol level to represent what is known. There is no reason to suppose the two languages are identical, or even that what is stored constitutes a declarative language of any sort. Moreover, there are clear intuitive cases where simply storing what you have been told would be a bad thing to do.

Consider indexicals, for example, that is, terms like "I," "you," "here," and "now," that might appear in an interaction language. If a fact about the world you are told is that *"There is a treasure buried here"* for instance, it would be a bad idea to absorb this information by storing the sentence verbatim. Two weeks from now, when you decide to go looking for the treasure, it is likely no longer true that it is located "here." You need to resolve the "here" at the time you are told the fact into a description of a location that can be used in different contexts. If later the question *"Is there a treasure here?"* is asked, we would want to resolve the "here" differently. We need to distinguish between how information is communicated to or retrieved from the system and how it is represented for long-term storage.

In this book, we will not emphasize indexicals like those above (although they are mentioned in an exercise). There is an important type of indexical that we *will* want to examine in considerable detail, however, and that is one that refers to the current state of knowledge.

Suppose, for example, we have a system that is attempting to solve a murder mystery, and that all it knows so far is that Tom and Dick were at the scene of the crime (and perhaps others). If the system is told that *"The murder was not committed by anyone you currently know to have been present,"* the system learns that the murderer was neither Tom nor Dick. It is this "currently" that makes the expression indexical. As the knowledge of the system changes, so will what this expression refers to, just as "here" did. Suppose the system later finds out that Harry was also at the scene and was in fact the murderer. If it is now asked *"Was the murder committed by someone you currently know to have been present?"* the correct answer is *yes*, despite what it was told before. As we will see in Chapter 3, it is extremely useful to imagine an interaction language that can use indexicals like these to request information or provide new information. But it will require us to distinguish clearly between an interaction language and any language used at the symbol level to represent facts for later use.

1.4 The rest of the book

The remaining chapters of the book are divided into two broad sections: Chapters 2 to 8 cover the basics in sequential order; Chapters 9 to 14 cover advanced research-oriented topics.

Introduction

- In Chapter 2, we start with a simple interaction language, a dialect of the language of first-order logic (with which we assume familiarity). However, there are good reasons to insist on some specific representational features, such as standard names and a special treatment of equality. With these, the semantic specification of the language ends up being clearer and more manageable than classical accounts. This will be especially significant when we incorporate epistemic features.
- In Chapter 3, we extend the first-order interaction language to include an epistemic operator, resulting in a language we call $\mathcal{KL}$. This involves being clear about what we mean by an epistemic state, distinct from a world state. This will allow us, among other things, to distinguish between questions about the world (e.g. the birds that do not fly) and questions about what is known (e.g. the birds that are known not to fly).
- Since what we mean by "knowledge" is so crucial to the enterprise here, in Chapter 4, we examine properties of knowledge in detail as reflected in the semantics of the language $\mathcal{KL}$. Among other things, we examine the interplay between quantifiers and knowledge, as well as the status of knowledge about knowledge.
- In Chapter 5, we define the **TELL** and **ASK** operations for the interaction language $\mathcal{KL}$. This provides a clear knowledge-level specification of the service to be provided by a knowledge representation system. We also include in this chapter a detailed example of the kind of questions and assertions that can be handled by our definition.
- In Chapter 6, we examine the relationship between the two views of knowledge mentioned above: knowledge in a KB, and knowledge in an abstract epistemic state. In other words, we look at the relationship between the symbol-level and knowledge-level views of knowledge. As it turns out, the correspondence between the two, as required by the semantics of the language $\mathcal{KL}$, is not exact.
- In Chapter 7, we prove that, despite the results of Chapter 6, it is possible to produce a symbol-level implementation of the interaction operations **TELL** and **ASK**, based on ordinary first-order reasoning. In particular, we show that the result of a **TELL** operation on a finitely represented state can itself always be properly represented, even if the sentence contains (indexical) references to what is currently known.
- In Chapter 8, we introduce a new concept called only-knowing that captures in a purely logical setting what is behind the **TELL** and **ASK** operations. The idea is to formalize using a new epistemic operator the assertion that a sentence is not only known, but all that is known.
- In Chapter 9, we relate only-knowing to Autoepistemic Logic, a special brand of so-called nonmonotonic logic which has been studied extensively in the literature. We are able to fully reconstruct Autoepistemic Logic using only-knowing and, in addition, extend it since we are using a more expressive language.

- In Chapter 10, we consider a proof theory for the logic of only-knowing. We show soundness and completeness in the propositional case and discuss why it is incomplete in the first-order case. Nevertheless, the axiom system allows us to obtain nontrivial first-order derivations using examples from Autoepistemic Logic.
- In Chapter 11, motivated by the fact that only-knowing is not that interesting when used as part of a query, we introduce a more focussed version, namely that of only-knowing something *about* a certain subject matter. We study, in a propositional setting, how **ASK** and **TELL** need to be adapted, how the new concept relates to Autoepistemic Logic, and how it gives rise to certain forms of reasoning about relevance.
- In Chapter 12, we begin our excursion into limited belief and a more tractable form of reasoning. Here we lay out the basic concepts of our approach, first in the propositional case and then moving to first-order. The main idea is to switch from a traditional two-valued semantics to a four-valued one. To keep matters as simple as possible, nested beliefs are first ignored altogether.
- In Chapter 13, we drop all these restrictions. We give new definitions of **TELL** and **ASK** based on the four-valued semantics and provide symbol-level implementations for them in a way quite similar to Chapter 7. Most importantly, we show that in many cases these routines now become tractable or at least decidable.
- In Chapter 14, we consider knowledge bases for reasoning about knowledge and action. For that, we amalgamate the logic of only-knowing with the situation calculus, a popular formalism in AI for reasoning about action. Besides a semantics which naturally extends the one we have developed for only-knowing, we provide a sound and complete second-order proof theory.

There are different ways of approaching this book. The first eight chapters are the core, but the remaining ones can be read more or less independently of each other. Those interested in default reasoning or nonmonotonic reasoning should read Chapter 9; those interested in proof systems and questions involving the logic of only-knowing should read Chapter 10; those interested in the issue of relevance and how it applies to default reasoning should read Chapter 11; those interested in tractable logical reasoning and the problem of logical omniscience should read Chapters 12 and 13 (in that order); finally, those interested in how knowledge relates to action (including perceptual action), for robotic applications, for example, should read Chapter 14.

1.5 Bibliographic notes

Much of the material in this chapter is shared with a forthcoming textbook on knowledge representation [10]. For a collection of readings on knowledge representation, see [9]. For a more philosophical discussion on knowledge and belief see [45, 12, 38], and [41] on the difference between the two. The notes at the end of Chapters 3 and 4 discuss attempts to formalize these notions. A general discussion of propositions, declarative sentences, and sentence tokens, as bearers of truth values, including the role played by indexicals, can be found in [4]. The connection between knowledge and commonsense is discussed in [101], one of the first papers on AI. The intentional stance is presented in [21], and critically examined in [22]. On Leibniz' views about thinking as a form of calculation see, for example, [28], vol. 3, p. 422. The Knowledge Representation Hypothesis is from Brian Smith's doctoral thesis [134], the Prologue of which appears in [9]. Procedural representations of knowledge are discussed in [143], and the criticism of AI by Hubert Dreyfus can be found in [26]. Zenon Pylyshyn discusses cognitive penetrability in [117], making a strong case for propositional representations to account for human-level competence. For Newell's knowledge and symbol levels, as well as the **TELL** and **ASK** functional interface, see the notes in Chapters 5 and 6. For general references on logic and entailment, see the notes in Chapter 2. Why reasoning needs to diverge from logic is discussed in [17] and [84]. For a review of the research in knowledge representation and reasoning in terms of this divergence, see [83]. For references on default (and logically unsound) reasoning, see the notes in Chapter 9.

1.6 Exercises

1. Consider a task requiring knowledge like baking a cake. Examine a recipe and state what needs to be known to follow the recipe.
2. In considering the distinction between knowledge and belief in this book, we take the view that belief is fundamental, and that knowledge is simply belief where the outside world happens to be cooperating (the belief is true, is arrived at by appropriate means, is held for the right reasons, and so on.). Describe an interpretation of the terms where knowledge is taken to be basic, and belief is understood in terms of it.
3. Explain in what sense reacting to a loud noise is and is not cognitively penetrable.
4. It has become fashionable to attempt to achieve intelligent behaviour in AI systems without using propositional representations. Speculate on what such a system should do when reading a book on South American geography.
5. Describe some ways in which the first-hand knowledge we have of some topic goes beyond what we are able to write down in a language. What accounts for our inability to express this knowledge?

2 A First-Order Logical Language

In this chapter, we will examine the properties of a first-order logical language that is suitable as a starting point at least for communicating with a KB about some application domain. As discussed in the previous chapter, we assume some familiarity with classical logical languages, propositional and quantificational, as discussed in any number of introductory logic texts (see the bibliographic notes). Here we concentrate mainly on the differences between our dialect of first-order logic and a standard one.

2.1 Why first-order logic?

We said in the previous chapter that the only feature of an interaction language that really mattered is that we had a clear and unambiguous account of what it meant for expressions in the language to be *true* or *false*. So why use a dialect of first-order logic for knowledge representation? It seems at first glance that this language is more suitable for expressing facts about *mathematical domains* such as the domain of numbers, sets, groups, and so on. This is why, after all, the language was invented by Frege at the turn of the century, and continues to be its main application in logical circles. These mathematical concepts, it might be thought, have very little in common with the typically vague and imprecise concepts underlying commonsense reasoning. Furthermore, quantification appears to be necessary only for stating facts about infinite domains, whereas many of the applications of knowledge representation concentrate on finite collections of objects.

The answer to these objections is best seen by considering how one might use a first-order language to express commonsense knowledge.

Each of the expressions of a first-order language in Figure 2.1 is accompanied by a gloss in English, interpreting the predicate, constant, and function symbols in the obvious way. Following these in each case is a question about what is being said.

Even though the intended domain here involves only a small collection of very simple objects, all of the facilities of full first-order logic with equality are being used. Consider the quantification in Example 4, for instance. If we are willing to assume that there are only finitely many blocks, can we not do without this universal quantifier? In one sense the answer is *yes*: we could simply state *of the blocks in the box* that they are light, as in:

$Light(block\_b) \land Light(block\_e)$.

The disjunction of Example 1 can be eliminated analogously by stating which of the two disjuncts is true

$In(block\_b, box)$,

---

1. *In(block_a, box)* ∨ *In(block_b, box)*
   Either block A or B is in the box.
   But which one?
2. ¬*In(block_c, box)*
   Block C is not in the box.
   But where is it?
3. ∃*x.In(x, box)*
   Something is in the box.
   But what is it?
4. ∀*x.In(x, box)* ⊃ *Light(x)*
   Everything in the box is light (in weight).
   But what are the things in the box?
5. *heaviest_block ≠ block_a*
   The heaviest block is not block A.
   But which block is the heaviest block?
6. *heaviest_block = favourite(john)*
   The heaviest block is also John's favourite.
   But what block is this?

Figure 2.1: Expressing knowledge in first-order logic

and similar considerations apply to the other examples.

The problem with this approach is simply that it requires us to know more than we may know in order to express what we want to express. We would need to know, for example, what blocks are in the box before we could say that they are all light. In other words, we would need to know (among other things) the answers to the questions listed after each example in order to express what the example expresses. In some applications, this knowledge will be available and it will be possible to *list directly* the properties of the objects in question without appeal to much in the way of logical notation:

> *On(block_a, table)*   *Heavy(block_a)*
> *In(block_b, box)*   *Light(block_b)*
> ...
> *heaviest_block = block_d*
> *favourite(john) = block_d.*

In this case, a simple language of something like

> < *object,attribute,value* >

triples would be sufficient.

But in cases where knowledge arrives incrementally, first-order logic (or English, or German for that matter) gives us much more: it allows us to say what we want to say without having to say more than we know. That is, from the point of view of knowledge representation, the expressiveness of first-order logic lies in what it allows us to leave *unsaid*.[1] Stated differently, the logical facilities of first-order logic with equality allow us to express knowledge that is *incomplete* in the sense that it does not completely pin down the facts about the situation being represented.[2] Thus what we are allowing by using a first-order language as our interaction language, is a system that can have, from a functional standpoint, incomplete knowledge of its application domain: it will be possible for the system to know that one of a set things must hold without also knowing which. This power is one of the hallmarks of knowledge representation formalisms and perhaps the major source of their complexity (conceptual and computational).

2.2 Why standard names?

Given the desirability of using disjunction, negation, equality and the rest of the baggage of first-order logic, the next question is: why not stop there? This is, after all, the place where "classical" logic (as described in text books) ends. What is the point of what we will call standard names?

Consider the expression

Teaches(*cs_100*, *best_friend*(*george*)).

This can be interpreted as saying that the best friend of George teaches a course called CS100. But if we asked "Who teaches CS100?" and were told only that it was the best friend of George, we would probably feel cheated. This *describes* the individual, but not necessarily in enough detail to *identify* him. The same could be said of an even more vague description like "it's a person with brown hair."

Given that we have the capability using first-order logic of expressing knowledge about the best friend of George without necessarily identifying him, a natural question we should consider is what it would mean to identify him. An obvious place to look is at assertions of equality. We might use something like

best_friend(*george*) = *father*(*bill*),

but this doesn't seem to say who he is unless we already know who the father of Bill is. If we have

father(*bill*) = *mister_smith*,

then again the question arises as to who is Mr. Smith.

So it seems that we have two options when it comes to knowing who somebody is. The first is simply to say that we cannot *identify* individuals directly using expressions of

---

[1] This is not strictly true, since there are cases in knowledge representation where we also want to deal with mathematical or infinite domains. For example, we may want to state general facts that apply to all points in time, or to all situations, all events, or whatever.

[2] This notion of incompleteness will be defined precisely later.

a first-order logic with equality. In this case, we would say that although we might know a collection of sentences of the form

$$Teaches(cs\_100, t), \quad (t_i = t_j), \quad (t_j \neq t_k),$$

none of these terms would be considered "special" in any way. Whether or not the system knows who any of the teachers are is something that cannot be determined by examining the known sentences. In fact, if this (admittedly nebulous) property should exist at all, it would be as a result of *non-linguistic* information that the system has acquired. From the point of view of a purely linguistic functional interface, we would never talk about the system knowing who or what (or when or where) something is, but only about the character of the terms t such that the system knows something expressed using t.

The second option is to say that it is a very useful concept to be able to distinguish between knowing that somebody or something must have a certain property and knowing who that individual is, *even for a system whose information is limited to a linguistic interface*. To do so, we need to introduce conventions into our language that go beyond those of standard first-order logic.

Perhaps the simplest mechanism for doing this is to imagine the space of all terms as partitioned into equivalence classes, where terms t_1 and t_2 are considered equivalent if $(t_1 = t_2)$ is true. Now imagine naming all potential equivalence classes using $^\#1, ^\#2, ^\#3$, and so on. We call these terms <u>standard names</u>. Then, we can simply *decide* as a linguistic convention that we will consider a term to be identified just in case we can name which equivalence class it belongs to.

So, for example, if we know that $best\_friend(george) = {^\#27}$ then we will say that we know who the best friend of George is. Similarly, if we do not know something of this form for say *mister_smith*, then regardless of what else we may know about him, and specifically what other terms we know to be equal to it, we will say that we do not know who Mr. Smith is.

By convention, the question of who $^\#27$ is does not arise. The term $^\#27$, unlike say $best\_friend(george)$ is intended to carry absolutely *no useful domain-dependent information* except that it is distinct from $^\#26$, and all the others. So while the ordinary equalities partition the terms into equivalence classes, the standard names anchor these classes and distinguish them from each other. As such they play a role like the *unique identifiers* of database formalisms (such as object identification numbers, like social security numbers).

If there is any doubt about the identity of an individual, we should not assign it a standard name. Fortunately, the language of first-order logic allows us to express knowledge without committing ourselves to the identity of the individuals involved. In fact, we may decide never to use standard names at all and stick to statements of equality and inequality between ordinary terms. In general, however, a term can be assigned to a standard name when we wish to express that it is distinct from all other terms that have been assigned standard names and when we do not wish to pursue further its identity.

2.3 The syntax of the language $\mathcal{L}$

We are now ready to describe the dialect of first-order logic called $\mathcal{L}$ that we will be using and the terminology and notation that goes with it.

The expressions of $\mathcal{L}$ are built up as sequences of symbols taken from the following two distinct sets:
1. the *logical symbols* consist of the following distinct sets:
    - a countably infinite supply of (individual) *variables*, written as x, y, or z, possibly with subscripts or superscripts.
    - a countably infinite supply of *standard names*, $^\#1$, $^\#2$, and so on, written schematically as n possibly with subscripts or superscripts.
    - the *equality* symbol, written $=$.
    - the usual logical connectives $\exists$, $\vee$, $\neg$ and punctuation $(,), ,$.
2. the *non-logical symbols* consist of two distinct sets:
    - the *predicate* symbols, written schematically as P, Q or R possibly with subscripts or superscripts, and intended to be domain specific properties and relations like *Person, Heavy, Teaches*.
    - the *function* symbols, written schematically as f, g or h possibly with subscripts or superscripts, and intended to denote mappings from individuals to individuals like *best_friend* or *father*. In case the mapping has no arguments, the function symbol is called a *constant* and is written schematically as b or c possibly with subscripts or superscripts; these are intended to denote individuals in the domain like *george, block_a*, or *heaviest_block*.

Each predicate or function symbol is assumed to have an *arity*, that is, a number indicating how many arguments it takes.

The logical symbols are the part of the alphabet of $\mathcal{L}$ that will have a fixed interpretation and use; the non-logical symbols, on the other hand, are the domain-specific elements of the vocabulary. Note that the equality symbol is taken to be domain-independent and is not considered to be a predicate symbol.

We are now ready to describe the expressions of $\mathcal{L}$. There are two types: <u>terms</u>, which are used to describe individuals in the application domain, and well-formed formulas or <u>wffs</u> which describe relations, properties or conditions in the application domain. We use schema variables t and u possibly with subscripts or superscripts to range over terms, and use Greek schema variables like α, β or γ possibly with subscripts or superscripts to range over wffs. We will use capitalized variables to range over *sets* of syntactic expressions. For example, G would be a set of function symbols, and Γ, a set of wffs.

Terms fall into three syntactic categories

1. variables,
2. standard names,
3. *function applications*, written as $f(t_1, \ldots, t_k)$ where the t_i are terms, and k is the arity of f. If the function symbol is a constant, it can be written without parentheses, as c instead of $c()$.

A term that contains no variables is called a *ground term*, and a ground term containing only a single function symbol is called a *primitive term*. In other words, a primitive term is of the form $f(n_1, \ldots, n_k)$, where $k \geq 0$, that is, a function application all of whose arguments are standard names.

The wffs of $\mathcal{L}$ are divided syntactically into atomic and non-atomic cases. The atomic formulas or *atoms* of $\mathcal{L}$ are of the form $P(t_1, \ldots, t_k)$, where P is a predicate symbol, the t_i are terms, and k is the arity of P. As with function applications, a *ground atom* is one without variables, and a *primitive atom* is a ground one where every t_i is a standard name. Thus, primitive expressions, both terms and atoms, contain a single non-logical symbol.

In general, a wff is one of the following:

1. an atom,
2. $(t_1 = t_2)$,
3. $\neg \alpha$,
4. $(\alpha \vee \beta)$,
5. $\exists x \alpha$.

As is the custom, we omit parentheses when the context is clear, use square parentheses or periods to increase readability, and freely write the usual abbreviations:

$$\forall x \alpha, (\alpha \wedge \beta), (\alpha \supset \beta), (\alpha \equiv \beta).$$

We also need the usual notion of a free or bound occurrence of a variable in a wff. The rigorous specification of these notions can be found in logic books, but informally, an occurrence of x in a wff is *bound* if it is located within a subwff of the form $\exists x. \alpha$ (in which case we say that it is within the *scope* of that quantifier $\exists$), and *free* otherwise. We use the notation α^x_t to mean the wff that results from textually replacing all free occurrences of the variable x in the wff α by the term t. Typically, the term t here will be a standard name. If $\vec{x}$ and $\vec{t}$ are sequences of variables and terms respectively and of the same size, by $\alpha^{\vec{x}}_{\vec{t}}$ we mean the wff that results by simultaneously replacing each free x_i by its corresponding t_i.

A First-Order Logical Language

One last syntactic notion: a *sentence* of $\mathcal{L}$ is a wff with no free variables. This is the most important syntactic category. Sentences and sentences alone will receive a truth value, can be believed, and so represent knowledge. In fact, we can think of $\mathcal{L}$ just as its set of sentences, with the rest of the syntactic machinery merely playing a supporting role.

2.4 Domain of quantification

Before discussing in detail how we will interpret the sentences of $\mathcal{L}$, we need to discuss an assumption that we make regarding the domain of quantification. This assumption, although not necessary, greatly simplifies the semantic specification of $\mathcal{L}$ and the technical analysis to follow.

The assumption is this: the application domain is considered to be isomorphic to the set of standard names. In other words, we assume that there is always a distinct object in the domain for each standard name (and so the domain must be infinite) and that each object in the domain has a distinct name (and so the domain must be countable).

So what does this rule out? First of all, unlike classical logic, we rule out domains with only finitely many individuals. This is not to say that predicates are required to be infinite; every predicate will be allowed to have a finite extension. But the sum total of *all* individuals in the domain must be infinite. If we wish to deal with finite domains, we use predicates (like *Object* or some such) and relativize what needs to be said to instances of these predicates. Another way of thinking about this is to say that the set of integers (or strings or some such) is always included in the domain of discourse even in otherwise "small" situations.

Conversely, we rule out domains where there are more objects than standard names. We do not want inaccessible individuals, that is individuals that have properties according to the predicate and function symbols, but cannot be referred to by name. If the domain is countable, this is not a real problem since we could have assigned the names differently to cover all the domain elements.

But if we had imagined a domain containing (say) the set of real numbers, this may seem to present a more serious difficulty. It is somewhat illusory, however: it is a well known result of ordinary classical logic that any satisfiable set of sentences is satisfiable in a *countable* domain.[3] What this means is that although we may be thinking of an uncountable domain like the reals, any collection of sentences of ordinary first-order logic that rules out the countable domains is guaranteed to be inconsistent! The real numbers might indeed be *compatible* with what we are talking about, but countable domains must always be too. So in the logic of $\mathcal{L}$, we simply take this one step further and imagine the

---

3 This is true for first-order logic, but not for higher-order logics.

domain as always being countable.

Note that although we assume the domain to be isomorphic to the set of standard names, we do not make this assumption for *constants*. As in ordinary classical logic, two constants may indeed refer to the same individual, and there may be individuals that are not named by any constant.[4] So, as we will see below, for the part of $\mathcal{L}$ that does not use standard names or equality, everything will be the same as in ordinary classical logic.

The main consequences of this assumption are:

- although it is sometimes desirable to talk about what individual in the domain a term refers to under some interpretation, it will never be *necessary* to do; instead we can talk about finding a *co-referring* standard name, since every individual has a unique name.
- it will similarly be possible to understand quantification *substitutionally*. For example, $\exists x.P(x)$ will be true just in case $P(n)$ is true for some standard name n, since every individual in the domain has a name.

It is these two assumptions that greatly simplify the semantic specification of $\mathcal{L}$.

2.5 World state

In the classical interpretation of sentences of first-order logic due to Tarski, one specifies a domain of discourse and appropriate functions and relations for the function and predicate symbols. Using these, a set of rules specify the truth value of every sentence. This is done by first considering the more general case where terms and wffs can have free variables, and using an assignment of domain elements to these variables.

In our case, the rules are much simpler. All the variability in the interpretation of sentences reduces to the understanding of the function and predicate symbols. We will still want to know the truth value of every sentence, of course, but this will be completely determined in a straightforward way, more like the way it is done in classical *propositional* logic.

Recall that in propositional logic, one specifies an interpretation by fixing an assignment to the atomic sentences, after which the truth value of the non-atomic sentences (disjunctions and negations) is recursively defined. In our case, we require two things to specify an interpretation: a truth value for each of the primitive atoms, and a standard name for each of the primitive terms. We call such an assignment a <u>world state</u> (or world, for short) and let W name the set of all world states. We use this term since it is these assignments that tell us the way the world is, relative to the language $\mathcal{L}$.

For example, suppose we only care about one function symbol *best_friend* and one

---

4 In this sense, what has been called the domain closure and unique name assumptions do not apply.

A First-Order Logical Language

predicate symbol *Person*. What we need to completely describe the way things are (that is, a world state) in this language is to say who the people are, and who is the best friend of whom. To say who the people are, we need only specify which sentences of the form *Person*(n) are true, since every individual is assumed to have a unique standard name. Similarly, we can handle best friends by specifying for each term of the form *best_friend*(n), the standard name of the best friend of the individual named n.[5] From this specification, as we will see below, the truth value of every sentence will be determined.

So what we have is that for any $w \in W$,

- $w[f(n_1, \ldots, n_k)]$ is a standard name, taken to be a specification of the primitive's unique co-referring standard name;
- $w[P(n_1, \ldots, n_k)]$ is either 0 or 1, taken to be a specification of the primitive's truth value, where 1 indicates truth.

In each case, we will say that w provides the <u>value</u> of the primitive expression.

2.6 Term and formula semantic evaluation

The generalization of term evaluation to non-primitive terms is straightforward. Suppose we want the value of the term $f(g(n), c)$ with respect to some w. If the value of the primitive terms $g(n)$ and c are n_1 and n_2 respectively, then the value of $f(g(n), c)$ is the value of the primitive term $f(n_1, n_2)$. In other words, to determine a co-referring standard name for a term, we recursively substitute co-referring standard names for the arguments, stopping at primitives.

More formally, the value of a ground term t at world state w, which we write as $w(t)$, is defined by

1. $w(n) = n$;
2. $w(f(t_1, \ldots, t_k)) = w[f(n_1, \ldots, n_k)]$, where $n_i = w(t_i)$.

We will never need to evaluate terms with variables.

We are now ready to state precisely what it means for a sentence α be true in a world state w, which we write as $w \models \alpha$. Informally, we proceed as follows: for atomic sentences and equalities, we evaluate the arguments and check the answers; for disjunctions and negations, we proceed as with ordinary propositional logic; for existential quantification, we use the fact that every individual has a name and consider each substitution instances of the wff in question. More precisely, we have:

1. $w \models P(t_1, \ldots, t_k)$ iff $w[P(n_1, \ldots, n_k)] = 1$, where $n_i = w(t_i)$;

---

[5] In case n happens not to be a person, we can assign it any other name that is also not a person, for example n itself, since we only care about people.

2. $w \models (t_1 = t_2)$ iff $w(t_1)$ is the same name as $w(t_2)$;
3. $w \models \neg\alpha$ iff it is not the case that $w \models \alpha$;
4. $w \models \alpha \vee \beta$ iff $w \models \alpha$ or $w \models \beta$;
5. $w \models \exists x\alpha$ iff for some name n, $w \models \alpha^x_n$.

Again there is no need to talk about the truth or falsity of wffs with free variables, even to deal with existential quantifiers.

2.7 Satisfiability, implication and validity

The semantic specification above completely determines which sentences of $\mathcal{L}$ are true and which are false given values for the primitives. Clearly, it is the non-logical symbols that carry the semantic burden here. The standard names, for example, are the fixed reference points in terms of which the predicate and function symbols are characterized. Each primitive expression deals with a single non-logical symbol and specifies one independent aspect of its meaning, namely its value for the given arguments. There is a world state for each possible value for each possible sequence of arguments for each primitive expression.

Although many or most of these world states will not be of interest to us, they determine the complete range of what can be true according to $\mathcal{L}$. We say that a set of sentences Γ is _satisfiable_ just in case there is some world state w such that $w \models \alpha$ for every α in Γ. In other words, Γ is satisfiable if there is some way the primitives could be assigned to make the sentences in Γ true. In this case, we will say that w satisfies Γ.

Although all primitives can be assigned freely and independently to their values, the semantic rules make certain sets unsatisfiable, even simple ones such as,

$$\{P(c), \neg P(n), (n = c)\}.$$

The semantic rules of $\mathcal{L}$ are such that if any two elements of this set are true, then the last one _must_ be false.

To focus on what _must_ hold according to the semantic rules of $\mathcal{L}$, we say that a sentence α is _logically implied_ by a set of sentences Γ, which we write $\Gamma \models \alpha$ just in case the set $\Gamma \cup \{\neg\alpha\}$ is unsatisfiable. Stated differently, Γ logically implies α if the truth of Γ forces α to be true also.

To preview what is to come, the reason logical implication is so important is this. We imagine a knowledge-based system acquiring information about some world state w in an incremental fashion. At any given point, it will have at its disposal not w itself, but only information about w in the form of a collection of sentences Γ, corresponding to what it has been told. If it now has to make a decision about what holds in w, for example to answer a question, it must use Γ since this alone represents what it knows. The trouble is that although w will satisfy Γ, in general many very different world states will as well. However, if $\Gamma \models \alpha$, then it is perfectly safe to conclude that α is also true in w since, according to the definition of entailment, any world state satisfying Γ (such as w) also satisfies α. So the information represented by Γ includes the fact that all of its implications are true in the intended world state.

Of particular concern to us is when the information represented by Γ is *finite*, that is, when Γ consists of the sentences $\{\alpha_1, \ldots, \alpha_k\}$. In this case, we can capture the notion of logical implication using a sentence of the language: it is easy to see that $\Gamma \models \alpha$ iff the sentence

$$(\neg\alpha_1 \vee \neg\alpha_2 \vee \ldots \vee \neg\alpha_k \vee \alpha)$$

comes out true at every world state. We call a sentence α *valid*, which we write as $\models \alpha$, if it is satisfied by every world state. With respect to finitely specifiable conditions, it is therefore the valid sentences of a logical language that determine what is *required* by the language. They also determine what is *allowed* by the language, since $\{\alpha_1, \ldots, \alpha_k\}$ is satisfiable iff the sentence $(\neg\alpha_1 \vee \ldots \vee \neg\alpha_k)$ is not valid. If the notion of a sentence is the culmination of the syntax of a language, the notion of validity is the culmination of its semantics. It is indeed often the case that a logic is simply identified with its set of valid sentences.

2.8 Properties of logic $\mathcal{L}$

We now examine some of the properties of the language $\mathcal{L}$ in preparation for the generalization to $\mathcal{KL}$ to follow. We assume some familiarity with a classical first-order language, and concentrate primarily on the unique properties of $\mathcal{L}$.

First of all, we establish that except perhaps for equality and standard names, $\mathcal{L}$ behaves identically with ordinary first-order languages:

Theorem 2.8.1: *Suppose α does not contain standard names or equality. Then $\models \alpha$ iff α is a valid sentence of ordinary first-order logic.*

Proof: First, suppose that $\models \alpha$, and suppose to the contrary that there is a Tarskian interpretation $\langle D, \Phi \rangle$ that satisfies $\neg\alpha$.[6] By the Skolem-Löwenheim theorem, we may assume without loss of generality that the domain D is countably infinite. Let $\{d_1, d_2, \ldots\}$ be an ordering of this set. Define a world state w where $w[P(^\#i_1, \ldots, ^\#i_k)] = 1$ iff the tuple $\langle d_{i_1}, \ldots, d_{i_k}\rangle$ is in $\Phi(p)$, and $w[f(^\#i_1, \ldots, ^\#i_k)] = {^\#i}$ iff $\Phi(f)(d_{i_1}, \ldots, d_{i_k}) = d_i$. It is not

---

6 That is, D is any non-empty set, and Φ is a mapping from function symbols to functions over D and from predicate symbols to relations over D, both of the appropriate arity.

hard to show for any sentence β without standard names or equality that β is true wrt w iff the interpretation $\langle D, \Phi \rangle$ satisfies β. (We leave the proof as an exercise.) Consequently, $w \models \neg\alpha$, contradicting the assumption that $\models \alpha$.

Similarly, suppose that $\not\models \alpha$, that is, that for some w we have that $w \models \neg\alpha$. Define a Tarskian interpretation $\langle D, \Phi \rangle$ as follows: D is the set of standard names, and $\Phi(p)$ is the set of tuples $\langle n_1, \ldots, n_k \rangle$ such that $w[P(n_1, \ldots, n_k)] = 1$, and $\Phi(f)(n_1, \ldots, n_k)$ is the name n such that $w[f(n_1, \ldots, n_k)] = n$. As above, we leave it as an exercise to show that w and $\langle D, \Phi \rangle$ must agree on the truth value of α, and so α is not valid in the classical sense either. ∎

We also get:

Corollary 2.8.2: *If Γ is a set of sentences without names or equality, then Γ is satisfiable iff it is satisfiable in ordinary first-order logic.*

The proof is analogous to that of the theorem. Thus, although the specification of $\mathcal{L}$ was much simpler than the traditional Tarskian account, the result is no different, when restricted to wffs of ordinary first-order logic without equality.

What can be say about equality in $\mathcal{L}$, that is, how is it different from standard first-order theories of equality (assuming we restrict our attention once again to wffs without standard names)? The main difference is that the following wffs are all valid in $\mathcal{L}$:

1. $\neg \exists x_1 \forall y (y = x_1)$,
2. $\neg \exists x_1 \exists x_2 \forall y (y = x_1) \lor (y = x_2)$,
3. $\neg \exists x_1 \exists x_2 \exists x_3 \forall y (y = x_1) \lor (y = x_2) \lor (y = x_3)$,

and so on. The i-th sentence in this enumeration says that there are not i individuals such that every individual is one of them; in other words, there are more than i individuals. If we call this set of sentences Δ, then Δ as a whole guarantees that there are *infinitely* many individuals, since no (finite) i will be sufficient.

This is clearly not a property of ordinary first-order logic where finite domains are allowed. It suggests that in $\mathcal{L}$ we should avoid talking about the domain as a whole, and always restrict our attention to individuals of a certain type. In other words, we should rarely write $\forall x.\alpha$, since after all, very little of interest will be true of *everything* (including numbers, people, bits of rock, events, and so on.) Instead, we should use a form of *typed* quantification and write something like

$$\forall x (P(x) \supset \alpha),$$

relativizing what we want to say to some predicate P. In fact, sentences like those of Δ that only contain *logical* symbols (including equality and standard names) have a special property:

Theorem 2.8.3: *If α contains only logical symbols, then either it or its negation is valid.*

Proof: Let w_1 and w_2 be any world states. Since the sentence α does not contain function or predicate symbols, by induction, $w_1 \models \alpha$ iff $w_2 \models \alpha$. Thus if α is satisfiable, it must be valid, and the theorem follows. ∎

If we think in terms of the information carried by a sentence, sentences with only logical symbols do not really express information about the world at all; they are either logically true or logically false.

To capture the precise relationship between our treatment and the ordinary treatment of equality, we need to quickly review the latter. In the usual treatment, equality is a regular predicate, but is specified to be an equivalence relation that allows substitution in arguments. So let Γ be the following sentences:

- reflexivity: $\forall x(x = x)$;
- symmetry: $\forall x \forall y (x = y) \supset (y = x)$;
- transitivity: $\forall x \forall y \forall z ((x = y) \land (y = z)) \supset (x = z)$;
- substitution of equals for functions: for any function symbol f,
$\forall x_1 \ldots \forall x_k \forall y_1 \ldots \forall y_k ((x_1 = y_1) \land \ldots \land (x_k = y_k)) \supset$
$$f(x_1, \ldots, x_k) = f(y_1, \ldots, y_k);$$
- substitution of equals for predicates: for any predicate symbol P,
$\forall x_1 \ldots \forall x_k \forall y_1 \ldots \forall y_k ((x_1 = y_1) \land \ldots \land (x_k = y_k)) \supset$
$$P(x_1, \ldots, x_k) \equiv P(y_1, \ldots, y_k).$$

Then, what we have is the following:

Theorem 2.8.4: *Let α be a sentence without standard names. Then α is valid iff in the classical account of first-order logic, $\Delta \cup \Gamma$ implies α.*

Proof: Left as an exercise. ∎

This theorem ensures that in $\mathcal{L}$ we get all the standard properties of equality, and moreover, what we get beyond the standard properties are the sentences of Δ, making the domain of quantification infinite.

Finally, we turn our attention to standard names. What can we say about them? Clearly, for each name n, the wff $(n = n)$ is valid, as we would expect from any well-behaved theory of equality. Less conventionally, perhaps, is that for any pair of distinct names n_1

and n_2, the wff $(n_1 \neq n_2)$ is also valid. So it is built into the logic that each standard name is equal to exactly one name, itself.

There is one more interesting property of standard names and that is, roughly, that they have no other special logical properties! In other words, the *only* feature that characterizes standard names is that they are different from each other. As a first approximation, we might want to say that if $\alpha_{\#1}^x$ is valid, then so must be $\alpha_{\#2}^x$, since there is nothing special about $^\#1$. But this is not quite right: imagine α is the wff $(x \neq {}^\#2)$; then $\alpha_{\#1}^x$ is valid, but $\alpha_{\#2}^x$ is not.

However, we can capture precisely what we want to say by talking about a consistent renaming of all the standard names in a sentence. First some notation: let * be a bijection from standard names to standard names. For any term t or wff α, we let t^* or α^* indicate the expression resulting from simultaneously replacing in t or α every name by its mapping under *. Then we get the following:

Theorem 2.8.5: *Let * be a bijection from standard names to standard names. Then α is valid iff α^* is valid.*

Proof: Here we prove the theorem only for the special case where α contains no function symbols. The more general case is left as an exercise.

To begin, let us define for any world state w, w^* to be the world state that is like w except that for any primitive wff α, $w^*[\alpha] = w[\alpha^*]$. We now show that for any α without function symbols, $w^* \models \alpha$ iff $w \models \alpha^*$. The proof is by induction over the depth of the formation tree of α.[7] The property clearly holds for atomic formulas $P(n_1, \ldots, n_k)$ by definition of w^*. For equalities, we have that $w^* \models (n_1 = n_2)$ iff n_1 and n_2 are the same iff (since * is a bijection) n_1^* and n_2^* are the same iff $w \models (n_1^* = n_2^*)$. For negations, we have that $w^* \models \neg\alpha$ iff it is not the case that $w^* \models \alpha$ iff (by induction) it is not the case that $w \models \alpha^*$ iff $w \models \neg\alpha^*$. For disjunctions, we have that $w^* \models (\alpha \vee \beta)$ iff $w^* \models \alpha$ or $w^* \models \beta$ iff (by induction) $w \models \alpha^*$ or $w \models \beta^*$ iff $w \models (\alpha \vee \beta)^*$. Finally, for existentials, we have that $w^* \models \exists x.\alpha$ iff for some name n, $w^* \models \alpha_n^x$ iff (by induction) for some name n, $w \models (\alpha_n^x)^*$ iff for some name n, $w \models (\alpha)_n^{*x}$ iff $w \models (\exists x.\alpha)^*$.

Now using this property, if for some w we have that $w \models \neg\alpha^*$, then for w^* we have that $w^* \models \neg\alpha$, and if for some w we have that $w \models \neg\alpha$, then for w^* we have that $w^* \models \neg\alpha^*$. So α must be valid iff α^* is. ∎

As a corollary to this we get:

---

[7] Since this type of induction proof is used so often throughout the book, we include this first one in full detail.

Corollary 2.8.6: *Let α be a formula with free variables $x_1, \ldots, x_k$ and let * be a bijection that leaves the names in α unchanged. Then for any standard names $n_1, \ldots, n_k$,*

$$\models \alpha_{n_1 \cdots n_k}^{x_1 \cdots x_k} \quad \text{iff} \quad \models \alpha_{n_1* \cdots n_k*}^{x_1 \cdots x_k}.$$

As a special case, we get:

Corollary 2.8.7: *Let α have a single free variable x and let n_1 and n_2 be names not appearing in α. Then, $\alpha_{n_1}^x$ is valid iff $\alpha_{n_2}^x$ is valid.*

Proof: Consider the bijection that swaps names n_1 and n_2 and leaves all other names unchanged. Then $(\alpha_{n_1}^x)^*$ is $\alpha_{n_2}^x$ and the result follows from the theorem. ∎

This establishes that names that do not appear in a wff can be used interchangeably. As a consequence we get:

Corollary 2.8.8: *Let α have a single free variable x and let n be a standard name not appearing in α. Let $n_1, \ldots, n_k$ be the all the standard names appearing in α. If α_n^x is valid and all the $\alpha_{n_i}^x$ are valid, then so is $\forall x \alpha$.*

Proof: To show a universal is valid, we need only show that all of its substitution instances are valid. For standard names appearing in α, the validity is assumed; for names not appearing in α, the same argument as the previous corollary can be used with the name n. ∎

Among other things, this shows under what conditions we are entitled to infer the validity of a universal from the validity of a finite collection of substitution instances. This will play a role in the axiomatization of $\mathcal{L}$ below.

To conclude this section on the logical properties of $\mathcal{L}$, we mention one additional minor difference between $\mathcal{L}$ and ordinary first-order logic. This difference would not be apparent looking only at validity, since it concerns infinite sets of sentences. Ordinary first-order logic is *compact*, that is, a set of sentences is satisfiable iff all of its finite subsets are. But this is not true of $\mathcal{L}$: there is a set of sentences of $\mathcal{L}$ that is unsatisfiable, but all of whose proper subsets are satisfiable:

$$\{\exists x P(x), \neg P(^\#1), \neg P(^\#2), \neg P(^\#3), \ldots\}.$$

The reason for the difference is that in $\mathcal{L}$ we can *name* every domain element (using an infinite collection of sentences). In ordinary logic, the set would be satisfiable since there would be the possibility of domain elements that are not named by any term. This difference is indeed minor since it requires an infinite set of sentences to exhibit it; the finite case is completely characterized by the above theorems.

2.9 Why a proof theory for $\mathcal{L}$?

We now turn to the development of a proof theory for $\mathcal{L}$. Before doing so, however, it is worth considering *why* we care about proof theories at all in this context, since the motivation here is perhaps non-standard.

First of all, we do *not* use a proof theory because we care about the structure of sound arguments, from premises to conclusions. Nor do we use a proof theory as the basis for a computational procedure for knowledge representation purposes. In fact, there is absolutely no reason to believe that the proof theory we will present is closer in any way to a realistic computational realization of a decision procedure for $\mathcal{L}$.

Rather, we examine a proof theory for one reason only: it gives us another revealing look at the valid sentences of $\mathcal{L}$, from a very different perspective. So far, we have defined the valid sentences in terms of truth: the valid sentences are those that come out true in all world states. With a proof theory, the picture we have is of a class of sentences defined by a closure operation: we start with a basic set (the axioms), then apply operations (the rules of inferences) on elements of the set until no new members are introduced.

A good analogy here is with the idea of a formal language, that is, a set of strings taken from some alphabet. We might describe a language as being all strings of the form

$$\{a^n b \mid n \geq 0\}.$$

We might also choose to describe the language as that which is produced by the following grammar:

$$S \rightarrow aS$$
$$S \rightarrow b$$

The two descriptions of this simple language are complementary, and each has its utility in certain contexts. The grammatical description is most like a proof theory since it describes the language by a closure operation: the language is the least set of strings such that b is in the language, and if S is in the language, then so is S with a prepended.

With this analogy we see clearly why this proof theory should not be understood procedurally. There is a clear separation between having a grammar for a language and having a *recognizer*. In fact, an efficient recognizer may or may not use the grammar explicitly. Similarly, having a proof theory is distinct from having a program that can prove (or even recognize) theorems. Such a theorem-prover may or may not use the proof theory, since it is the valid sentences that count, not the particular axioms and rules of inference.

2.10 Universal generalization

Most of the proof theory described below is standard. The main difference involves standard names. This shows up clearly in the treatment of the rule of universal generalization

Universal generalization is the rule that allows universals to be concluded from arguments involving "arbitrary values." The standard way to phrase this rule is using open wffs, as in

From α, infer $\forall x.\alpha$.

The reason that this rule is *sound*, that is, that the conclusion is valid given that the premise is valid, involves two cases: if α does not have x free, then clearly $\forall x.\alpha$ is valid; if it does have x free, then α is talking about some particular value of x. So the argument goes: if α is valid, then there is nothing special about x, and so the universal must be valid also.

In the case of $\mathcal{L}$, we do not need to appeal to wffs with free variables since we can use standard names. A first step might be to have a rule with an infinite set of premises like this:

From $\alpha^x_{\#_1}, \alpha^x_{\#_2}, \alpha^x_{\#_3}, \ldots$, infer $\forall x.\alpha$.

According to our semantics, this rule is clearly sound: if α is valid for all standard names replacing x, then the universal must be valid also.

However, there is a finitary version of this rule that does the trick because of Corollary 2.8.8 discussed earlier: From $\alpha^x_{n_1}, \ldots, \alpha^x_{n_k}$ where the n_i range over all the standard names in α and at least one standard name not in α, infer $\forall x.\alpha$. Thus, to infer a universal, we need only look at a finite number of arguments, one for each name in the wff, and one extra name. The soundness of this strategy is immediate from the corollary.

2.11 The proof theory

Except for the rule of universal generalization, the proof theory for $\mathcal{L}$ is not very surprising. First, we have the following axioms, for any formula α, β, or γ, any variable x, and any closed term t:

1. $\alpha \supset (\beta \supset \alpha)$
2. $(\alpha \supset (\beta \supset \gamma)) \supset ((\alpha \supset \beta) \supset (\alpha \supset \gamma))$
3. $(\neg\beta \supset \neg\alpha) \supset ((\neg\beta \supset \alpha) \supset \beta)$
4. $\forall x(\alpha \supset \beta) \supset (\alpha \supset \forall x\beta)$, provided that x does not occur free in α
5. $\forall x\alpha \supset \alpha^x_t$
6. $(n = n) \wedge (n \neq m)$, for any distinct n, m

The rules of inference are *modus ponens* and *universal generalization*, as discussed in the previous section:

1. From α and $(\alpha \supset \beta)$, infer β.
2. From $\alpha_{n_1}^x, \ldots, \alpha_{n_k}^x$, infer $\forall x \alpha$, provided the n_i range over all names in α and at least one not in α.

As usual, we say that α is a *theorem* of $\mathcal{L}$, which we write as $\vdash \alpha$, iff there is a sequence of wffs $\alpha_1, \alpha_2, \ldots, \alpha_k$, where $\alpha_k = \alpha$ and each α_i in the sequence is either an instance of an axiom, or follows from earlier sentences in the sequence by one of the two rules of inference. If Γ is any set of sentences, we say that Γ *derives* α, written as $\Gamma \vdash \alpha$ iff Γ contains sentences $\gamma_1, \ldots, \gamma_k$, where $k \geq 0$ and such that $\vdash ((\gamma_1 \wedge \ldots \wedge \gamma_k) \supset \alpha)$. Finally, we say that Γ is *inconsistent* if it contains sentences $\gamma_1, \gamma_2, \ldots, \gamma_k$, where $k > 0$ and such that $\{\gamma_1, \gamma_2, \ldots, \gamma_{k-1}\} \vdash \neg \gamma_k$.

The first three axioms above are typical ones that are used (with the rule of modus ponens) to characterize propositional logic. In fact, they could be replaced by any combination of axioms and rules that correctly captures ordinary propositional logic. We therefore simply state the following without proof:

Theorem 2.11.1: *A sentence α is a theorem of ordinary propositional logic iff it can be derived using just the first three axioms and the rule of modus ponens.*

The next two axioms (and the rule of universal generalization) are the typical way quantifiers are formalized in a proof theory (although as noted above, universal generalization is handled differently here). Finally, the last axiom is the one and only addition that is necessary to handle equality; the usual formalization is much more complex. Note that all that is needed to capture the properties of standard names (as distinct from other terms) are the axiom of equality and the rule of generalization.

The most important property of this proof theory (following our discussion above of its role) is that it correctly matches the semantic characterization given earlier:

Theorem 2.11.2: $\models \alpha$ *iff* $\vdash \alpha$.

Proof: The proof has two parts: *soundness* involves establishing that everything derivable is valid; *completeness* involves showing that any valid sentence is derivable.

The proof of the former is easy, and proceeds by induction on the length of the derivation of α: establish (case by case) that all instances of axioms are valid, and then show that each rule of inference preserves validity (using Corollary 2.8.8). The details are left as an exercise.

A First-Order Logical Language

The proof of the latter is more challenging. The usual way to show that a sentence that is not derivable is not valid, is to show that any finite consistent set of sentences is satisfiable. This is sufficient since if a sentence α is not derivable, then $\{\neg\alpha\}$ must be consistent, and so $\{\neg\alpha\}$ would be satisfiable, in which case α would not be valid. We will not prove here that finite consistent sets are indeed satisfiable, since the details of the proof can be reconstructed from the proof for the more general language $\mathcal{KL}$ to follow. The basic structure of the argument, however, is to show how the set can be extended to an infinite superset that remains consistent and that contains for every sentence, either the sentence or its negation. From this set (that also has other properties), a satisfying world state w is constructed directly: for any primitive term t and primitive atom α, $w[t] = n$ iff $t = n$ is an element of the set, and $w[\alpha] = 1$ iff α is an element of the set. This style of completeness proof is called a *Henkin proof*. ∎

Thus the valid sentences are the same as those that are derivable. We obtain as an easy corollary:

Corollary 2.11.3: *If Γ is a finite set of sentences, then $\Gamma \models \alpha$ iff $\Gamma \vdash \alpha$.*

We leave it as an exercise to show that this corollary need not hold when Γ is infinite.

2.12 Example derivation

The whole point of introducing a proof theory is to provide a different perspective on the working of the semantics of $\mathcal{L}$. Consider, for example, the fact that equals can be substituted for equals as in

$$\forall y \forall x.(x = y) \supset (f(x) = f(y)).$$

We can prove this sentence valid as follows: Let w be any world state, and n and m be any standard names. If $w \models (n = m)$, then they must be the same standard name, and $f(n)$ and $f(m)$ must be the same terms. Thus $w \models (f(n) = f(m))$ also. Since this works for any pair of names and any world state, the universal must be valid. Using the proof theory of $\mathcal{L}$, we will show that the above sentence is derivable, which gives a different argument for its validity.

To show a derivation, we will list a sequence of sentences, one per line, followed by a justification. If the justification is of the form **Ax**, this means that the current line is an axiom; if it is of the form **UG**, this means that the current line is derivable from the preceding line and perhaps some earlier ones by universal generalization; if it is of the form **MP**, this means that the current line is some β and that there is a previous α such

1.	$^{\#}1 = {^{\#}1}$	Ax
2.	$\forall x(x = x)$	UG
3.	$f(^{\#}1) = f(^{\#}1)$	MP
4.	$(^{\#}1 = {^{\#}1} \supset f(^{\#}1) = f(^{\#}1))$	MP
5.	$^{\#}2 \neq {^{\#}1}$	Ax
6.	$(^{\#}2 = {^{\#}1} \supset f(^{\#}2) = f(^{\#}1))$	MP
7.	$\forall x(x = {^{\#}1} \supset f(x) = f(^{\#}1))$	UG
8.	$\forall y \forall x(x = y \supset f(x) = f(y))$	UG

Figure 2.2: A sample derivation in $\mathcal{L}$

that $(\alpha \supset \beta)$ is either an axiom or on an earlier line. The derivation for the substitutivity property is in Figure 2.2.

Note how at the end, universal generalization is used twice, once for each universally quantified variable. The first application on line 7, has α being the formula

$$(x = {^{\#}1}) \supset (f(x) = f(^{\#}1)).$$

This uses one standard name, $^{\#}1$, and so to apply generalization, we need to prove two instances of α, $\alpha^x_{\#1}$ and for some n not in α, α^x_n. The former is line 4, and the latter is line 6, where n is $^{\#}2$.

The final application of universal generalization is on line 8 for the formula

$$\forall x(x = y \supset f(x) = f(y)).$$

This uses no standard names, and so all we need is an instance with the variable y replaced by any standard name. This occurs on line 7 with y replaced by $^{\#}1$.

A similar strategy would be used to prove a sentence of the form $\forall x \forall y \forall z. \beta$, where β has no standard names. In this case, 3 new standard names would be used, call them $^{\#}1$, $^{\#}2$, $^{\#}3$. Then, to conclude the universal for z, it would be necessary to prove 3 formulas: $\beta^{x\,y\,z}_{\#1\,\#2\,\#1}$, $\beta^{x\,y\,z}_{\#1\,\#2\,\#2}$, and $\beta^{x\,y\,z}_{\#1\,\#2\,\#3}$. To conclude the universal for y, 2 previous formulas are required: $\forall z. \beta^{x\,y}_{\#1\,\#1}$ and $\forall z. \beta^{x\,y}_{\#1\,\#2}$. To conclude the final sentence, only 1 previous sentence is used: $\forall y \forall z. \beta^x_{\#1}$. In general, to prove a sentence with no standard names but k universal variables, k new standard names must be introduced, and a total of $k!$ previous sentences must be established.

As a final example, we show that a standard property of first-order logics holds for $\mathcal{L}$:

Theorem 2.12.1:

$$\vdash \forall x(\alpha \supset \beta) \supset ((\forall x \alpha) \supset (\forall x \beta)).$$

A First-Order Logical Language

Proof: The proof proceeds by deriving a universal:

$$\vdash \forall x[\forall x(\alpha \supset \beta) \supset ((\forall x\alpha) \supset \beta)].$$

To derive this universal we need to get,

$$\vdash [\forall x(\alpha \supset \beta) \supset ((\forall x\alpha) \supset \beta_n^x)],$$

for all the names n appearing in α or β, and an additional one. To derive this, we use the fact that

$$\vdash \forall x(\alpha \supset \beta) \supset (\alpha_n^x \supset \beta_n^x)$$

and

$$\vdash (\forall x\alpha \supset \alpha_n^x),$$

which are axioms, and put these two together using ordinary properties of propositional logic. Now with the above universal in hand, we get by *modus ponens* by distributing over the universal:

$$\vdash \forall x(\alpha \supset \beta) \supset \forall x((\forall x\alpha) \supset \beta).$$

Then distributing once more over the universal, we get:

$$\vdash \forall x(\alpha \supset \beta) \supset ((\forall x\alpha) \supset (\forall x\beta)),$$

which completes the proof. ∎

Thus the proof here again depends on the number of standard names in the sentence, unlike the case in ordinary logics.

2.13 Bibliographic notes

There are many excellent introductions to classical first-order logic among which [107] and [31]. The non-modal parts of [53] also offer a very clear and succinct presentation. Our use of standard names was inspired by a similar construct in a textbook by Smullyan [136]. There they were called "parameters," and this was also the name used in the first presentation of $\mathcal{L}$ in [80]. Standard names also owe much to the idea of unique identifiers (sometimes called object identifiers) in database management, for which, see [3], for example. The presentation of first-order logic here is non-standard in that it concentrates on the truth of sentences, not on the denotation of terms. This approach has been called a truth-value semantics by Leblanc [78]. Denotation and reference has been a major preoccupation of logicians, especially in modal contexts, attempting to capture the meaning of natural language noun phrases. See the references at the end of chapters 3 and 4 regarding terms and their denotations. On the substitutional interpretation of quantification, see Leblanc's paper above as well as [79], and for a more critical discussion, [63]. Logic, and first-order logic especially, has acquired a position of prominence in Knowledge Representation. For why all of first-order logic with equality is needed, see [108]; for why we should stop there, see [103].

2.14 Exercises

1. Complete the proof of Theorem 2.8.1.
2. Prove Theorem 2.8.4.
3. Let $*$ be a bijection from standard names to standard names as in Theorem 2.8.5. Suppose that w_1 and w_2 are world states that satisfy $(w_1[t])^* = w_2[t^*]$ for every primitive term t. Prove by induction that for every closed term t, $(w_1(t))^* = w_2(t^*)$.
4. Prove Theorem 2.8.5 for the case where α may contain function symbols. Hint: define w^* so that on primitive terms t, $w^*[t]$ equals $(w[t^*])^{*-1}$ (and therefore, that $(w^*[t])^* = w[t^*]$), and redo the induction using the result of the previous exercise.
5. Show that $\vdash \forall x \forall y (x = y \supset y = x)$.
6. Show that $\vdash \forall x(\alpha \wedge \beta) \equiv (\forall x \alpha) \wedge (\forall x \beta)$.
7. Show that $\vdash \exists x(t = x)$. Hint: Use the fact that $\forall x (^\#1 \neq x) \supset (^\#1 \neq {}^\#1)$ is an axiom. Then apply contra-positives, generalization, and specialization.
8. Show that $\vdash \exists x((\exists x \alpha) \supset \alpha)$. Hint: Show that
$$\vdash \forall x((\exists x \alpha) \wedge \neg \alpha) \supset (\exists x \alpha) \wedge (\forall x \neg \alpha),$$
then apply contrapositives.
9. Prove that the logic of $\mathcal{L}$ is sound.
10. When discussing rules of inference, some logic textbooks use the term "truth-preserving": a rule is truth-preserving if whenever the premises of the rule are true, the conclusion is also true. For example, *modus ponens* is truth-preserving. Show that our version of universal generalization is *not* truth-preserving, but is "validity-preserving". Explain why truth-preserving rules are not needed for a logic to be sound.
11. Prove that Corollary 2.11.3 fails when Γ is infinite. Hint: consider what is implied by the set of sentences $\{P(^\#1), P(^\#2), P(^\#3), \ldots\}$. Show a consistent set of sentences that is unsatisfiable.
12. In some logic textbooks, derivability is defined directly by something like: $\Gamma \vdash \alpha$ iff there is a sequence of wffs $\alpha_1, \alpha_2, \ldots, \alpha_k$, where $\alpha_k = \alpha$ and each α_i in the sequence is either an instance of a logical axiom, a member of Γ, or follows from earlier sentences in the sequence by one of the two rules of inference. (In this account, the theorems of the language would then be defined as the sentences derivable from the empty set of

premises.) Give an example of a Γ and an α where this definition and ours diverge. Comment on why our definition of derivability is more suitable for our semantics.

13. Extend the semantic description of the language to incorporate complex predicates:

    (a) Every predicate symbol P is a predicate.

    (b) If α is a wff then $\lambda(x_1, \ldots, x_k)\alpha$ is also a predicate (of arity k).

14. Extend the semantic description of the language to incorporate definite descriptions:

    If α is a wff then $\iota x.\alpha$ is a term.

    You will probably want to introduce a new special standard name $^\#0$ as the equivalence class for non-referring terms.

3 An Epistemic Logical Language

In this chapter, we introduce a new logical language called $\mathcal{KL}$ that goes beyond the first-order language considered in the previous chapter. Like $\mathcal{L}$, $\mathcal{KL}$ is intended as a language for communicating with a KB, but unlike $\mathcal{L}$, in $\mathcal{KL}$ we can talk of what is or is not *known*, in addition to what is or is not true in the world. We begin by considering why a simple first-order language like $\mathcal{L}$ is insufficient by itself. It turns out that it is precisely the *incomplete knowledge* expressible using $\mathcal{L}$ that compels us to go beyond $\mathcal{L}$. We briefly consider two other strategies for dealing with this incomplete knowledge, before settling on $\mathcal{KL}$ as the cleanest and most general approach. We then discuss, first informally, and then formally, the semantics of $\mathcal{KL}$.

3.1 Why not just use $\mathcal{L}$?

Given that we imagine a KB as representing knowledge about the world as expressed in a language like $\mathcal{L}$, why would we ever want to go beyond $\mathcal{L}$? To see the reason most clearly, we will put aside temporarily the idea of a functional interface and imagine that a KB consists simply of a finite set of sentences from $\mathcal{L}$. In our examples, we will mostly use a single two-place predicate *Teach*, where the sentence *Teach*(t_1, t_2) is intended to be true if the person referred to by t_1 teaches the person referred to by t_2 in some course. Instead of writing standard names like #17 for arguments, we will adopt the following convention: for this chapter, proper names starting with a "*t*" like *tina* or *tom* are to be understood not as constants, but as standard names which will be used as the *t*eacher argument; proper names starting with an "*s*" like *sara* or *sam* are to be understood as standard names which will be used as the *s*tudent argument. This is only for readability.

So, for example, we could have a KB consisting of the two sentences

{*Teach*(*ted*, *sue*), (*Teach*(*tina*, *sue*) ∨ *Teach*(*tara*, *sue*))}.

Note that this KB has incomplete knowledge in that it knows that one of Tina or Tara teaches Sue, but does not know which. We cannot simply ask the KB to produce a list of Sue's teachers, for instance, since it does not know who they all are. On the other hand, the system *should* know that Sue has a teacher other than Ted. In addition, it should realize that it does not know who this other teacher is. Consequently, we should be able to ask

Does Sue have a teacher who is not yet known to be her teacher?

and expect to get the answer *yes*. In other words, the system should realize that its list of Sue's teachers is currently incomplete. The reason we need to go beyond $\mathcal{L}$ is that there is no way to express this question as a sentence of $\mathcal{L}$.

3.2 Known vs. potential instances

Before going into ways of dealing with this issue, let us be clear about what we mean by saying that somebody is a known teacher. For a standard name n and a predicate P, we say that n is a <u>known instance</u> of P if the sentence $P(n)$ is known to be true; we say that n is a <u>potential instance</u> of P if the sentence $\neg P(n)$ is not known to be true. This can obviously be generalized to predicates with additional arguments or even to arbitrary open formulas.

The main point is that saying that somebody is a known teacher is not just talking about the way the world is, like saying that somebody is a teacher; it is a property of the KB in that it says that a certain sentence is known to be true. Thus we can distinguish between the following three sets of individuals: the known teachers, the actual teachers, and the potential teachers, where the first and last category depend on the state of the KB, and the middle category depends on the state of the world. For a KB whose knowledge is *accurate*, we would expect

Known instances $\subseteq$ Actual instances

Actual instances $\subseteq$ Potential instances.

For a KB whose knowledge was also *complete*, we would expect the reverse as well, so that all three sets would be the same.

Thus, the known and potential instances bound from below and above respectively the actual instances of a predicate. As more knowledge is acquired using $\mathcal{L}$, these bounds can become tighter. For example, the sentence

$$\forall x [Teacher(x) \supset (x = {}^{\#}1) \vee (x = {}^{\#}7) \vee (x = {}^{\#}9)]$$

serves to narrow the set of potential instances of the predicate to the three individuals named. It does not provide any new known instances, but rules out all but the three named. To tighten the bounds from below, the obvious way is to name an instance, as in $Teacher({}^{\#}1)$. But more generally, we can describe the set of instances using wffs like

$$\exists x [Teacher(x) \wedge x \neq {}^{\#}1] Teacher({}^{\#}3) \vee Teacher({}^{\#}7) Teacher(best\_friend({}^{\#}8)).$$

None of these directly result in more known instances, but they serve to augment what is known about the actual ones.

3.3 Three approaches to incomplete knowledge

We said above that we needed to go beyond $\mathcal{L}$ because it was impossible to express in $\mathcal{L}$ the question about whether there were any teachers that were not yet known teachers of Sue. But perhaps we were too quick in making that assessment. What about

$$\exists x Teach(x, sue) \wedge \neg Known\_teach(x, sue),$$

An Epistemic Logical Language

that is, why not use *Known_teach* as a predicate. In the above example, Ted would be both a teacher and a known teacher of Sue, but although one of Tina and Tara is a teacher of Sue, *neither* would be a known teacher.

The trouble with this approach concerns the relation between the two predicates *Teach* and *Known_teach*. The two predicates are clearly not independent. Observe that if we found out that Tom was a teacher of Sue, we would immediately want to conclude that Tom was a known teacher of Sue. Since this holds for any individual, it appears that the sentence

$$\forall x. Teach(x, sue) \supset Known\_teach(x, sue)$$

must be true. Unfortunately it is not, since we would then get that

$$Known\_teach(tina, sue) \lor Known\_teach(tara, sue),$$

which is false since neither is a known teacher.

In a nutshell, the reason we should not have a predicate *Known_teach* is that it is not a property of the world of teachers, but of the knowledge about that world. To find out if somebody is a known teacher, it is not sufficient to look carefully at the set of teachers in the world; it depends crucially on everything else that is known.

A second approach to dealing with this issue involves using $\mathcal{L}$ but with a *3-valued logic* instead of the current 2-valued one. Instead of sentences being merely true or false, we would allow them to take on the value *unknown*. For example, the sentence *Teach(ted, sue)* would be true, but the sentence *Teach(tina, sue)* would be unknown.

The problem with this approach is how to specify the semantics of $\mathcal{L}$. In particular, the *unknown* truth value does not seem to behave like the other two. For example, in the 2-valued logic $\mathcal{L}$, the truth value of a sentence $(\alpha \lor \beta)$ is a direct function of the truth values of α and β: if either is true then the disjunction is true, and otherwise it is false. But what would the truth table be for a 3-valued logic? Suppose α and β are both unknown; the only reasonable conclusion is that the disjunction $(\alpha \lor \beta)$ should be unknown as well. For example, if *Teach(tom, sam)* and *Teach(tom, sara)* are both unknown, then so must be (*Teach(tom, sam)* $\lor$ *Teach(tom, sara)*).

Unfortunately, this does not work. In the above example, the truth value for both *Teach(tina, sue)* and *Teach(tara, sue)* is unknown, yet their disjunction is clearly known to be true (since it is one of the sentences in the KB). As with *Known_teach* above, the problem is that we cannot assign an appropriate truth value to a sentence without taking into account the totality of what is known. Again, whether or not a sentence is considered unknown is not a property of the world of teachers, but of what is known about that world.

In summary, to talk about the known teachers, it appears that we need to be able to use sentences in two distinct ways: we need to be able to say that a sentence is true or false (in the world), and we need to be able to say that a sentence is known or unknown (by the KB). To handle the former, we use a sentence of $\mathcal{L}$ directly; this then precludes using it for the latter.

The third approach, and the one we will be using throughout, is to augment the language $\mathcal{L}$ so that for every sentence α, there is another sentence that can be read as "α is known." Then, instead of saying that α is known to be true, we would say that the sentence "α is known" is true; instead of saying α is not known to be true, we would say that "α is known" is false. By extending the language in this way, we only have to talk about which sentences are true or false as before, even though we care about which are known or unknown. It is this extension to the language that constitutes $\mathcal{KL}$.

3.4 The language $\mathcal{KL}$

Syntactically, the language $\mathcal{KL}$ is the same as $\mathcal{L}$ except that it has one extra logical symbol, ***K***, and one extra formation rule for wffs:

If α is a formula, then ***K***α is a formula too.

Informally, ***K***α should be read as "α is currently known to be true."

Before looking at the semantics of $\mathcal{KL}$, it is worth examining these sentences informally. We can distinguish between two main types of sentences in $\mathcal{KL}$. First, the *objective* sentences of $\mathcal{KL}$ are those that are also sentences of $\mathcal{L}$. These are sentences whose truth value depends only on the state of the world; they say nothing about what is or is not known. The second category of sentence are the *subjective* sentences, which are those where every function or predicate symbol appears within the scope of a ***K*** operator. These are sentences whose truth value depends only on what is known; they say nothing about the state of the world.[1] Of course, there are also mixed sentences that are neither purely subjective nor objective, as in

$$P(^\#1) \land \neg \boldsymbol{K} Q(^\#1).$$

The truth value here depends on both the state of the world and the epistemic state.

For example, the objective sentence $\neg\textit{Teach}(\textit{tina}, \textit{sue})$ is true or false depending on whether Tom teaches Sue; this fact may or may not be known. Similarly, the subjective sentence $\neg\boldsymbol{K}\textit{Teach}(\textit{tina}, \textit{sue})$ says that it is not known that Tina teaches Sue. This says nothing about the world of teachers, in that Tina may or may not actually teach Sue; it is purely an assertion about the KB. Finally a mixed sentence like

$$\textit{Teach}(\textit{tara}, \textit{sue}) \land \neg\boldsymbol{K}\textit{Teach}(\textit{tara}, \textit{sue})$$

---

[1] Recall that we do not assume that something known is necessarily true in the world.

talks both about the world state and the epistemic state: it says that Tara actually teaches Sue even though this is not currently known by the system.

It is worth noting that sentences of $\mathcal{L}$ that contain no predicate or function symbols like $\forall x.(x = x)$ are strictly speaking both objective and subjective, according to the above definition. As shown in Theorem 2.8.3, these special sentences do not depend on either the state of the world or on what is known, and so are either logically true (valid) or logically false (unsatisfiable).

One very important distinction that can be made by $\mathcal{KL}$ and that will come up repeatedly is that between the following two subjective sentences:

$K\exists x.Teach(x, sam)$ and $\exists x.KTeach(x, sam)$

The first says of a particular sentence of $\mathcal{L}$, that it is known to be true. The second sentence says that for some value of x, a certain sentence involving x is known to be true. The first says that it is known that Sam has a teacher; the second says that there is an x for which it is known that x teaches Sam, that is, Sam has a *known* teacher.

The difference between the two would show up, for example, when all that was known was the sentence $(Teach(tom, sam) \vee Teach(tara, sam))$. In this epistemic state, the first sentence would be true since it is known that somebody teaches Sam. But the second sentence would be false since nobody is known to teach Sam. The first sentence merely requires the existence of a teacher to be known, but the second requires the KB to know *who* the teacher is. So, for example, if what was known was $Teach(tom, sam)$, then both sentences would be true.

A final feature of $\mathcal{KL}$ worth noting is that a K operator may appear within the scope of other K operators. For example, the sentence $K\neg KTeach(tom, sam)$ says that it is known that Tom is not known to teach Sam. The knowledge that is expressed here is not objective since it uses a K operator. This type of subjective knowledge is usually called *meta-knowledge*. The most useful application of meta-knowledge is when the object of belief is neither objective nor subjective. Consider, for example, the sentence

$K[\exists x.Teach(x, sue) \wedge KTeach(x, sam)]$.

This sentence expresses knowledge that is both about the world and the epistemic state: what is known is that Sue has a teacher (world) who is among the known teachers of Sam (knowledge). This is a much stronger claim than

$K[\exists x.Teach(x, sue) \wedge Teach(x, sam)]$,

where what is known is that Sue has a teacher among the teachers of Sam, since the known teachers of Sam are usually a much smaller set than the teachers of Sam. One of the most powerful and useful features of $\mathcal{KL}$ is that it allows us to express this variety of meta-knowledge, knowledge about the relationship between the world and the epistemic state.

3.5 Possible worlds

In the previous section, we saw informally that the truth value of sentences of $\mathcal{KL}$ depended on both a world state and an epistemic state. Before defining the latter precisely in Section 3.8, it is worth spending some time reviewing the general idea that will be used, which is that of *possible worlds*.

The notion of a possible world goes back to the philosopher Leibniz in the 18th century, but its technical development is mainly due to Kripke beginning in 1959. Its application to knowledge is primarily due to Hintikka starting in 1962. The main idea is actually already implicit in the way we have treated the semantics of $\mathcal{L}$: although there is only one world (about which we are interested in making assertions, having knowledge and so on), there are many different *ways* the world can be, only one of which is the way it actually is. Each of these different ways is what is called a possible world. So to say there are two possible worlds is *not* to say that there are two realities (with two individuals corresponding to Socrates and so on), but that *the* world can be two different ways. A possible world is *actual* if that is the way the world really is.

To see why this notion is useful, consider the following two sentences:

1. If I put my hand into a fire, it will feel hot.
2. If I put my hand into a fire, it will feel cold.

Intuitively, we would like to claim that the first sentence is true and the second one is false. But why should that be? Let us assume (for the sake of argument) that I will never put my hand into a fire. In this case, the antecedent of the conditional is false, and so both sentences would be equally true. So although these sentences use future tenses, they cannot be understood as simple claims about the future.

The usual explanation for how we should understand these sentences and why the truth values are different is that we have to consider a possible world where I do put my hand into fire. In other words, we imagine a (so called counter-factual) possible world that is exactly like the actual one *except* that at some point, I put in my hand into a fire. The claim in the first sentence is that *in this possible world* it feels hot (correct), and in the second, that it feels cold (incorrect).

The difficulty in the previous example is being precise about what it means for a possible world to be exactly like reality except for a few changes. This is a partial description of a new way things could be, and it is often not clear exactly what is being described. For example, in the possible world where I put my hand into a fire, we may assume that the laws of physics continue to apply (otherwise they would not be laws). But clearly there is more to it than just a difference in hand motions. If I burn my hand, this will be a start of a chain of consequences with potentially far-reaching implications. Moreover, in a possible world where I am *willing* to put my hand into a fire, I obviously am very different from the way I really am. And what are the consequences of those differences? Would I live in the same city, have the same job, friends, family, and so on?

For our purposes, rather than describing possible worlds as being minimal changes to other possible worlds, we can describe them directly in terms of what sentences are true, treating any two possible worlds that satisfy the same sentences as equivalent. For the part of the world that is purely objective, we can think of a possible world as modeled by what we have called a *world state*: a specification of the values for every primitive expression.[2]

3.6 Objective knowledge in possible worlds

The relationship between knowledge and possible worlds is this: We imagine that what a knower cares about is the way the world is, that is, the possible world that is actual. At any given point, the knower will not have determined this in full detail, but perhaps some possibilities will have been ruled out. As more information is acquired, more and more possible worlds can be eliminated. Eventually, the knower may have eliminated all but a single possible world, which would then be taken to be *the* way things really are. But this final state of complete knowledge may never be achieved, and in general, incomplete knowledge will force the agent to deal with a set of possibilities.

An epistemic state, then, can be modeled by the set of possible worlds that have not been ruled out by the knower as being the actual one. For purely objective knowledge, we can think of an epistemic state as a set of world states. Consider Figure 3.1, for example.

This picture illustrates an epistemic state called e_5 where all but three possible worlds have been eliminated, w_1, w_2, and w_3. We assume that these three worlds assign a truth value to the primitive atoms as illustrated.[3] So we are imagining in this case that the knower has decided that the real world must be in one of the three world states illustrated.

What does the knower believe in this situation? The idea of the possible-world understanding of knowledge (due to Hintikka) is that what is known for sure is what would be true regardless of which possible world turns out to be the correct one. That is, we take a conservative view and say that what is known is exactly what is true in *all* the world states that make up the epistemic state. This has the effect of guaranteeing that as long as the real world is among these alternatives, what is known will be true in reality. In the figure, *Teach*(*ted*, *sue*) would be known, since it comes out true in all three worlds. The

---

2 The situation will be complicated by the fact that we also want to treat knowledge as part of a possible world; but for the moment, we limit ourselves to purely objective knowledge.
3 For concreteness, we may assume that they assign all other primitive atoms to false, and all primitive terms to the standard name #1. Nothing hinges on this assumption.

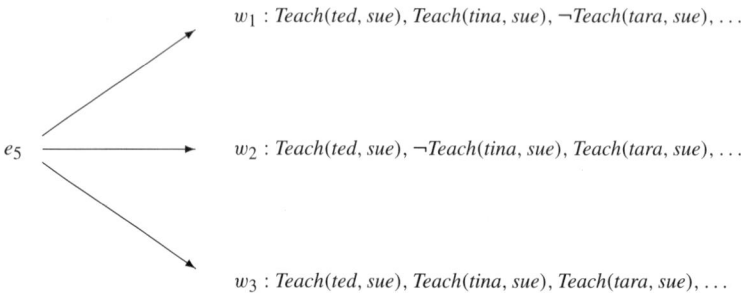

Figure 3.1: An epistemic state modeled as three world states

disjunction

$$(Teach(tina, sue) \lor Teach(tara, sue))$$

would also be known since at least one disjunct is true in each world state. On the other hand, neither *Teach(tina, sue)* nor *Teach(tara, sue)* is known since in either case, there is a world state where it comes out false.

With this possible-world understanding of knowledge, and unlike the 3-valued approach discussed earlier, we can see how two sentences can be unknown while their disjunction is known. Moreover, we can see what it would mean to have *complete* knowledge of the world: this corresponds to a case where the epistemic state can be modeled by a single world state. With complete knowledge, everything not known to be true is known to be false.

It need not be the case that this knowledge is *accurate*, however. To show whether or not the real world is among those in the epistemic state, we need to augment our diagrams to show which world state is actual. Thus, we will introduce a new label for a world state immediately beside the epistemic state as in Figure 3.2 as a way of indicating the real state of the world. In this figure, the knowledge of the world is indeed accurate, so everything known is true. When knowledge is both accurate and complete, the epistemic state would be modeled by the set consisting of just the real world state.

An Epistemic Logical Language

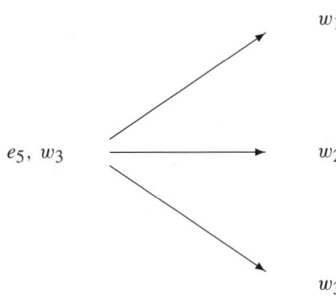

Figure 3.2: Including the actual world state

3.7 Meta-knowledge and some simplifications

So far, our possible-world account applies only to objective knowledge: each epistemic state is characterized by the (objective) world states that have not been ruled out. But as we said earlier, $\mathcal{KL}$ allows for the possibility of knowledge about the epistemic state as well.

To handle this, the simplest way is to imagine that we must deal with an enlarged notion of possible world, let us call it a *possible universe*, that consists of both a world state and an epistemic state. At any given point, only one world state and only one epistemic state will be actual. These correspond to the way the world really is and to what is really known (which is the left side of the diagrams). However, there are other possible ways the universe could be: other sentences could be true, and other sentences could be known (which is the right side of the diagrams).

To handle meta-knowledge, we assume that the agent is interested in determining both the real state of the world *and* the real state of knowledge. As before, at any point, only some of these possible universes will have been ruled out. Thus we now imagine an epistemic state as involving a set of possible universes, each consisting of both a world state and an epistemic state. Ignoring the circularity in this for a moment, the picture we have is more like that of Figure 3.3. The difference is that on the right of the diagram, instead of a list of world states, we have a list of pairs consisting of an epistemic state and a world state.

The interpretation of this diagram is this: we imagine the actual universe as being in epistemic state e_5 and world state w_3. Moreover, e_5 is an epistemic state that rules out all

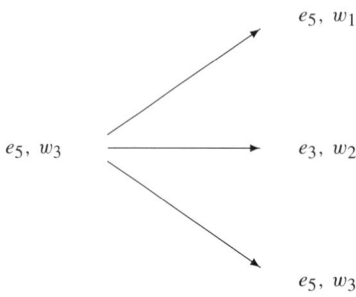

Figure 3.3: Knowledge about epistemic states

but three possibilities: in the first, the world is in w_1 and the knower is in state e_5; in the second, the world is in state w_2 and the knower in state e_3; in the final, the world is in state w_3 and the knower is in state e_5. We can already see that this knowledge is accurate since the real universe is one of the three possibilities. This does not yet complete the specification, however, since we have not yet described epistemic state e_3. It could, for example, introduce new world states and still further epistemic states requiring additional elaboration.

But without this extra complication, we can already see how meta-knowledge will be handled in simple cases. An objective sentence ϕ is considered known if it comes out true in each alternative possible universe. So for epistemic state e_5, this involves world states w_1, w_2, and w_3. Now a subjective sentence like $\boldsymbol{K}\psi$ is analogously considered known in epistemic state e_5 if it comes out true in each alternative possible universe. Thus, we would require $\boldsymbol{K}\psi$ to be true in both epistemic state e_5 and e_3, since as far as the knower is concerned, either could be the real epistemic state. So the principle is the same in both cases: to find out if an arbitrary sentence α is known, test if α is true in all of the alternative possible universes, by using the world state for the objective parts, or recursively, the epistemic state for the subjective part.

But what exactly is an epistemic state in this enlarged view? It cannot simply be a *set* of possible universes since that would require in the above example epistemic state e_5 to contain itself, among other things.

A general and very elegant way of handling this circularity was first proposed by Kripke. Instead of thinking of universes as pairs of world states and epistemic states, we can think of them as atomic indices and use two additional relations:

An Epistemic Logical Language

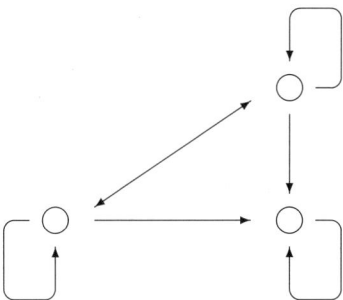

Figure 3.4: Epistemic state understood as an accessibility graph

- a relation that tells us for each index, the value of the primitive expressions at that index; this is the world state part;
- a relation (called the *accessibility* relation) that tells us for each index, what other indices are considered to be possible; this is the epistemic state part.

Ignoring the world state part, then, what we have is a set of points and a binary relation over them, which can be most clearly illustrated using a graph as in Figure 3.4. In this graph, we have three indices (corresponding to three possible universes). The arrows indicate the accessibility relation. For example, from the leftmost universe (which corresponds to $\langle e_5, w_3 \rangle$ from before), all three universes are possible; from the topmost universe (which corresponds to $\langle e_5, w_1 \rangle$ from before), again all three are possible, so the epistemic state is the same; in the bottom one, (which corresponds to $\langle e_3, w_2 \rangle$ from before), the epistemic state is different, and only a single universe is considered possible. Thus, to find out what is known with respect to any of these indices, we need only find out what is true at all the accessible indices.

While this mechanism of accessibility relations is powerful and elegant, it is too general for our needs. This is because we are willing to make a simplifying assumption about subjective knowledge:

Assumption *Purely subjective meta-knowledge is both complete and accurate.*

One way of thinking about this is that we assume that a knower, by introspection, can determine his/her true internal subjective state. To say that this subjective knowledge is complete is to say that there is no doubt in the knower about what is or is not known; to say that this subjective knowledge is accurate is to say that what the knower believes about

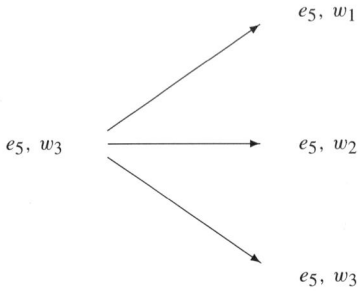

Figure 3.5: Complete and accurate meta-knowledge

this internal state is always correct.[4]

We can represent this diagrammatically as in Figure 3.5. The key observation about this picture is that while there is uncertainty about the real world state in e_5, there is no uncertainty about the epistemic state: there is a single epistemic state in the list of alternatives (this is the completeness part of the assumption) and it is indeed the correct epistemic state e_5 (this is the accuracy part).

While this simplification is not necessary and we could have continued under the more general setting of accessibility relations, it does allow us to avoid much of the complexity since there is always exactly one epistemic state accessible from another. Thus, we can treat an epistemic state simply as a *set* of world states as before, keeping the fixed epistemic part in the background. As we will see, this still leaves room for interesting meta-knowledge that is not purely subjective, and certainly sufficient richness to keep us occupied.

3.8 The semantics of $\mathcal{KL}$

With the preliminaries out of the way, we now turn to formally specifying the conditions under which sentences of $\mathcal{KL}$ are considered to be true or false (and thus, indirectly, known or unknown). As discussed above, a sentence is considered to be true or false in a universe

---

[4] Actually, the assumption as stated here is not quite right. As discussed in the next chapter, we assume that subjective meta-knowledge is accurate only as long as the epistemic state is consistent. But this is a detail that need not concern us yet.

An Epistemic Logical Language

consisting of both a world state and an epistemic state. World states are modeled exactly as they were in $\mathcal{L}$, namely as functions from primitive expressions to their values, and epistemic states are modeled as sets of world states. We write $e, w \models \alpha$ to indicate that α is true in world state w and epistemic state e. We proceed recursively as follows: all cases except sentences dominated by K operators are handled as in $\mathcal{L}$, and $K\alpha$ is true if α is true at every universe whose epistemic state is e and whose world state is a member of e. In detail, we have

1. $e, w \models P(t_1, \ldots, t_k)$ iff $w[P(n_1, \ldots n_k)] = 1$, where $n_i = w(t_i)$;
2. $e, w \models (t_1 = t_2)$ iff $w(t_1)$ is the same name as $w(t_2)$;
3. $e, w \models \neg\alpha$ iff it is not the case that $e, w \models \alpha$;
4. $e, w \models \alpha \vee \beta$ iff $e, w \models \alpha$ or $e, w \models \beta$;
5. $e, w \models \exists x.\alpha$ iff for some name n, $e, w \models \alpha_n^x$;
6. $e, w \models K\alpha$ iff for every $w' \in e$, $e, w' \models \alpha$.

Except for the last clause, the definition is the same as it was with $\mathcal{L}$, with an extra parameter e carried around. In the last clause, we consider the truth of α at a range of alternative world states w', but in keeping with our assumption, the epistemic state in all these alternatives remains fixed at e.

As before, we say that a set of sentences Γ is *satisfiable* just in case there is some world state w and an epistemic state e such that $e, w \models \alpha$ for every α in Γ, in which case we say that w and e satisfy Γ. We say that α is *valid* if it is satisfied by every world and epistemic state. Finally, we say that a sentence α is *logically implied* by a set of sentences Γ, which we write $\Gamma \models \alpha$, iff the set $\Gamma \cup \{\neg\alpha\}$ is unsatisfiable.

As a notational matter, we will use the Greek letters σ and τ to range only over subjective sentences, and ϕ and ψ to range only over objective sentences. We will often write $e \models \sigma$ and $w \models \phi$ for subjective and objective sentences respectively.

In a sense we are done. The rest of the book can be thought of as an exploration of the properties of this semantic definition.

3.9 Bibliographic notes

The original idea of possible worlds goes back to Leibniz (see [28]), although the first satisfactory mathematical treatment is due to Kripke [62]. For the use of possible worlds in interpreting counterfactual conditionals, see [89] and [138]. The first to apply Kripke's possible-world model to the formalization of knowledge was Hintikka [50]. Excellent general textbooks on modal logic and possible-world semantics are [53] and [14]. The modal system we are using in this book would be called weak-S5 in the terminology of [53] or

K45 in the terminology of [14]. Much of the research effort in the area of modal logic concerns variant modal systems for different applications. The book by Fagin, Halpern, Moses and Vardi [32] offers a modern treatment of modal logic for knowledge. The authors present and discuss in depth a number of variant modal systems, many of them considerably more complex than $\mathcal{KL}$, although they restrict their attention to purely propositional languages. See [53] to get a glimpse of why general first-order modal logic is so troublesome. Much of the difficulty is caused by wanting to allow different domains of discourse in different possible worlds, corresponding to the intuition that what does or does not exist may vary from world to world. But this intuition seems to be fraught with difficulty [118, 55]. The approach we take here is that properties of objects may indeed change from world to world, including perhaps having a physical presence of some sort, but that there is only one fixed universal set of objects (possibly without physical presence) to begin with. See [52] and the references therein for a discussion of such existence assumptions.

3.10 Exercises

1. Prove that if an epistemic state contains more than one world state, then the knowledge is incomplete (that is, some sentence is neither known to be true nor known to be false).
2. Show that the truth value of a sentence where no predicate or function symbol appears within the scope of a ***K*** does not depend on the epistemic state. Similarly, show that the truth value of a sentence where no predicate or function symbol appears outside the scope of a ***K*** does not depend on the world state.

4 Logical Properties of Knowledge

In this chapter, we will undertake an analysis of the logical properties of knowledge. Since it is the semantics of the language $\mathcal{KL}$ that determines for us what it means for something to be known, we will undertake this analysis by examining closely the logical properties of $\mathcal{KL}$ itself.

We begin by showing how knowledge and truth behave similarly and differently, for objective and subjective knowledge. We then do the same for knowledge and validity. Next, we consider the issue of known individuals and how this relates to knowledge of universals. Then, we show that we have circumscribed the basic characteristics of knowledge by building an axiom system for $\mathcal{KL}$ and proving it sound and complete. Finally, given the simplifying assumption we have made about meta-knowledge, we consider the question as to whether the language itself can be simplified by eliminating all expressions of meta-knowledge.

This chapter contains two non-trivial theorems: Theorem 4.5.1 and Theorem 4.6.2. We have included the proofs of these inline as we feel that it is important to master the mathematical techniques involved in the analysis of knowledge. These techniques are further developed in exercises at the end of the chapter. Theorems or lemmas that do not contain proofs should also be thought of as exercises, with only those of special interest listed explicitly at the end.

4.1 Knowledge and truth

There are many parallels between the notions of knowledge and truth as they appear in $\mathcal{KL}$. First, it is worth noting that they are distinct notions, that is, that something can be true and not known and vice versa:

Theorem 4.1.1: *There are sentences α such that*
 1. $\{\alpha \wedge \neg \boldsymbol{K}\alpha\}$ *is satisfiable;*
 2. $\{\neg\alpha \wedge \boldsymbol{K}\alpha\}$ *is satisfiable.*

Proof: Let α be any primitive sentence, and choose w and w' so that $w \models \alpha$ and $w' \models \neg\alpha$. Then the first sentence above is satisfied when the world state is w and the epistemic state is $\{w, w'\}$; the second sentence above is satisfied when the world state is w' and the epistemic state is $\{w\}$. ∎

Thus, we have that for some α, $\not\models (\alpha \supset K\alpha)$ and $\not\models (K\alpha \supset \alpha)$. In particular, there is no requirement that what is known be true, and so, as we have noted before, the term "belief" may be more appropriate.

There is, however, a class of sentences for which there is a correspondence between knowledge and truth: the subjective sentences. In all cases, the truth of a subjective sentence implies that the sentence is also known to be true. This is shown by induction over subjective sentences, where the base case consists of equalities between standard names, sentences of the form $K\alpha$, and their negations. So we begin with:

Lemma 4.1.2: $\models (n_1 = n_2) \supset K(n_1 = n_2)$ and $\models (n_1 \neq n_2) \supset K(n_1 \neq n_2)$.

Proof: The truth value of $(n_1 = n_2)$ does not depend on the choice of world or epistemic state, but only on whether the two names are the same or not. Consequently, if it is true (or false), it will be true (or false) for every world state in the epistemic state. Thus it will be known to be true. ∎

Lemma 4.1.3: $\models K\alpha \supset KK\alpha$ and $\models \neg K\alpha \supset K\neg K\alpha$.

Proof: The truth value of $K\alpha$ does not depend on the world state in question but only on the epistemic state. Thus, if it is true (or false) for some pair w and e, it will also be true for w' and e for every $w' \in e$. Thus it will be known to be true. ∎

Combining these two and using induction, we obtain:

Theorem 4.1.4: *For any subjective sentence σ,* $\models (\sigma \supset K\sigma)$.

Thus any true sentence about what is or is not known is known to be true. Another way of putting this is to say that the concept of knowledge we are dealing with is such that there is never any reason to tell a knowledge base a fact about itself; it already has *complete* knowledge about such matters.

But what about the converse? Is it the case that any subjective sentence that is known to be true is indeed true? In the previous chapter, we assumed informally that meta-knowledge was *accurate*, that is, that subjective meta-knowledge was indeed true. Here, however, we must be more precise and admit that this is not exactly right.

The complication involves the epistemic state where the knowledge is *inconsistent*. This state is modeled by the empty set of world states, meaning that all possibilities have been ruled out. In this state *all* sentences are known, and so it cannot be the case that the

Logical Properties of Knowledge

known subjective sentences are all true. For example, for any primitive sentence ϕ, the sentence $\neg K\phi$ will be false, but believed to be true. In other words, the set of sentences $\{K\neg K\phi, K\phi\}$ is satisfiable.

However, for consistent epistemic states, we do have that any subjective meta-knowledge is true. Since a consistent epistemic state can be thought of as one where at least one sentence is not believed, we get:

Theorem 4.1.5: *For any α and any subjective σ,* $\models (\neg K\alpha \supset (K\sigma \supset \sigma))$.

The proof of this is similar to the proof of the previous theorem.

4.2 Knowledge and validity

There is, as it turns out, also a parallel between knowledge and validity (or provability). Consider the case of an objective sentence: an objective sentence is known iff it comes out true in all world states contained in the epistemic state; it is valid, on the other hand, if it comes out true in *all* states. Thus for objective sentences, validity coincides with a special case of knowledge, namely where the epistemic state contains all world states. But more importantly, generalizing from this observation, we have:

Theorem 4.2.1: *If* $\models \alpha$ *then* $\models K\alpha$.

Thus, as we saw above, although true sentences need not be known in general, *valid* sentences will always be known. Thus, it is important to distinguish in $\mathcal{KL}$ between the following two claims:

1. $\models (\alpha \supset K\alpha)$;
2. If $\models \alpha$ then $\models K\alpha$.

Only the second one is correct.

We will consider the converse of the above theorem in a moment. First, let us consider another property of validity: closure under rules of inference. Knowledge has this property as well, for exactly the same reason:

Theorem 4.2.2: $\models K\alpha \wedge K(\alpha \supset \beta) \supset K\beta$.

This says that knowledge is closed under *modus ponens*. For universal generalization, we will take the infinitary version of the rule and say that if all (infinitely many) instances of a formula are known, then so is the universal version of the formula:

Theorem 4.2.3: $\models \forall x K\alpha \supset K\forall x\alpha$.

The finitary version of this principle will be discussed later.

Taken together, the previous three theorems allow us to consider knowledge as some sort of provability operation. That is, recalling the derivability operation in $\mathcal{L}$, one possible characterization is the following:

1. All of the axioms are derivable.
2. If α and $(\alpha \supset \beta)$ are derivable, then so is β.
3. If α_n^x is derivable for every name n, then so is $\forall x\alpha$.

The three theorems above allow us to replace "derivable" in the above by "known to be true", as well as moving from $\mathcal{L}$ to $\mathcal{KL}$.

But a better way to look at these theorems is that they are special cases of the principle that knowledge is closed under logical consequence:

Theorem 4.2.4: *Let e be any epistemic state. Let Γ be any set of sentences such that for every $\gamma \in \Gamma$, $e \models K\gamma$. Further, suppose that $\Gamma \models \alpha$. Then $e \models K\alpha$.*

In other words, if (some of) what is known logically implies α, then α must be known as well. This property is sometimes referred to as *logical omniscience* since it says that a knowledge base is always "aware" of all of the logical consequences of what it knows. It is as if the knowledge base were always able to instantly do logical reasoning over everything that it knows, and therefore believe that these sentences must be true also.[1] We will collectively refer to these four theorems by the name of the most general one, Theorem 4.2.4.

Finally, let us return to the converse of Theorem 4.2.1. Is it the case, that sentences that are always believed must be logically valid? The answer is no. For example, although we do not require a knowledge base to be accurate, it turns out that a knowledge base will always *believe* that it is accurate, in the following sense:

Theorem 4.2.5: $\models K(K\alpha \supset \alpha)$

Proof: There are two cases. Suppose $K\alpha$ is true for some e. Then $K(K\alpha \supset \alpha)$ must be true by Theorem 4.2.4, since $\{\alpha\}$ logically implies $(\beta \supset \alpha)$ for any β. On the other hand, if $\neg K\alpha$ is true, then $K\neg K\alpha$ must be true by Theorem 4.1.4, and so $K(K\alpha \supset \alpha)$ is again

---

[1] This is not a very realistic assumption for real agents (with finite resources), but it is one that makes the characterization of knowledge much simpler. We will take up the topic of relaxing this assumption later in Chapters 12 and 13.

Logical Properties of Knowledge

true by Theorem 4.2.4. So for any e, $K(K\alpha \supset \alpha)$ must be true, and the theorem follows. ∎

Thus a knowledge base always believes that if it believes something, then it must be true, even though this principle is not valid. We can think of this as saying that a knowledge base is always *confident* of what it knows. It does not allow for the possibility that something it believes is false. Or put another way, this says that all the knowledge base has to go on are its beliefs, all of which are equally reliable, and so it has no *reason* (that is, no belief) to doubt anything that it believes.

If "know" is not the appropriate term here (since what is known in our sense is not required to be true), neither is "believe," at least in the sense of allowing for the fact that you might be mistaken. Perhaps a more accurate term would be "is absolutely sure of" which would not require truth, but would preclude doubts.[2]

4.3 Known individuals

As we observed before, there is more to understanding what is known than simply identifying the sentences known to be true. We also want to be able to distinguish between epistemic states where it is known that Sue has a teacher and epistemic states where the identity of that teacher is known. We can adapt the standard philosophical jargon of *de dicto* and *de re* knowledge to describe the situation. If Sue is known to have a teacher *de dicto*, this means that the sentence saying that Sue has a teacher (that is, the simple existential) is known to be true. If Sue is known to have a teacher *de re*, this means that there is some individual who is known to be a teacher of Sue. Formally, the two conditions would be expressed as follows:

de dicto: $K\exists x\, Teach(x, sue)$ is true at e iff for every $w \in e$, there is a name n such that $w \models Teach(n, sue)$.

de re: $\exists x\, K Teach(x, sue)$ is true at e iff there is a name n such that for every $w \in e$, $w \models Teach(n, sue)$.

Much of the richness and complexity of $\mathcal{KL}$ is a direct result of this difference, which semantically, reduces to an order of quantifiers.

The first property to observe about this distinction is that *de re* knowledge implies *de dicto*, but not vice versa.

Theorem 4.3.1: $\models (\exists x K\alpha \supset K\exists x\alpha)$ but $\not\models (K\exists x\alpha \supset \exists x K\alpha)$

---

[2] This is still not right because of logical omniscience. A more accurate gloss for $K\alpha$ would be "α follows logically from what the system is absolutely sure of," although even this is not quite right because of introspection and meta-knowledge.

This follows immediately from the semantic characterization given above. What does imply *de re* knowledge, however, is knowledge involving particular standard names:

Theorem 4.3.2: $\models KP(n) \supset \exists x KP(x)$.

But this does not work if there is uncertainty about the identity of the individual:

Theorem 4.3.3: *Let n_1 and n_2 be distinct. Then*

$$\not\models K(P(n_1) \lor P(n_2)) \supset \exists x KP(x).$$

Proof: Let w be such that $w \models P(n)$ iff $n = n_1$, and w' be such that $w \models P(n)$ iff $n = n_2$. Let $e = \{w, w'\}$. Then e satisfies the left hand side, but not the right hand side. ∎

Of course, by Theorem 4.2.4, we would still have *de dicto* knowledge here, since an existential follows from the disjunction.

Similarly, there will be no *de re* knowledge if the uncertainty about the identity of the individual is because of a non-standard name:

Theorem 4.3.4: *Suppose t is a primitive term. Then,* $\not\models (KP(t) \supset \exists x KP(x))$.

Proof: Let w and w' be as above except that $w[t] = n_1$ and $w'[t] = n_2$. Then again e satisfies the left hand side, but not the right hand side. ∎

In all of the above cases, what is at issue is the existence of a *fixed* individual for each of the world states making up the epistemic state. The language $\mathcal{L}$ allows us to express properties of individuals without fixing the identity of the individual in question. The fact that $(KP(t) \supset \exists x KP(x))$ is not valid in $\mathcal{KL}$ means that the sentence

$$\forall x. KP(x) \supset KP(t)$$

is not valid either. This, in turn, is an instance of $(\forall x \alpha \supset \alpha_t^x)$ which in general cannot be valid either. This is very different from the situation in the logic of $\mathcal{L}$, since this last sentence was in fact an *axiom* in the proof theory of $\mathcal{L}$, often called the *axiom of specialization*.

So why exactly does the axiom of specialization fail in $\mathcal{KL}$? Consider this example. Suppose the following sentence is true:

$$\forall x. K\text{Teacher}(x) \lor K\neg\text{Teacher}(x).$$

That is, for every individual x, either x is known to be a teacher or known not to be a teacher. In other words, the knowledge base has an opinion about every individual, one way or another. This can happen, for instance, when it is known that $^\#3$ is the one and only teacher. Now consider the sentence

$$K\textit{Teacher}(t) \lor K\neg\textit{Teacher}(t),$$

where t is the term *best_friend(mother(sam))*. Clearly the first sentence can be true without the second one being true, when the identity of the best friend of the mother of Sam is unknown. In particular, even if we know that $^\#3$ is the one and only teacher, it does not follow that we know whether or not t is a teacher, since we may not know if $(t = {^\#3})$ is true or not.

This is an example of the failure of the axiom of specialization and it is due to the fact that the identity of non-standard terms may be unknown. In fact, we can show that the axiom of specialization *does* hold provided that the replacement for the variable never places a function symbol within the scope of a K. First we need this lemma:

Lemma 4.3.5: *Let t be any term, w any world state, e any epistemic state. Suppose that α is a formula with at most a single free variable x, and that none of the free occurrences of x in α are within the scope of a K. Assume that $w(t)$ is n. Then*

$$e, w \models \alpha_n^x \text{ iff } e, w \models \alpha_t^x.$$

Proof: The proof is by induction on the length of α. If α is an atomic sentence, this clearly holds. If α is an equality, it also holds by induction on the structure of the terms in the equality. For negations and conjunctions, the lemma holds by induction. If α is of the form $\forall y\beta$, then there are two cases: if y is the same as x, then the lemma holds trivially since x does not appear free in α; if y is distinct from x, then $e, w \models (\forall y\beta)_n^x$ iff $e, w \models \forall y(\beta_n^x)$ iff $e, w \models (\beta_n^x)_{n'}^y$ for every n', iff $e, w \models (\beta_{n'}^y)_n^x$ for every n', iff (by induction) $e, w \models (\beta_{n'}^y)_t^x$ for every n', iff $e, w \models (\beta_t^x)_{n'}^y$ for every n', iff $e, w \models \forall y(\beta_t^x)$ iff $e, w \models (\forall y\beta)_t^x$. Finally, if α is of the form $K\beta$, then the lemma holds trivially since x does not occur freely within the scope of a K. ∎

With this lemma, we then get:

Theorem 4.3.6: $\models (\forall x\alpha \supset \alpha_t^x)$, *provided that when replacing x by t, no function symbol is introduced within the scope of a K.*

Proof: If the term t is a standard name, this follows immediately from the semantics of universal quantification. Otherwise, it must be the case that no free occurrence of x in α is within the scope of a K. Suppose that for some w and e, $e, w \models \forall x\alpha$. Then we must have, $e, w \models \alpha_n^x$ for every n. In particular, consider the n which is $w(t)$. By the above lemma, we must have $e, w \models \alpha_t^x$. ∎

Thus, specialization holds as long as the term t is a standard name or x does not appear free in α within the scope of a **K**. Other sound restrictions of the axiom of specialization are also considered in the exercises.

As a simple consequence of this theorem we get that equals can be substituted for equals provided that this does not involve placing a non-standard term within the scope of a **K**:

Theorem 4.3.7: *Suppose that t and t' are terms, and that α has at most a single free variable x. Further assume that neither α_t^x nor $\alpha_{t'}^x$ introduces a function symbol within the scope of a **K**. Then,*

$$\models (t = t' \supset \alpha_t^x \equiv \alpha_{t'}^x).$$

Proof: First observe that for any pair of names, n and n', we have

$$\models (n = n' \supset \alpha_n^x \equiv \alpha_{n'}^n).$$

since equality between names holds iff the names are the same. Thus we get that

$$\models \forall y \forall y'(y = y' \supset \alpha_y^x \equiv \alpha_{y'}^x).$$

The theorem then follows immediately from Theorem 4.3.6. ∎

4.4 An axiom system for $\mathcal{KL}$

Having examined various properties of $\mathcal{KL}$, we are now ready to turn to an axiomatization of the logic. As in the case of $\mathcal{L}$, the principal reason for doing this is to provide a simple but very different picture of the valid sentences, phrased in term of an initial set (the axioms) and closure conditions (the rules of inference). As it turns out, the rules of inference we need for $\mathcal{KL}$ are just those of $\mathcal{L}$: *modus ponens* and universal generalization. So we need only list the axioms, which are in Figure 4.1. The definition of theorem, derivability, consistency, and inconsistency, are the same as they were in $\mathcal{L}$. Again we use the notation $\vdash \alpha$ to say that α is a theorem, and $\Gamma \vdash \alpha$ to say that Γ derives α.

As in $\mathcal{L}$, it is fairly easy to establish soundness:

Theorem 4.4.1: *If a sentence of $\mathcal{KL}$ is derivable, then it is valid.*

The proof of soundness, as usual, is by induction on the length of the derivation. The basis of the induction proof depends on the validity of the above axioms, all of which

1. Axioms of $\mathcal{L}$:
   All instances of the axioms of $\mathcal{L}$, but with the proviso on the axiom of specialization that no function symbol is introduced within the scope of a $\boldsymbol{K}$.
2. Knowledge of axioms:
   $\boldsymbol{K}\alpha$, where α is an instance of an axiom of $\mathcal{L}$, again with the proviso on specialization;
3. Knowledge closed under modus ponens:
   $\boldsymbol{K}(\alpha \supset \beta) \supset (\boldsymbol{K}\alpha \supset \boldsymbol{K}\beta)$;
4. Knowledge closed under universal generalization:
   $\forall x \boldsymbol{K}\alpha \supset \boldsymbol{K}\forall x\alpha$;
5. Complete knowledge of subjective truths:
   $(\sigma \supset \boldsymbol{K}\sigma)$, where σ is subjective.

Figure 4.1: Axioms for $\mathcal{KL}$

were established in the previous section. Note, for example, that without the proviso on the axiom of specialization, the system would be *unsound*, in that it would be possible to derive non-valid sentences. The rule of *modus ponens* clearly preserves validity. So all we need to establish is that the *finitary* version of universal generalization works, that is, that if α_n^x is valid for every name in α and at least one not in α, then $\forall x\alpha$ is valid too. In the case of $\mathcal{L}$, this was Corollary 2.8.8 of Theorem 2.8.5; here we need a similar corollary for a generalized theorem:

Theorem 4.4.2: *Let * be a bijection from names to names. For any term t or wff α, let t^* or α^* be the result of simultaneously replacing in t or α every name by its mapping under *. Then α is valid iff α^* is valid.*

Proof: Similar to the proof of Theorem 2.8.5. We need to define w^* as before, and here we also need to define e^* as $\{w^* \mid w \in e\}$. ∎

Corollary 4.4.3: *Let α have a single free variable x and let n be a standard name not appearing in α. Let $n_1, \ldots, n_k$ be the all the standard names appearing in α. If α_n^x is valid and all the $\alpha_{n_i}^x$ are valid, then so is $\forall x\alpha$.*

Proof: The same argument as that of Corollary 2.8.8. ∎

Establishing the *completeness* of this axiom system, that is, that the above axioms are sufficient to generate *all* the valid sentences is much more challenging, as it requires examining the properties of $\mathcal{KL}$ in fine detail. Before doing so, it is worth looking at some simple derivations.

To help in the presentation of derivations, we will use the following property of the proof theory of $\mathcal{KL}$:

Theorem 4.4.4: *If $\vdash \alpha$, then $\vdash \mathbf{K}\alpha$.*

Proof: The proof is by induction on the length of the derivation of α. First suppose that α is an axiom of $\mathcal{KL}$. There are two cases: if it is an instance of an axiom of $\mathcal{L}$ (with proviso), then $\mathbf{K}\alpha$ is also an axiom of $\mathcal{KL}$ and so is derivable; all other axioms of $\mathcal{KL}$ are subjective, and so if σ is any other axiom, $(\sigma \supset \mathbf{K}\sigma)$ is also an axiom, in which case $\mathbf{K}\sigma$ is again derivable by *modus ponens*. If, on the other hand, α follows from some earlier derivable β and $(\beta \supset \alpha)$, then by induction, $\mathbf{K}\beta$ and $\mathbf{K}(\beta \supset \alpha)$ must also be derivable, and so $\mathbf{K}\alpha$ is derivable by *modus ponens* and the axiom of closure of knowledge under *modus ponens*. Finally, if α is of the form $\forall x \beta$, and is derivable from $\beta_{n_1}^x$ to $\beta_{n_k}^x$ by universal generalization, then by induction, $\mathbf{K}\beta_{n_1}^x$ to $\mathbf{K}\beta_{n_k}^x$ are also derivable, in which case, $\forall x \mathbf{K}\beta$ follows from universal generalization, and then $\mathbf{K}\forall x\beta$, by *modus ponens* and the axiom of closure of knowledge under universal generalization. ∎

So although as an axiom we only state that the axioms of $\mathcal{L}$ are known, this theorem shows that any derivable sentence of $\mathcal{KL}$ is also known. This means that the proof theory behaves as if there was an additional rule of inference (sometimes called *knowledge generalization*) which says: from α, infer $\mathbf{K}\alpha$.

We will use the same notation for derivations as we did with $\mathcal{L}$, with two additions. First, a justification marked $\mathcal{L}$ means that the current line is derivable from the previous one (and perhaps earlier ones too), as a theorem of $\mathcal{L}$ alone. In other words, we will not go into any detail involving sub-derivations that use only the axioms of $\mathcal{L}$. Second, a justification **KG** means that the current line is formed by putting a $\mathbf{K}$ in front of an earlier line (appealing to the above theorem).

Figure 4.2 contains a derivation of $\mathbf{K}(\mathbf{K}\alpha \supset \alpha)$, whose validity was proven directly in the previous section. The last step is derived using properties of $\mathcal{L}$, from steps 3 and 8: if $(\beta \supset \gamma)$ and $(\neg \beta \supset \gamma)$ are both derivable, then so is γ.

As a second example, consider the fact that subjective knowledge must be accurate when knowledge is consistent, that is, that

$$\vdash \neg \mathbf{K}\alpha \supset (\mathbf{K}\sigma \supset \sigma).$$

This is easily shown (and left as an exercise) given the following:

Theorem 4.4.5: $\vdash \neg \mathbf{K}\alpha \supset (\mathbf{K}\beta \supset \neg \mathbf{K}\neg\beta)$.

Proof: See Figure 4.3. ∎

Logical Properties of Knowledge

1.	$\alpha \supset (K\alpha \supset \alpha)$	$\mathcal{L}$
2.	$K(\alpha \supset (K\alpha \supset \alpha))$	KG
3.	$K\alpha \supset K(K\alpha \supset \alpha)$	MP
4.	$\neg K\alpha \supset (K\alpha \supset \alpha)$	$\mathcal{L}$
5.	$K(\neg K\alpha \supset (K\alpha \supset \alpha))$	KG
6.	$K\neg K\alpha \supset K(K\alpha \supset \alpha)$	MP
7.	$\neg K\alpha \supset K\neg K\alpha$	Ax
8.	$\neg K\alpha \supset K(K\alpha \supset \alpha)$	MP
9.	$K(K\alpha \supset \alpha)$	$\mathcal{L}$

Figure 4.2: A derivation in $\mathcal{KL}$

1.	$\beta \supset (\neg\beta \supset \alpha)$	$\mathcal{L}$
2.	$K(\beta \supset (\neg\beta \supset \alpha))$	KG
3.	$K\beta \supset (K\neg\beta \supset K\alpha)$	MP
4.	$\neg K\alpha \supset (K\beta \supset \neg K\neg\beta)$	$\mathcal{L}$

Figure 4.3: Derivation of not knowing a wff and its negation

So, as long as there is a single sentence α that is not known, there is no sentence β such that both it and its negation are known. The proviso is necessary since it is possible for every sentence to be known.

As a third example, we derive $(\forall x \forall y (x = y) \supset K(x = y))$ in Figure 4.4. This shows that all equalities (among standard names) are known. A similar derivation can be used to show that inequalities are also known. Note that the first two lines here use the fact that equality sentences that do not use function symbols are subjective, and hence known. So this does *not* permit the derivation of $((t_1 = t_2) \supset K(t_1 = t_2))$ for non-standard terms t_i, since the axiom of specialization with proviso cannot put a function symbol within the scope of a K.

4.5 A Completeness proof

We now turn our attention to the completeness of the axiomatization of $\mathcal{KL}$:

Theorem 4.5.1: *If a sentence of $\mathcal{KL}$ is valid, then it is derivable.*

1.	$(^\#1 = {}^\#1) \supset K(^\#1 = {}^\#1)$	Ax
2.	$(^\#1 = {}^\#2) \supset K(^\#1 = {}^\#2)$	Ax
3.	$\forall y (^\#1 = y) \supset K(^\#1 = y)$	UG
4.	$\forall x \forall y (x = y) \supset K(x = y)$	UG

Figure 4.4: Derivation of knowing equality of names

As we discussed in the case of $\mathcal{L}$, we can prove this by showing that any consistent sentence is satisfiable. This is sufficient since if a sentence α is valid, $\neg \alpha$ is unsatisfiable, and so, $\neg \alpha$ would be inconsistent, and therefore, $\neg \neg \alpha$ derivable, and consequently α derivable as well, since (as it is easy to show) $(\neg \neg \alpha \supset \alpha)$ is derivable in $\mathcal{KL}$ (and $\mathcal{L}$).

To show that every consistent sentence is satisfiable, we proceed in two stages: first we show that every finite consistent set of sentences can be extended to what we will call a T-set; then we show that every T-set can be satisfied.

To define a T-set, we start with the notion of a maximally consistent set: a set of sentences is <u>maximally consistent</u> iff it is consistent and any proper superset is inconsistent. The following are properties of maximally consistent sets that derive directly from properties of ordinary first-order logic and we will not prove here:

Lemma 4.5.2:

1. *Every consistent set can be extended to a maximally consistent set. That is, for every consistent Γ, there is a maximally consistent Γ' such that $\Gamma \subseteq \Gamma'$.*
2. *If Γ is maximally consistent then $\neg \alpha \in \Gamma$ iff $\alpha \notin \Gamma$.*
3. *If Γ is maximally consistent then $(\alpha \wedge \beta) \in \Gamma$ iff $\alpha \in \Gamma$ and $\beta \in \Gamma$.*
4. *If Γ is maximally consistent and $\Gamma \vdash \alpha$, then $\alpha \in \Gamma$.*

Note that we are *not* claiming for a maximally consistent Γ that if $\exists x \alpha \in \Gamma$, that for some n, $\alpha_n^x \in \Gamma$. In fact, the set

$$\{\exists x P(x), \neg P(^\#1), \neg P(^\#2), \ldots\}$$

is consistent (and can be extended to a maximally consistent set) since there is no contradiction for any finite subset of the set. Similarly, the infinite set

$$\{\neg K \forall x P(x), KP(^\#1), KP(^\#2), \ldots\}$$

is consistent as is

$$\{(t \neq {}^\#1), (t \neq {}^\#2), (t \neq {}^\#3), \ldots\}.$$

Logical Properties of Knowledge

However, none of these sets are satisfiable, and so a T-set must go beyond maximal consistency if it is to be satisfiable.

To handle these cases, we first define the concept of an *E-form*: the E-forms with respect to a variable x are the least set of wffs with only x free such that:

1. If α is a formula with just a free variable x, then $(\exists x \alpha \supset \alpha)$ is an E-form with respect to x;
2. If t is a closed term, then $(t = x)$ is an E-form with respect to x;
3. If α is any sentence and β is an E-form with respect to x, then so is the formula $(\neg K\alpha \supset \neg K(\beta \supset \alpha))$.

A substitution α_n^x, where α is an E-form with respect to x and n is any standard name is called an *instance* of the E-form. Now we define a *T-set* to be a maximally consistent set that contains at least one instance of every E-form.

Notice how T-sets rule out cases like the above: for example, if a T-set contains $\exists x P(x)$, then for some n it must also contain $(\exists x P(x) \supset P(n))$, and thus, it must also contain $P(n)$ by Lemma 4.5.2. That is, if a T-set contains $\exists x \alpha$, it must also contain a witness to this existential.

4.5.1 Part 1

In this subsection, we prove

Theorem 4.5.3: *Every finite consistent set can be extended to a T-set.*

We begin by showing that the existential closure of every E-form is derivable in $\mathcal{KL}$:

Lemma 4.5.4: *If α is an E-form wrt x, then $\vdash \exists x \alpha$.*

Proof: The proof is by induction on the composition of the E-form. If the E-form is one of the two base cases, then the lemma holds by virtue of properties of $\mathcal{L}$, and were given as exercises in Chapter 2. Otherwise assume that α is any sentence, β is an E-form wrt x, and that by induction, $\exists x \beta$ is derivable. It is easy to show that $(\forall x K\gamma \supset K\exists x \gamma)$ is derivable for any γ, and in particular,

$$\forall x K(\beta \supset \alpha) \supset K\exists x(\beta \supset \alpha)$$

is derivable. By properties of $\mathcal{L}$,

$$K\exists x(\beta \supset \alpha) \supset K((\exists x \beta) \supset \alpha)$$

is also derivable, since x does not appear free in α. However, $\vdash K\exists x\beta$ since $\vdash \exists x\beta$, by Theorem 4.4.4. So putting all these together, we get that

$$\vdash \forall x K(\beta \supset \alpha) \supset K\alpha,$$

and so

$$\vdash \neg K\alpha \supset \exists x \neg K(\beta \supset \alpha).$$

Finally, using properties of $\mathcal{L}$, we can move the existential to the front, and get that

$$\vdash \exists x. \neg K\alpha \supset \neg K(\beta \supset \alpha),$$

since x does not occur free in α. ∎

Next we have that

Lemma 4.5.5: *If Γ is a finite consistent set of sentences, and β is an E-form, then there is a name n such that $\Gamma \cup \{\beta_n^x\}$ is consistent.*

Proof: Suppose not. Let γ be the conjunction of sentences in Γ. Then,

$$\vdash (\beta_n^x \supset \neg \gamma),$$

for every n, and so,

$$\vdash \forall x (\beta \supset \neg \gamma).$$

But, by the previous lemma, $\vdash \exists x\beta$. Therefore, since x does not occur free in γ, we get $\vdash \neg \gamma$, contradicting the consistency of Γ. ∎

We can now prove the theorem of this subsection:

Proof: Suppose all sentences of $\mathcal{KL}$ are enumerated by $\alpha_1, \alpha_2, \alpha_3, \ldots$ and that all E-forms are enumerated by $\beta_1, \beta_2, \beta_3, \ldots$. We will first define a sequence of finite sets of sentences, $\Gamma_0, \Gamma_1, \Gamma_2, \ldots$ and show that each must be consistent. First, let Γ_0 be the given finite consistent set of sentences. Now assume that Γ_i has been defined and is consistent. Let α be α_i if $\Gamma_i \cup \{\alpha_i\}$ is consistent, and $\neg \alpha_i$, otherwise; then $\Gamma_i \cup \{\alpha\}$ is consistent. Let β be the instance of the E-form β_i that is consistent with $\Gamma_i \cup \{\alpha\}$, promised by the previous lemma, and let Γ_{i+1} be $\Gamma_i \cup \{\alpha, \beta\}$. This set must be consistent also. Finally, let Γ be the union of all Γ_i. This set is maximally consistent and also contains an instance of every E-form. ∎

This shows that any finite consistent set can be extended to a T-set.

4.5.2 Part 2

What remains to be shown is this:

Logical Properties of Knowledge

Theorem 4.5.6: *Every T-set can be satisfied.*

In fact, what we will show is that a T-set completely determines an epistemic and world state, that is, that for each T-set Γ, there is an e and w such that
$$\Gamma = \{\gamma \mid e, w \models \gamma\}.$$
As we will show, from the fact that a T-set is maximally consistent, negations and conjunctions are handled properly; because a T-set also has an instance of every E-form, existentials are also accounted for. So all we really need to do is establish that the K operator is treated properly.

In what follows, we let $\Re(\Gamma)$ be the set of all T-sets Γ' such that for every α, if $K\alpha \in \Gamma$, then $\alpha \in \Gamma'$.

First we define a mapping from T-sets to world states: for any Γ that is a T-set, w_Γ is the world state w such that for any primitive, $w[\phi] = 1$ iff $\phi \in \Gamma$, and $w[t] = n$ iff $(t = n) \in \Gamma$. From the properties of T-sets, we get by induction:

Lemma 4.5.7: *If Γ is a T-set, then for any objective ϕ, $\phi \in \Gamma$ iff $w_\Gamma \models \phi$.*

Note that this handles the completeness for the objective part of $\mathcal{L}$. To handle the rest of $\mathcal{KL}$, first we show:

Lemma 4.5.8: *If Γ is a T-set, and $\neg K\alpha \in \Gamma$, then for some $\Gamma' \in \Re(\Gamma)$, $\neg \alpha \in \Gamma$.*

Proof: We will show that there must be a Γ' that has these properties: it contains $\neg\alpha$, it contains an instance of every E-form, it contains every γ such that $K\gamma \in \Gamma$, and it is consistent. It is then immediate that this Γ' can be extended to a maximally consistent set, which is therefore a member of $\Re(\Gamma)$.

First observe, that since $\neg K\alpha \in \Gamma$, and Γ is a T-set, every E-form β has an instance β_n^x such that $\neg K(\beta_n^x \supset \alpha) \in \Gamma$. Let $\beta_1, \beta_2, \ldots$ and so on, be all such instances, and let Γ' be this set, together with $\neg\alpha$ and all γ such that $K\gamma \in \Gamma$. What remains is to show that this Γ' is consistent.

Observe that for any finite subset $\{\beta_1, \ldots, \beta_k\}$ of the instances of E-forms in Γ', we have the sentence
$$\neg K(\beta_1 \supset (\beta_2 \supset \ldots \supset \alpha)\ldots)$$
in Γ. This is by induction on the size of the subset, using the closure property of T-sets. Now suppose to the contrary that Γ' is inconsistent. Then for some γ such that $K\gamma \in \Gamma$, and some finite set of β_i as above, we have that
$$\vdash (\gamma \supset (\beta_1 \supset (\beta_2 \supset \ldots \supset \alpha)\ldots)),$$

and thus,

$$\vdash (K\gamma \supset K(\beta_1 \supset (\beta_2 \supset \ldots \supset \alpha)\ldots)).$$

Since $K\gamma \in \Gamma$, this would imply that

$$K(\beta_1 \supset (\beta_2 \supset \ldots \supset \alpha)\ldots)$$

was in Γ too, contradicting the consistency of Γ itself. ∎

Next, we associate an epistemic state to each T-set as follows: for any Γ that is a T-set, let e_Γ be defined as

$$\{w_{\Gamma'} \mid \Gamma' \in \Re(\Gamma)\}.$$

Now we are ready to prove the theorem of this subsection. Specifically, we prove that if Γ is a T-set, and $w = w_\Gamma$ and $e = e_\Gamma$ then $\alpha \in \Gamma$ iff $e, w \models \alpha$, and consequently Γ is satisfied by this w and e.

Proof: The proof is by induction on the length of α. If α is a atomic sentence or an equality, the theorem holds by Lemma 4.5.7. The theorem holds for negations and conjunctions by induction, and for existential quantification by induction and the properties of T-sets. Finally, consider the case of $K\alpha$. If $K\alpha \in \Gamma$, then for every every $\Gamma' \in \Re(\Gamma)$, $\alpha \in \Gamma'$; thus, by induction, for every $w' \in e$, $e, w' \models \alpha$, and so $e \models K\alpha$. Conversely, if $K\alpha \notin \Gamma$, then $\neg K\alpha \in \Gamma$, and so by Lemma 4.5.8, for some $\Gamma' \in \Re(\Gamma)$, $\alpha \notin \Gamma'$; thus, by induction, for some $w' \in e$, $e, w' \models \neg \alpha$, and so $e \models \neg K\alpha$. ∎

This ends the completeness proof.

4.5.3 Variant systems

It is appropriate at this stage to consider some simple variants of $\mathcal{KL}$ and see how the axiomatization and the proof of completeness would have to be modified to deal with them.

Perhaps the simplest variant would be one where knowledge was required to be *consistent*. Currently, the set consisting of all sentences of the form $K\alpha$ is satisfiable, but only by the epistemic state that is the empty set of world states. Semantically, to make sure that knowledge is consistent, we need only require an epistemic state to be non-empty. To capture this property axiomatically, we simply change the axiom stating that subjective knowledge is complete to one stating that it is both complete and accurate:

subjective knowledge is complete and accurate: $(\sigma \equiv K\sigma)$.

This is clearly sound for the new semantics. To show that it is complete, we need only show that for any T-set Γ, the set $\Re(\Gamma)$ is non-empty. To see why it must be, observe that Γ cannot contain every $K\gamma$ since it would have to contain $K\neg K\alpha$, and then $\neg K\alpha$ by the above axiom, violating consistency. Thus it must contain, $\neg K\gamma$ for some γ, and then by Lemma 4.5.8, $\Re(\Gamma)$ is non-empty.

Another simple variant of $\mathcal{KL}$ would require all knowledge to be *accurate*: we only look at pairs $\langle e, w \rangle$ such that $w \in e$. In terms of the proof theory, this can be handled by adding the axiom

knowledge is accurate: $(K\alpha \supset \alpha)$

To see why this is sufficient, we need only show that for any T-set Γ we have that $w_\Gamma \in e_\Gamma$. In fact, we get a stronger property, namely that $\Gamma \in \Re(\Gamma)$, as a direct consequence of the above axiom and T-set closure.

A final variant that is less plausible in general is that the knowledge is *complete*. As we said earlier, this is modeled semantically by having an epistemic state consisting of a single world state. In the proof theory, we would add the following axiom

knowledge is complete: $(\neg K\alpha \supset K\neg\alpha.)$

To see why this is sufficient, we need only show that for any T-set Γ, $\Re(\Gamma)$ consists of a singleton set, which we leave as an exercise.

4.6 Reducibility

Having looked at a proof theory for $\mathcal{KL}$ and a few simple variants, we now turn our attention to a very different logical property of knowledge having to do with meta-knowledge. This will also constitute the first time there is clear difference between the quantifier-free subset of $\mathcal{KL}$ and the full version. We will use the term *propositional* subset to mean the subset of $\mathcal{KL}$ without quantifiers.

If we look at the semantic and axiomatic accounts of $\mathcal{KL}$, it might appear that the simplifying assumption made regarding meta-knowledge makes the whole notion dispensable. Assuming that knowledge is consistent, for example, we have that both $KK\alpha \equiv K\alpha$ and $K\neg K\alpha \equiv \neg K\alpha$ are valid. This means that we can always reduce strings of K operators and negations down to at most a *single* K operator. So the question we wish to address in this section is this: can we generalize this idea and eliminate *all* nesting of K operators, without losing expressive power? In other words, is it possible to take any sentence and find an equivalent one where the K operator only dominates objective sentences? If we can, this would mean that meta-knowledge offers essentially nothing over objective knowledge.

As it turns out, the answer to the question is *yes* for the propositional subset of $\mathcal{KL}$, and *no* for the full language. First the propositional case:

Theorem 4.6.1: *For any α in the propositional part of $\mathcal{KL}$, there is an α', where α' has no* *nesting of **K** operators and* $\models (\alpha \equiv \alpha')$.

Proof: The proof is based on induction on the depth of nesting of **K** operators in α, but here we will merely present it in outline. Assume that α has a subformula $\boldsymbol{K\beta}$ where β uses **K** operators. First, we observe that because of the usual properties of the propositional part of $\mathcal{L}$, we can put β into a logically equivalent conjunctive normal form (CNF) β' where β' is a conjunction of disjunctions of extended literals, where an extended literal is a (possibly negated) sentence that is either objective or the form $\boldsymbol{K\gamma}$. So we have that $\boldsymbol{K\beta}$ is equivalent to $\boldsymbol{K\beta'}$. Next, we use the fact that both of these are valid:

$$\boldsymbol{K}(\beta_1 \wedge \beta_2) \equiv (\boldsymbol{K}\beta_1 \wedge \boldsymbol{K}\beta_2)$$

and

$$\boldsymbol{K}(\phi \vee \boldsymbol{K}\gamma_1 \vee \neg \boldsymbol{K}\gamma_2) \equiv (\boldsymbol{K}\phi \vee \boldsymbol{K}\gamma_1 \vee \neg \boldsymbol{K}\gamma_2).$$

(See the exercises.) By applying this repeatedly, we get that $\boldsymbol{K\beta'}$ is equivalent to $\boldsymbol{K\beta''}$ where the latter has reduced the level of nesting by one. By applying this repeatedly to α, we eliminate all nesting of **K** operators. ∎

So this theorem shows that talk of meta-knowledge in the propositional part of $\mathcal{KL}$ can be replaced by completely equivalent talk about objective knowledge. If there is anything *new* to meta-knowledge, it is in its interaction with the quantifiers. Note that the above proof fails for the full version of $\mathcal{KL}$ because there is no way to distribute the **K** over some version of a CNF: although we can move **K** operators inwards when we have something like $\boldsymbol{K}\forall x\alpha$, we cannot do so for sentences like $\boldsymbol{K}\exists x\alpha$.

In the full quantified version of $\mathcal{KL}$, we will show that there are indeed sentences with nested **K** operators that cannot be rephrased in terms of objective knowledge. In particular, the sentence

$$\boldsymbol{K}\exists x[P(x) \wedge \neg \boldsymbol{K}P(x)],$$

which we will call λ, cannot be so reduced:

Theorem 4.6.2: *For any α, if* $\models (\alpha \equiv \lambda)$, *then α has nested **K** operators.*

The proof proceeds by constructing two epistemic states e_1 and e_2 that agree on all objective knowledge but disagree on λ. This is sufficient, since any proposed α without nested **K** operators cannot be equivalent to λ, since although e_1 and e_2 will assign the same truth value to α, they will assign different truth values to λ.

We construct e_1 and e_2 as follows. Let Ω be some infinite set of standard names containing $^\#1$ whose complement is also infinite.[3] Let Φ be the set of objective sentences consisting of $\{(t = {^\#1})\}$ for every primitive term t, $\{\neg\phi\}$ for every primitive sentence ϕ whose predicate letter is not P, and finally $\{P(n)\}$ for every $n \in \Omega$. Let e_1 be $\{w \mid w \models \Phi\}$. Let $\overline{w}$ be the (unique) element of e_1 such that for every $n \notin \Omega$, $w \models \neg P(n)$. Finally, let e_2 be $e_1 \setminus \{\overline{w}\}$. This gives us that for all $w \in e_1$ and for all $n \in \Omega$, $w \models P(n)$; worlds in e_2 have this property also, and in addition, because $\overline{w} \notin e_2$, they also each have $w \models P(n)$ for some $n \notin \Omega$.

The first thing to observe is that e_1 and e_2 disagree on λ. Specifically,

$$e_1 \models \neg K \exists x[P(x) \wedge \neg KP(x)]$$

but

$$e_2 \models K \exists x[P(x) \wedge \neg KP(x)].$$

This is because they do agree on the *known* instances of P, in that

$$e_1 \models KP(n) \text{ iff } e_2 \models KP(n) \text{ iff } n \in \Omega,$$

and so the presence of $\overline{w}$ in e_1 makes λ false, since $\overline{w}$ satisfies $P(n)$ only for the known instances of P.

To complete the proof, we need only show that e_1 and e_2 agree on all objective knowledge and hence on all sentences without nested K operators. Showing that if $e_1 \models K\phi$ then $e_2 \models K\phi$ is trivial, since $e_2 \subset e_1$; the converse will take some work.

First we prove the following:

Lemma 4.6.3: *Let n_1 and n_2 be distinct names that are not members of Ω. Then for any ϕ, $\overline{w} \models \phi$ iff $\overline{w} \models \phi^*$, where ϕ^* is ϕ with n_1 and n_2 interchanged.*

Proof: By induction on the structure of ϕ, given that $\overline{w}$ is defined in a way that treats the two names exactly the same. ∎

Next, assume that the names that are not in Ω are enumerated as $m_1, m_2, m_3, \ldots$, and define a corresponding sequence of worlds $w_1, w_2, w_3, \ldots$, as follows: w_i is the unique element of e_2 such that $w_i \models P(n)$ iff $n \in \Omega$ or $n = m_i$. Then we get the following:

Lemma 4.6.4: *Let n be any name in Ω other than $^\#1$. Then for any ϕ, $w_i \models \phi$ iff $w_i \models \phi^*$, where ϕ^* is the result of interchanging n and m_i in ϕ.*

Proof: By induction on the structure of ϕ, given that w_i is defined in a way that treats the two names exactly the same. (The proviso regarding $^\#1$ is necessary because we have

---

3 An example is: $\{^\#1, {^\#3}, {^\#5}, \ldots\}$.

made it be the value of all primitive terms.) ∎

Using these two lemmas, we obtain:

Lemma 4.6.5: *Suppose $m_i \notin \Omega$. Then for any ϕ which does not mention m_i, $\overline{w} \models \phi$ iff $w_i \models \phi$.*

Proof: The proof is by induction on ϕ. The only tricky case is for existentials.
In one direction, if $\overline{w} \models \exists x \phi$, then $\overline{w} \models \phi_n^x$ for some n. There are two cases: if $n \neq m_i$, we get that $w_i \models \phi_n^x$ by induction, and so $w_i \models \exists x \phi$; however, if $n = m_i$, then by the first lemma above, if we let n' be some distinct name that does not appear in ϕ, and such that $n' \notin \Omega$, we get that $\overline{w} \models \phi_{n'}^x$. Then by induction, $w_i \models \phi_{n'}^x$, and so $w_i \models \exists x \phi$.
In the other direction, if $w_i \models \exists x \phi$, then $w_i \models \phi_n^x$ for some n. Again, there are two cases: if $n \neq m_i$, we get that $\overline{w} \models \exists x \phi$ as above; however, if $n = m_i$, then by the second lemma above, if we choose any n' not mentioned in ϕ such that $n' \in \Omega$ and $n' \neq {}^{\#}1$, we get that $w_i \models \phi_{n'}^x$. Then by induction, we get that $\overline{w} \models \phi_{n'}^x$, and so $\overline{w} \models \exists x \phi$. ∎

Now we can finish the proof of the theorem. If $e_1 \models \neg K\phi$, then for some $w \in e_1$ we have $w \models \neg \phi$. If $w \in e_2$, we are done; otherwise, $w = \overline{w}$, and so choose some $m_i \notin \Omega$ that does not appear in ϕ. By the lemma above, $w_i \models \neg \phi$, where $w_i \in e_2$. Either way, for some $w \in e_2$ we have $w \models \neg \phi$, and so $e_2 \models \neg K\phi$.

In the end, what this theorem shows is that the knowledge expressed by λ cannot be expressed in terms of objective knowledge, even allowing that subjective knowledge is complete and accurate. What the sentence λ expresses is that the KB knows that it has incomplete knowledge about P: there is an individual with property P not currently known to have that property. The above theorem shows that this is a form of knowledge that goes beyond mere objective knowledge.

This completes our purely logical analysis of $\mathcal{KL}$. In the chapters to follow, we will apply $\mathcal{KL}$ to the ask of interacting with a knowledge base.

4.7 Bibliographic notes

The properties of $\mathcal{KL}$ were first presented in [80], and then in [82]. Many of the properties discussed here will come up again in later chapters. Logical omniscience was first discussed by Hintikka [50] and, because it appears to have direct bearing on computational issues, has received considerable attention since then. See Chapters 12 and 13 and the references there for a more thorough discussion of this issue and a model of knowledge

Logical Properties of Knowledge

without logical omniscience. Other properties of the propositional subset of $\mathcal{KL}$ and numerous variants can be found in [32]. Turning to the quantificational aspects, the *de dicto / de re* distinction is a major one in quantified modal logics. See [53] for an introduction to the issue. A more philosophical discussion can be found in [94]. The philosopher Quine, among others, has maintained that quantifying into a modal context is fundamentally incoherent [118], although his arguments (concerning, for instance, confusion of identity) require *de re* belief without using standard names. In this context, our standard names are often called "rigid designators" in that that they designate the same individual in every possible world. Note that we use these as logical names [129], that is, as terms in our logical language, without claiming them to be anything like proper names found in natural languages. See [95, 64] for a discussion of these issues. The Henkin-style completeness proof presented here (including the use of E-forms) is adapted from [53].

4.8 Exercises

1. State whether or not each of the following properties of knowledge holds in general, and if not, whether it holds when the sentence α is subjective and when knowledge is consistent:
   (a) $\models (\alpha \supset K\alpha)$;
   (b) $\models (K\alpha \supset \alpha)$;
   (c) if $\models \alpha$ then $\models K\alpha$;
   (d) if $\models K\alpha$ then $\models \alpha$;
   (e) $\models K(\alpha \supset K\alpha)$;
   (f) $\models K(K\alpha \supset \alpha)$.

2. Show that $\models K\alpha \wedge K(\alpha \supset \beta) \supset K\beta$ and $\models \forall x K\alpha \supset K\forall x\alpha$.

3. Divide the set of subjective sentences into three categories: positive, negative, and mixed. Show for the positive case, we have that $(\sigma \equiv K\sigma)$, even when the knowledge is inconsistent.

4. Use Theorem 4.2.5 to show that a knowledge base will always believe that either it does not believe α or it does not believe $\neg\alpha$. Thus, a knowledge base always believes it is consistent.

5. Show that $\vdash K\alpha \supset K(\alpha \vee \beta)$.

6. Use Theorem 4.4.5 to show $\vdash \neg K\alpha \supset (K\sigma \supset \sigma)$.

7. Show that $\{(t_1 = t_2), K(t_1 \neq t_2)\}$ is satisfiable.

8. Show that Theorem 4.3.7 is false without the proviso on introducing a function symbol

within the scope of a K.

9. Show a derivation of a non-valid sentence that could happen if there were no proviso on the axiom of specialization.

10. Show that the axiom of specialization is valid without proviso when the value of the term t being substituted is correctly known. That is, show that
$$\models \exists y(y = t) \wedge K(y = t)) \supset (\forall x \alpha \supset \alpha_t^x).$$
Show that this does not hold when the value of the term t is known but need not be correct. That is, show that
$$\not\models \exists y K(y = t) \supset (\forall x \alpha \supset \alpha_t^x).$$

11. Show that $(\neg K\alpha \supset K\neg\alpha)$ is sufficient to characterize complete knowledge. That is, show that the given axiomatization of $\mathcal{KL}$ with this axiom added is complete for a semantics where an epistemic state is required to be a singleton set.

12. Show that $\models [K(\alpha \vee \sigma) \equiv K\alpha \vee \sigma]$, when σ is subjective. This is a generalization of the property used in the proof of Theorem 4.6.1.

13. Show that the axiom $\forall x K\alpha \supset K\forall x\alpha$ can be replaced by $K\forall x(K\alpha \supset \alpha)$, in the sense that one is derivable from the other, given the other axioms.

5 The TELL and ASK Operations

In the previous chapter, we examined the properties of the language $\mathcal{KL}$ in detail to develop a clear understanding of when sentences in this language were true or false. In this chapter, we will use the language as a way of communicating with a knowledge base or KB. We will use $\mathcal{KL}$ both to find out what is known and to provide new knowledge. We begin by defining these interaction operations, and examining a few immediate properties. We then illustrate the use of these operations on a larger example KB, emphasizing the power of $\mathcal{KL}$ as an interaction language. In the next chapter, we will examine some of the deeper properties and implications of the definitions presented here.

5.1 Overview

After our somewhat lengthy excursion into the logical properties of $\mathcal{KL}$, it is perhaps worthwhile to briefly review where we stand. What we have, so far, is a logical language $\mathcal{KL}$ together with a precise specification of what it means for sentences in this language to be true or false, and what it means for sentences to be known or unknown, as a function of a given world and epistemic state. A world state here is modeled as a function from primitive expressions to their values, and an epistemic state is modeled as a set of world states.

What we intend to do with this language is use it as a way of interacting with the KB of a knowledge-based system. Roughly, we will find out if something is known by the KB by asking it a question formulated as a sentence in $\mathcal{KL}$. Similarly, we will make something known to the KB by telling it that some sentence of $\mathcal{KL}$ is true. Thus, we envision for now two operations to be performed on a KB:

1. **ASK**$[\alpha, e] \in \{yes, no\}$
   In an epistemic state e, we determine if α is known, by using an **ASK** operation. The result we expect is a simple answer, *yes* or *no*.
2. **TELL**$[\alpha, e] = e'$
   In an epistemic state e, we add information to the KB by performing a **TELL** operation. The result is a new epistemic state, e'.

Note that while the first argument to these operations is a symbolic structure (that is, a sentence of $\mathcal{KL}$), the second argument is an epistemic state, not some symbolic representation of one. For now at least, our approach to these operations will not depend on how what is known is actually represented in a KB.

If we think of a KB as an abstract data type, accessible only in terms of the operations it provides, we need one more operation to give us an initial epistemic state:

3. **INITIAL**[ ] = e_0
   The epistemic state before any **TELL** operations.

The idea is that we imagine the full lifetime of a KB as proceeding through a number of states $e_0, e_1, e_2, \ldots$ in sequence, where e_0 is as above, and for every $i > 0$, there are sentences α_i, such that $e_i = $ **TELL**$[\alpha_i, e_{i-1}]$. In any such state, we can use **ASK** to determine what is known.

These three operations together constitute a full functional interface to a KR system as described in Chapter 1. The user of the system has access to the KB only through this fairly narrow interface. The KR system builder's job is to implement these three operations somehow using whatever representational means are appropriate.

In the rest of this chapter, we examine the three operations in more detail.

5.2 The ASK operation

The purpose of the **ASK** operation is clear: ultimately, we want to find out from a knowledge base if a sentence α is true or not. The complication is that a KB will not always be able to answer this, since all it has to go on is what it knows. There are, in fact, four possible outcomes:

1. it might believe that α is true;
2. it might believe that α is false;
3. it might not know whether α is true or false;
4. it might be inconsistent, and believe both that α is true and that it is false.

This is not a result of anything like a multi-valued logic, but simply the result of the different possible epistemic states that can arise. In effect, for various α, each of the following sets of sentences are satisfiable (corresponding to the cases above):

1. $\{\boldsymbol{K}\alpha, \neg\boldsymbol{K}\neg\alpha\}$;
2. $\{\neg\boldsymbol{K}\alpha, \boldsymbol{K}\neg\alpha\}$;
3. $\{\neg\boldsymbol{K}\alpha, \neg\boldsymbol{K}\neg\alpha\}$;
4. $\{\boldsymbol{K}\alpha, \boldsymbol{K}\neg\alpha\}$;

At best then, **ASK** can do no more than tell us which of the four mutually exclusive alternatives holds for the current epistemic state.

There is, however, a slightly simpler interaction we can use. Instead of asking whether or not α is *true*, we could ask the KB whether or not α is *known*. In the first and fourth cases above, the KB would answer affirmatively, and in the second and third cases, it would answer negatively. If we then ask if $\neg\alpha$ is known, in the second and fourth case, it would answer affirmatively, and in the first and third, negatively. The net effect of this shift is

The TELL and ASK Operations

that we can still determine which of the four cases holds if we so desire, but by asking two questions: one for α and one for $\neg\alpha$.

Since this decoupling of the positive and negative cases leads to a simpler definition for **ASK**, we will use this convention. The translation to the other form of question is straightforward. With this in mind, we define <u>ASK</u> as follows:

Definition 5.2.1: For any sentence α of $\mathcal{KL}$ and any epistemic state e,

$$\text{ASK}[\alpha, e] = \begin{cases} yes & \text{if } e \models K\alpha \\ no & \text{otherwise} \end{cases}$$

So at a given epistemic state e, **ASK** returns *yes* not if α is true (which would require a world state), but if $K\alpha$ is true at e. Thus the semantics of $\mathcal{KL}$, which tells us what it means for a sentence like $K\alpha$ to be true, does all the work in the definition.

5.3 The initial epistemic state: e_0

As discussed in Chapter 3 and will become clearer when we look at **TELL**, an epistemic state is a set of world states that is progressively narrowed as information is acquired. In finding out that ϕ is true, we eliminate from the epistemic state the world states where ϕ is false. So the smaller the set of world states, the more complete the knowledge.

It is worth considering the properties of the most uninformed epistemic state, which we will call <u>e_0</u>. This state consists of *all* world states, in that nothing is known that could eliminate any possible world. Of course, all of the valid sentences will be known in e_0. Furthermore, these are the only objective sentences that are known:

Theorem 5.3.1: *If ϕ is objective, then* $\text{ASK}[\phi, e_0] = yes$ *iff* $\models \phi$.

So nothing is known about the world in e_0, in that any objective sentence that is false in some world state is not known to be true.

What other knowledge does e_0 have? Like any other epistemic state, it will know about its own subjective state. So for example, if $\neg\phi$ is satisfiable, then $e_0 \models \neg K\phi$, and so $e_0 \models K\neg K\phi$, which means that $\text{ASK}[\neg K\phi, e_0]$ is *yes*.

So like every other epistemic state, e_0 knows all the valid sentences as well as all the true subjective ones. Somewhat surprisingly, perhaps, still more is known:

Theorem 5.3.2: *If α is $[\exists x P(x) \supset \exists x. P(x) \wedge \neg K P(x)]$, then* $\text{ASK}[\alpha, e_0] = yes$.

Proof: First observe that $e_0 \models \neg KP(n)$, for every n. Thus, for every $w \in e$, if $w \models \exists x P(x)$, then $e_0, w \models \exists x[P(x) \land \neg KP(x)]$. From this it follows that $e_0 \models \boldsymbol{K\alpha}$. ∎

Note that this α is neither a valid sentence nor a true subjective one. What α says is that if there is a P, then it must be an unknown P. Intuitively, this sentence is known in e_0 because although it is not known in e_0 whether or not there are any instances of P, it is known that there no *known* instances, and consequently, any instance of P must be an unknown one.

So although e_0 has no purely objective knowledge, it does have non-trivial knowledge, something perhaps not immediately obvious given the logical analysis of $\mathcal{KL}$ in the previous chapter. In fact, we have:

Theorem 5.3.3: *The set $\{\alpha \mid \textbf{ASK}[\alpha, e_0] = yes\}$ is not recursively enumerable.*

Proof: It is a property of ordinary first-order logic (and consequently, of $\mathcal{L}$) that the set of all objective satisfiable sentences is not recursively enumerable. But observe that $\textbf{ASK}[\neg\boldsymbol{K}\neg\phi, e_0] = yes$ iff ϕ is satisfiable. ∎

This shows the considerable power assumed under our concept of knowledge, and is yet another reason for wanting to look at more limited versions in Chapters 12 and 13.

5.4 The monotonicity of knowledge

As described above, acquiring knowledge means moving from an epistemic state modeled by a set of world states to a subset of those world states. It follows that objective knowledge is preserved by this acquisition of knowledge:

Theorem 5.4.1: *If ϕ is objective and $\textbf{ASK}[\phi, e] = yes$, then for any epistemic state $e' \subseteq e$, $\textbf{ASK}[\phi, e'] = yes$.*

So as information is acquired, anything objective that was believed will continue to be believed. We refer to this property of knowledge as *objective monotonicity*.[1]

However, this monotonicity does *not* hold in general for arbitrary sentences of $\mathcal{KL}$. That is to say, there is no guarantee that once a sentence is believed it will continue to be believed as further information is acquired:

---

1 The idea of acquiring knowledge to *revise* existing objective knowledge is much more complex and is discussed in the bibliographic notes.

The TELL and ASK Operations

Theorem 5.4.2: *Knowledge is nonmonotonic: there is a sentence α and an epistemic state e such that $\text{ASK}[\alpha, e] = yes$, but $\text{ASK}[\alpha, e'] = no$, for some $e' \subset e$.*

Proof: Let α be the sentence $\neg KP(n)$, for some standard name n. Since $e_0 \models \alpha$, we have that $\text{ASK}[\alpha, e_0] = yes$. In other words, the initial epistemic state e_0 knows that it does not believe that $P(n)$. Now suppose we eliminate all world states where $P(n)$ is false: let e' be the set of all worlds states w such that $w \models P(n)$. Clearly, $e' \models KP(n)$, and in particular, $e' \models \neg K\alpha$, and so $\text{ASK}[\alpha, e'] = no$. ∎

So a knowledge base may know that it does not believe some sentence, but if it later comes to believe that sentence, it needs to revise its original subjective belief. Thus, a knowledge base will change its mind about certain things even though it is acquiring objective knowledge purely monotonically.

5.5 The TELL operation

It is the **TELL** operation that defines how knowledge is acquired. What we are after with **TELL** is perhaps the simplest form of information acquisition, where there is no revision of objective knowledge, no retraction, and no forgetting. We assert using **TELL** that a sentence is true; if this assertion is inconsistent with what is already known, we simply move into the inconsistent epistemic state, and leave it at that.

After some objective ϕ has been asserted, we would expect that in the resulting state, ϕ should be believed, as should anything previously known about the world. Moreover, we want to be in a state where this is *all* that is known. That is, we should not get new objective beliefs that are independent of ϕ and what was known before. Consequently, on being told ϕ in some state e, we want the *largest* (that is the least informed) epistemic state where ϕ is believed and which is a subset of e. When ϕ is objective, there is a unique epistemic state that has these properties:

$$\text{TELL}[\phi, e] = e \cap \{w \mid w \models \phi.\}$$

Thus, as we have been assuming, asserting ϕ simply means eliminating the world states where ϕ is false.

Consider now a non-objective sentence, α, which is $\exists x[P(x) \wedge \neg KP(x)]$. This sentence says that there is an unknown P, and would be true (at w and e) if the instances of P in w contained some individual that was not among the known instances in e. What could it mean for a KB to be *told* that this sentence is true? There are at least two possible interpretations:

- one might interpret this is as asserting that there is a *forever* unknown P. In other

words, there is a *P* that is not known to be *P* and will continue to be unknown, no matter what further knowledge is acquired.
- one might interpret this as asserting that there is a *currently* unknown *P*. In other words, there is a *P* that is not among the known *P*, in the current epistemic state. But as this state changes, this *P* may become known.

Clearly both readings have their uses: if we want to state properties about this and future states of knowledge, the first reading is more appropriate; if we want to state properties about the world, but in terms of the current epistemic state, the second reading is preferred. Our focus here is clearly on objective knowledge, and so we will take the second reading, although the issue of constraining (future) states of knowledge more generally will reappear in Chapter 8.

The impact of this decision is that on being told a sentence that contains K operators, we can interpret references to what is known to be about the current epistemic state. We do not have to worry about the epistemic state that will result after the update, or any other future state. Consequently, we can do what we did for objective assertions, except we now use the current epistemic state to deal with the non-objective parts. Thus, we define *TELL* as follows:

Definition 5.5.1: For any sentence α of $\mathcal{KL}$ and any epistemic state e,

$\textbf{TELL}[\alpha, e] = e \cap \{w \mid e, w \models \alpha\}$.

By interpreting K operators in α in terms of the given epistemic state (the e argument), we do not need a complex fixed-point construction, and we are guaranteed that the resulting epistemic state is always uniquely defined, as it was in the objective case.

For example, consider $\textbf{TELL}[\exists x. P(x) \land KQ(x), e]$. As an assertion this tells the system that there exists something that has property P. Moreover, this something is also known to have property Q. But known when? The answer we take here is that the individual is known to have property Q in the epistemic state e just before the assertion. In other words, we are asserting that there is an individual n such that $P(n)$ holds and such that $e \models KQ(n)$.

The reason we need to be careful about what epistemic state we use is that as a result of the assertion, we may be changing whether or not an individual is known to have some property. Its status before and after the assertion can be different. Because of this, it can happen that a sentence α is not believed after it has been asserted:

Theorem 5.5.2: *There is a sentence α and an epistemic state e such that*

$\textbf{ASK}[\alpha, \textbf{TELL}[\alpha, e]] = no$.

Proof: Let e be e_0 and let α be $[P(n) \wedge \exists x. P(x) \wedge \neg KP(x)]$, for some standard name n. If we let $e' = \textbf{TELL}[\alpha, e_0]$, then $e' = \{w \mid w \models P(n)\}$. But e' has no reason to believe that there is another P apart from n, and so $e' \models \neg K \exists x. P(x) \wedge \neg KP(x)$, from which it follows that $\textbf{ASK}[\alpha, e'] = no$. ∎

So by the above definition, a knowledge base can be told something and yet end up not believing it. This means that a **TELL** operation cannot be interpreted as an instruction to believe its sentence argument. Rather, it needs to be understood as an assertion that this argument is true (or more precisely, that it *was* true at the time of the assertion), and an instruction to draw conclusions about the world from this fact. Because of this, an assertion of a purely subjective sentence (which carries no information about the world) is either redundant or contradictory:

Theorem 5.5.3: *If σ is subjective, then* $\textbf{TELL}[\sigma, e] = e$ *or* $\textbf{TELL}[\sigma, e] = \{\}$.

It is not too surprising that we cannot tell the knowledge base anything about what it knows, since we have assumed that it already has complete knowledge of its own subjective state. In fact, this last result generalizes to any sentence whose truth value is known:

Theorem 5.5.4: *If* $\textbf{ASK}[\alpha, e] = yes$, *then* $\textbf{TELL}[\alpha, e] = e$; *if* $\textbf{ASK}[\neg \alpha, e] = yes$, *then* $\textbf{TELL}[\alpha, e] = \{\}$.

So, for example, if we start in e_0, we get that $\textbf{TELL}[KP(c) \vee KQ(c), e_0] = \{\}$. That is, if we tell the system that it either knows $P(c)$ or it knows $Q(c)$, this is inconsistent because it already knows that it does not know either. On the other hand, a similar assertion, $\textbf{TELL}[P(c) \vee Q(c), e_0]$ is fine, since the objective sentence here is unknown.

5.6 Closed world assertions

What is the purpose of non-objective assertions? The answer is that they allow us to express facts about the world that could otherwise not be made without knowing the contents of the knowledge base. A simple example of this is what has been called the *closed world assumption*. The idea, roughly, is to be able to tell a knowledge base in a certain epistemic state that its information about some part of the world is complete. In its simplest form, we would like to be able tell the system that it already knows every instance of a predicate P. For this, we use a sentence of $\mathcal{KL}$ like

$$\forall x[P(x) \supset KP(x)],$$

which we will call γ.[2] This sentence can be read as saying that every P is currently known, or equivalently, anything not currently known to be a P is not one. So telling the system γ is like telling it $\neg P(n)$, for every n such that $e \models \neg KP(n)$.

For example, suppose that
$$e = \{w \mid w \models P(^\#1) \text{ and } w \models P(^\#2)\}.$$
For this epistemic state, we have that $e \models K[P(^\#1) \wedge P(^\#2)]$. That is, in e, it is known that $P(^\#1)$ and $P(^\#2)$ are both true. But it is not known in e whether or not there are any other n such that $P(n)$ is true. In other words, there are world states in e where $^\#1$ and $^\#2$ are the only instances of P, and there are world states in e where there are others. So we have:
$$e \models \neg K\forall x[P(x) \equiv (x = {}^\#1) \vee (x = {}^\#2)]$$
and
$$e \models \neg K\neg\forall x[P(x) \equiv (x = {}^\#1) \vee (x = {}^\#2)].$$
However, if we now assert γ and get $e' = \text{TELL}[\gamma, e]$, then we have
$$e' \models K\forall x[P(x) \equiv (x = {}^\#1) \vee (x = {}^\#2)].$$
We have told the system that $^\#1$ and $^\#2$ are the *only* instances of P. The important point is that using γ we told it this without having to list the instances of P. We were able to do this because in e, we have the following:
$$e \models K\forall x[KP(x) \equiv (x = {}^\#1) \vee (x = {}^\#2)].$$
In other words, although the system does not know if there are any additional instances of P, it does know what all the *known* instances are. So by telling it γ, it is then able to determine that these known instances are the complete list.

As a second example, suppose that we have
$$e = \{w \mid w \models P(n), \text{ for every } n \neq {}^\#1\}.$$
In this epistemic state, $P(n)$ is known to be true for every n apart from $^\#1$. However, $P(^\#1)$ is undecided since there are world states in e where $P(^\#1)$ is true and others where it is false. If we now assign $e' = \text{TELL}[\gamma, e]$, we settle the issue:
$$e' \models K\forall x[P(x) \equiv x \neq {}^\#1].$$
In this case, we could not have listed explicitly all of the instances of P even if we had wanted to since there are infinitely many. But γ works as intended here as well.

As a final example, suppose we have
$$e = \{w \mid w \models P(^\#1) \text{ or } w \models P(^\#2)\}.$$
In this state, we have all world states that satisfy either $P(^\#1)$ or $P(^\#2)$. So an instance of P is known to exist, $e \models K\exists x P(x)$, although there are no known instances, since for every n, there is a world state in e satisfying $\neg P(n)$. If we now define $e' = \text{TELL}[\gamma, e]$, we get

---

[2] Note that the formula λ introduced on Page 74 is actually $K\neg\gamma$.

the result that e' is empty, the inconsistent epistemic state. The reason for this is that in e, γ is already known to be false. In other words, $e \models K\exists x[P(x) \land \neg KP(x)]$, because $e \models K\neg KP(^\#1)$ and $e \models K\neg KP(^\#2)$, but $e \models K[P(^\#1) \lor P(^\#2)]$. Thus, in this example, it is already known that there is an instance of P apart from the known ones, and any attempt to assert otherwise leads to an inconsistency.

So although we can express the closed world assumption as a sentence of $\mathcal{KL}$, we must be careful in how it is used. In certain states, the knowledge base will already know that it is missing an instance of the predicate in question and should not be told otherwise. However, we can prove that in all other cases, **TELL** will behave properly and result in a consistent epistemic state where the closed world assumption is known to be true. To show this, we need the following lemma:

Lemma 5.6.1: *Suppose e is an epistemic state where $e \models \neg K\neg\gamma$. Let $e' = $ **TELL**$[\gamma, e]$. Then for any w, $e, w \models \gamma$ iff $e', w \models \gamma$.*

Proof: We will show that for any n, $e \models KP(n)$ iff $e' \models KP(n)$, from which the conclusion follows. If $e \models KP(n)$, then $e' \models KP(n)$, since $e' \subseteq e$. Conversely, if $e \models \neg KP(n)$, then since $e \models \neg K\neg\gamma$, there is a $w \in e$ such that $e, w \models \gamma$, and thus for which, $w \models \neg P(n)$. Since $e, w \models \gamma$, we have that $w \in e'$, and so, $e' \models \neg KP(n)$. ∎

Then we get:

Theorem 5.6.2: *Suppose e is an epistemic state where $e \models \neg K\neg\gamma$. Then*

$$\mathbf{ASK}[\gamma, \mathbf{TELL}[\gamma, e]] = yes.$$

Proof: Let $e' = $ **TELL**$[\gamma, e]$. Suppose that $w \in e'$, and so $e, w \models \gamma$. From the lemma above, it follows that $e', w \models \gamma$. Since this applies to any $w \in e'$, we have that $e' \models K\gamma$, and hence **ASK**$[\gamma, e']=yes$. ∎

So this shows that as long as the closed world assumption γ is not known to be false, we can always assert that it is true, and end up in a consistent epistemic state where it is known to be true (quite unlike the example in Theorem 5.5.2). Moreover, we will see in Chapter 8 that the resulting epistemic state also has the desirable property that nothing else is known: in a precise sense, *all* that is known in this state is γ together with what was known before.

- *Teach*(*tom, sam*)
- *Teach*(*tina, sue*) ∧ [*Teach*(*tom, sue*) ∨ *Teach*(*ted, sue*)]
- ∃*x Teach*(*x, sara*)
- ∀*x*[*Teach*(*x, sandy*) ≡ (*x* = *ted*)]
- ∀*x*∀*y*[*Teach*(*x, y*) ⊃ (*y* = *sam*) ∨ (*y* = *sue*) ∨ (*y* = *sara*) ∨ (*y* = *sandy*)]

Figure 5.1: The Example KB

5.7 A detailed example

In this final section, we will consider a larger example of the use of **TELL** and **ASK**. This will allow us to explore how these operations can be used to probe what is known and add to it in a way that would be impossible if we only used $\mathcal{L}$. In the first subsection below, we will start with an initial epistemic state and ask a series of questions. In the second subsection, we will consider the effect of various assertions.

To keep the discussion as intuitive as possible, we will define the initial epistemic state as the set of all worlds satisfying a finite collection of sentences which can be thought of as a KB, a representation of what is known. As in earlier chapters, we will use a single predicate, *Teach*, and again use proper names that begin with a "*t*" for the first argument, the teacher, and proper names that begin with an "*s*" for the second argument, the student. As before, these should be understood as standard names. We will not use either function or constant symbols in any of the examples unless explicitly indicated.

The knowledge base we begin with appears in Figure 5.1. Notice that the first four sentences tell us what we know about each of the four students. The last sentence says that as far as the teaching relation is concerned, there are only four students. The starting epistemic state e is defined as the set of all world states satisfying this KB. Equivalently, $e = \textbf{TELL}[\text{KB}, e_0]$. Thus we have that $e \models K\phi$ (where ϕ is objective) iff (KB ⊃ ϕ) is valid.

5.7.1 Examples of ASK

To illustrate the operation of **ASK**, we consider various arguments of the form **ASK**[α, e], where e is the epistemic state above and α is one of the sentences below. For clarity, we will use three possible answers: TRUE means that α is answered *yes* (and ¬α, *no*), FALSE means that ¬α is answered *yes* (and α, *no*), and UNKNOWN, otherwise.

1. *Teach*(*tom, sam*)   TRUE

   This is the simplest type of question, and is clearly known to be true.

2. *Teach*(*tom, sandy*)   FALSE

   The sentence *Teach*(*tom, sandy*) is known to be false, since Ted is the only teacher of

Sandy, and because these are assumed to be standard names, they are distinct.

3. *Teach*(*tom, sue*) UNKNOWN
Neither this question nor its negation are known to be true. Tom may or may not teach Sue, according to what is known in *e*.

4. **K***Teach*(*tom, sue*) FALSE
In contrast to the previous question, here we are asking a subjective question about what is known. Such questions are always known to be true or to be false. In this case, it is known in *e* that Tom is not known to teach Sue. That is, the system realizes that the previous question was not known.

5. $\exists x\, Teach(x, sara)$ TRUE
Again this is a simple objective question that can be answered directly. It is known that Sara has at least one teacher, although no teacher has been named.

6. $\exists x\, \mathbf{K}Teach(x, sara)$ FALSE
Here the question is whether Sara has any known teachers. The system knows that although Sara has a teacher, none are as yet known. With this question and the preceding, we can distinguish between knowledge of the existence of a teacher and knowledge of the identity of that teacher.

7. $\exists x\, \mathbf{K}Teach(x, sue)$ TRUE
Unlike Sara, Sue does have a known teacher, namely Tina. Thus in her case, we know of both the existence and the identity of a teacher. Note that this question and the one before it is subjective and so would always be answered TRUE or FALSE.

8. $\exists x[Teach(x, sue) \land \neg \mathbf{K}Teach(x, sue)]$ TRUE
Having established in the previous question that Sue has a known teacher, here we are asking if she has a teacher apart from the known ones. In other words, is the list of teachers known for Sue incomplete? The answer is yes. There is only a single known teacher for Sue, Tina; but it is also known that one of Tom or Ted teaches her, and neither is a known teacher of Sue.

9. $\exists x[Teach(x, sandy) \land \neg \mathbf{K}Teach(x, sandy)]$ FALSE
If we ask the same question of Sandy, the answer is no. It is known that Ted is her teacher and her only one. Thus, Sandy has no teachers other than her single known teacher.

10. $\exists x[Teach(x, sam) \land \neg \mathbf{K}Teach(x, sam)]$ UNKNOWN
If we ask the same question about Sam, the answer is unknown. We know that Tom teaches Sam, but we have no other information. So Sam may or may not have a teacher apart from this known one. Note that this question (like the two preceding ones) is not a subjective sentence, and consequently can be believed to be true, believed to be false,

or here, neither.

11. $\exists y K \forall x [Teach(x, y) \supset KTeach(x, y)]$ TRUE

This is a generalized version of the preceding question posed subjectively. Do you know someone whose teachers are all known? That is, can we name somebody whose list of teachers is complete? The answer is yes: Sandy. Note that to verify this, we need to establish that the question is known to be true, which involves checking the truth of a sentence with K operators nested to depth three.

12. $\exists y(y \neq sam) \wedge \neg Kif[\forall xTeach(x, y) \supset KTeach(x, y)]$ FALSE

where $Kif\alpha$ is an abbreviation for $K\alpha \vee K\neg\alpha$.

We established in an earlier question that Sam may or may not have teachers other than the known ones. This question asks if there is an individual other than Sam for which this is true. In other words, is there anyone other than Sam for which you don't know if you are missing any teachers? The answer is no because we know that there are only three cases to consider apart from Sam: for Sandy, we know that we have all the teachers, and for Sue and Sara, we know that we are missing one.

13. $\exists y K \exists x [Teach(x, y) \wedge \exists z [(y \neq z) \wedge KTeach(x, z)]]$ TRUE

This question asks if there is an individual y known to have the property that one of her teachers x is known to teach somebody else z. For this to be true, we need to know who y is, we need not know who the x is, but for each such x, we must know who the z is.

The answer here is yes because of Sue. For every $w \in e$, we either have

$$e, w \models Teach(tom, sue) \wedge (sue \neq sam) \wedge KTeach(tom, sam)$$

or

$$e, w \models Teach(ted, sue) \wedge (sue \neq sandy) \wedge KTeach(ted, sandy).$$

So no matter if Tom or Ted teaches Sue, both of them are known to teach someone other than Sue. Consequently, Sue is known to have a teacher x who is known to teach somebody else, even though we do not know who that x is:

$$e \models K\exists x.Teach(x, sue) \wedge \exists z[(sue \neq z) \wedge KTeach(x, z)].$$

Note that by using nested K operators and quantifiers, we can insist on the individual z being known as a function of some other unknown individual x. This would fail, for example, if instead of knowing that Tom teaches Sam, all we knew was that Tom teaches someone other than Sue.

5.7.2 Examples of TELL

To illustrate the operation of **TELL**, we will consider some example assertions. In each case, we present a sentence α followed by notation $[e \rightarrow e']$, where e is either the initial epistemic state above or the result of a previous assertion, and where $e' = \textbf{TELL}[\alpha, e]$.

1. $\forall x[\textit{Teach}(x, \textit{sue}) \supset \textit{Teach}(x, \textit{sara})]$ \hfill $[e \to e_1]$

   This is an assertion of a sentence without $\textbf{\textit{K}}$ operators, and so the resulting state is simply the set of world states w that are in e and satisfy the assertion. In all such world states, we have that Tina teaches Sara, as does one of Tom or Ted. Consequently, these facts would be known in e_1.

2. $\forall x[\textbf{\textit{K}}\textit{Teach}(x, \textit{sue}) \supset \textit{Teach}(x, \textit{sara})]$ \hfill $[e \to e_2]$

   In this assertion, we do not say that every teacher of Sue teaches Sara, but only the currently known ones. In the initial state e, Tina is the only known teacher of Sue, so this assertion says nothing about either Tom or Ted. In both e_1 and e_2 we would have a single known teacher for Sara, namely Tina, but only in e_1 would we also know that there was an additional teacher apart from Tina. So if β is the sentence

   $$\exists x[\textit{Teach}(x, \textit{sara}) \land \neg \textbf{\textit{K}}\textit{Teach}(x, \textit{sara})]$$

   then $\textbf{ASK}[\beta, e]$ is *yes*, and $\textbf{ASK}[\beta, e_1]$ is *yes*, but since $e_2 \models \neg \textbf{\textit{K}}\beta \land \neg \textbf{\textit{K}}\neg\beta$, both $\textbf{ASK}[\beta, e_2]$ and $\textbf{ASK}[\neg\beta, e_2]$ would be *no*. In other words, we started out by knowing we were missing some of Sara's teachers, but after finding out that Tina is one of them, we no longer know whether or not we are still missing any.

3. $\forall x[\textit{Teach}(x, \textit{sara}) \supset \exists y \textbf{\textit{K}}\textit{Teach}(x, y)]$ \hfill $[e_2 \to e_3]$

   This starts in the state e_2 where all that is known about Sara is that Tina teaches her, and we assert that all of her teachers have a currently known student. This means that all of her teachers must be one of Tom, Tina, or Ted, since these are the only individuals with known students (for Sam, Sue and Sara, and Sandy, respectively) in e_2. The assertion, however, does not identify any new teachers for Sara so that the known teachers in e_3 are the same as the known teachers in e_2: just Tina.

   This is a good example of how a set can be bounded from below by the known instances and from above by the potential instances (that is, the individuals not known to be non-instances). In this case, Sara's teachers are bounded from below by {*tina*}, and from above by {*tom, tina, ted*}. To have complete knowledge of Sara's teachers, all that is needed is to settle the case of Tom and Ted.

4. $\forall x[\textit{Teach}(x, \textit{sara}) \supset \neg \exists y \textbf{\textit{K}}[\textit{Teach}(x, \textit{sue}) \lor \textit{Teach}(y, \textit{sue})]]$ \hfill $[e \to e_4]$

   The assertion starts from e again and says that anyone who teaches Sara cannot be one of two individuals such that it is known that one of them teaches Sue. After this assertion we have the same set of known teachers for Sara, but the assertion addresses the potential teachers. It clearly rules out Tina as a teacher (since for this x there is such a y, namely Tina herself). Taking the case of Tom, although he is not known to teach Sue in e, it is known that either Tom or Ted teaches Sue, so Tom is an x such that there is a y for which it is known that either x or y teaches Sue. Thus Tom cannot be a teacher of Sara. A similar argument rules out Ted. Thus, we have ruled out as potential teachers Tina, Tom and Ted.

The last assertion above is a good example of using a set of potential instances as candidates or suspects in an assertion. Consider an individual, Terry, about which absolutely nothing is known in e. Neither Tom nor Terry are known teachers of Sue, and both Terry and Tom are potential teachers of Sue. However, there is a difference between the two: Tom is a candidate teacher in the sense that it is known that one of Tom or Ted teaches Sue; nothing comparable is known about Terry. More generally, we might say that x is a <u>candidate</u> instance of P if x is not a known instance, but is a member of a candidate set for P, where a candidate set is a minimal set of individuals such that it is known that one of them is an instance of P. In the case of Sue's teachers, the set {Tom, Ted} is a candidate set, and so both are candidates. We know that one of Tom, Ted, and Terry must also be a teacher of Sue, but this is not a minimal set, and so Terry is not a candidate.

This notion of a candidate is useful to help form intuitions about default reasoning, which we explore in Chapter 9. It is often useful to be able to assume of certain individuals that they have a certain property unless they are known not to have it. For example, a person might be assumed to be innocent of a crime unless proven guilty. In some cases, however, there may be a set of suspects, where there is good reason to believe that one of them is guilty, although none of them are known to be guilty. Although any suspect might be innocent, they cannot all be assumed to be innocent. Although we may wish to assume innocence as broadly as possible, we may have to temporarily withhold that assumption for the candidate set.

5.8 Other operations

Having examined definitions for the knowledge-level operations of **TELL** and **ASK**, it is worth remembering that these are only two of many possible interaction operations we might consider. In this section, we briefly investigate two others: one involving definitions, and one involving wh-questions.

5.8.1 Definitions

The idea behind introducing <u>definitions</u> in a KB is this: we want to extend our vocabulary of predicate or function symbols in terms of the existing ones. For example, we might want to have a predicate symbol *FlyingBird*, which instead of being independent of all other predicate symbols, simply means the conjunction of *Bird* and *Fly*.

We could, of course, simply assert a universally quantified biconditional such as

$$\forall x (FlyingBird(x) \equiv Bird(x) \land Fly(x)).$$

The trouble with this is that it looks exactly like any other fact we might know about the predicate. For example, in some application we might know that all and only the birds in my cage fly:

$$\forall x (FlyingBird(x) \equiv Bird(x) \wedge InCage(x)).$$

Logically, the two facts are indistinguishable, but clearly the second one is not intended to say what the predicate *means*.

One simple way of handling definitional information is to imagine an epistemic state as having two components: $e = \langle e_a, e_d \rangle$, where e_a is the *assertional epistemic state* resulting from **TELL** operations as before, and e_d is the *definitional epistemic state* resulting from new **DEFINE** operations. So **TELL**$[\alpha, \langle e_a, e_d \rangle]$ would be defined to change e_a only, as specified earlier. Similarly, **DEFINE**$[P(\vec{x}), \phi[\vec{x}], \langle e_a, e_d \rangle]$ would be defined to change e_d only. The P here is an *n-ary* predicate symbol, and ϕ is a formula with n free variables, its definition.[3] For example, we could have

$$\textbf{DEFINE}[FlyingBird(x), (Bird(x) \wedge Fly(x)), e]$$

Having separated e_a and e_d, we can now define the **DEFINE** operation to be that of asserting the universally quantified biconditional over e_d (with no danger of confusion).[4]

With the epistemic state broken into two parts, we can consider asking questions about the definitions only. For example, we can define

$$\textbf{ASK-DEF}[\alpha, \langle e_a, e_d \rangle] = \begin{cases} yes & \text{if } e_d \models K\alpha \\ no & \text{otherwise} \end{cases}$$

This is just like ordinary **ASK** except that it only uses e_d: it determines whether or not α is known to be true *by definition*. For **ASK** itself, we want to use both definitions and assertions to determine what is known. Thus, we would redefine it as

$$\textbf{ASK}[\alpha, \langle e_a, e_d \rangle] = \begin{cases} yes & \text{if } (e_a \cap e_d) \models K\alpha \\ no & \text{otherwise} \end{cases}$$

So for example, if we **DEFINE** the predicate *FlyingBird* as above, then assert using **TELL** that *FlyingBird(tweety)* is true, then **ASK** will correctly confirm that *Fly(tweety)* is true.

5.8.2 Wh-questions

In addition to yes/no questions, any knowledge representation system will need to answer *wh-questions*, that is, questions beginning with words like "who," "what," "when," "where," and "how." For example, we might want to find out: who are the teachers of

---

3 There may be good reasons to restrict the language of ϕ as is often done in description logics, or even to extend it to allow, for example, recursive definitions.
4 We may also want to consider other forms of definitions that do not lead to biconditionals. For example, while it may not be possible to define a "natural kind" term like *Bird*, we may still wish to express necessary definitional properties, such as its being an animal.

Sara? In an epistemic state with incomplete knowledge about teachers, however, it is not clear what form of answer would be appropriate. The simplest solution is to take an operator like **WH-ASK**[*Teacher*(*x*, *sara*), *e*] to be a question about the *known* teachers of Sara. While this would mean that the question *Teacher*(*x*, *sara*) and *KTeacher*(*x*, *sara*) would get the same answer, the question ¬*K*¬*Teacher*(*x*, *sara*) could still be used to find out about *potential* teachers of Sara. The actual teachers of Sara, of course, lie between these two sets.

This suggests the following definition:

WH-ASK[$\alpha[\vec{x}], e$] = $\{\vec{n} \mid e \models K\alpha[\vec{n}]\}$.

The main problem with this definition is that the answer could be an infinite set of standard names, for example, when α is $(x \neq {}^{\#}1)$. We will see in Chapter 7 how even an infinite answer to a **WH-ASK** question can be finitely represented at the symbol level.

Simply returning a set of standard names for a wh-question may not be very illuminating, however. What we would ultimately like is an answer in *descriptive* terms, using meaningful constants and function symbols. This is not to suggest that we would be better off with

$\{\vec{t} \mid e \models K\alpha[\vec{t}]\}$

as the answer to a wh-question, since we would then have no way of knowing how many answers there were (because many of the terms *t* could be co-referential). A better idea is to define a new operator, **DESCRIBE**[*n*, *e*] which takes a standard name as argument and returns the set of terms known to be co-referential with *n*:

DESCRIBE[n, e] = $\{t \mid e \models K(t = n)\}$.

We will see again in Chapter 7 how this potentially infinite set of terms can be represented finitely.

5.9 Bibliographic notes

The idea of characterizing a knowledge representation and reasoning service in terms of tell and ask operations first appeared in [80] and [82]. It appeared subsequently in a variety of publications, most notably in a general AI textbook [130]. The idea was inspired by similar operations defining abstract data types like stacks and queues [96]. The **TELL** operation itself presents the simplest possible model of how an epistemic state changes, where contradicting information leads to an inconsistent state. For a more delicate approach, which may involve giving up some past beliefs to preserve consistency, see [38, 61], or any of the many papers in the area of belief revision. See also Chapter 11 and the papers referred to there, for ideas on how to formally characterize forgetting earlier information. The closed world assumption first appeared in [119], and was one of the motivations for the earliest forms of nonmonotonic reasoning, further discussed in Chapter 9. The distinction between assertions and definitions is discussed in [8] and appeared in the KRYPTON knowledge representation system [7]. The idea of using a limited language for defining predicates (or concepts) derives from early work on semantic networks (see [34], for example), and especially the KL-ONE system [11]. Some of this research then evolved into the subarea of description logics (see [23, 24], for instance). See [18] for a treatment of natural-kind concepts that do not admit necessary and sufficient definitions. As to wh-questions, returning more than just yes/no answers is of course the mainstay of database systems. A comparable story can be told for simple forms of knowledge bases, as in logic programming [97]. For a more general KB, perhaps the clearest account is that of Green's answer extraction [44, 114], although this still only applies when the KB is in a restricted syntactic form.

5.10 Exercises

1. Show that subjective knowledge is complete, in that if σ is subjective, then either **ASK**$[\sigma, e] = yes$ or **ASK**$[\neg\sigma, e] = yes$.

2. Show that **ASK**$[\forall x. P(x) \supset \neg KP(x), e_0] = yes$.

3. Prove that if ϕ and ψ are objective, the order in which they are asserted is unimportant:
   TELL$[\phi, \textbf{TELL}[\psi, e]] = \textbf{TELL}[\psi, \textbf{TELL}[\phi, e]]$.

4. Give an example epistemic state where an individual is known to have some property, but after an assertion, it not known to have that property.

5. Present a non-subjective sentence for which knowledge is nonmonotonic.

6. Consider the positive and negative subjective sentences of Exercise 3 of Chapter 4. Show that knowledge is monotonic for the positive ones.

7. Construct an example epistemic state (as a set of world states satisfying some property) where the assertion of γ is redundant because it is already known to be true.

8. Construct a variant of the example KB such that unlike Question 11,

   $\exists y K \forall x[Teach(x, y) \supset KTeach(x, y)]$

   comes out FALSE, but

   $K \exists y \forall x[Teach(x, y) \supset KTeach(x, y)]$

   comes out TRUE.

9. Describe or construct an epistemic state where the candidate instances of P are the same as the potential instances.

10. Consider the following generalization of candidate instances. Define an objective sen-

tence ϕ to be *explainable* in state e iff there is some objective ψ such that $\neg K \neg \psi$ and $K(\psi \supset \phi)$ are both true. If also $\models (\psi \supset \phi)$, we say that ϕ is trivially explainable. Show that if n is a candidate instance of P then $P(n)$ is non-trivially explainable, but that the converse is not true. Show that n is a potential instance of P iff $P(n)$ is explainable.

11. What is **ASK**$[\exists x[\neg \exists y(\textbf{\textit{K}}\textit{Teach}(x, y)) \wedge \textit{Teach}(x, sue)], e]$?

12. In **ASK** Number 13, show that we cannot move any other quantifiers outside the K operator without changing the meaning of the question.

13. Show that
$$e_3 \models \exists y_1 y_2 y_3 \textbf{\textit{K}} \forall x [\textit{Teach}(x, sara) \supset (x = y_1) \vee (x = y_2) \vee (x = y_3)].$$

14. Describe the result of **TELL**$[\forall x[\textit{Teach}(x, n) \supset \textbf{\textit{K}}\textit{Teach}(x, n)], e]$, for $n = $ *sue*, *sam*, *sandy*, *sara*.

15. Define a new interaction operator **HOW-MANY** which tells us how many instances of a formula are known to be true. In particular, it should take a formula with one free variable as argument, and return a pair of numbers $\langle i, j \rangle$, such that it is known that there are at least i and at most j true instances of the formula. Describe a KB where this operator would provide useful information, but **WH-ASK** would not.

6 Knowledge Bases as Representations of Epistemic States

In what we have seen so far, we have been thinking about knowledge in two very different ways:

- as characterized by a symbolic *knowledge base* or KB, that is, a collection of sentences about the world, where what is known is what can be inferred from the sentences;
- as characterized by an *epistemic state*, that is, a set of world states, where what is known is what is true in all of the world states.

While these two notions are clearly related, it turns out that there are also interesting differences between them. In this chapter we will explore in detail their relationship.

In the first section, we observe that many epistemic states are equivalent, in the sense that they satisfy exactly the same set of sentences. Choosing representatives for these equivalence classes of epistemic states simplifies the results to follow. In Section 6.2, we define what it means for an objective KB to represent what is known in an epistemic state. An epistemic state is defined to be representable if such a KB exists. In Section 6.3 we prove that there are epistemic states that are not representable in this sense, but in Section 6.4, we show that we can usually ignore these states, in the sense that any satisfiable sentence of $\mathcal{KL}$ is satisfied in a representable epistemic state. The KB in question may need to be infinite, however, and we prove in Section 6.5 that finitely representable epistemic states are not sufficient: there is a satisfiable sentence of $\mathcal{KL}$ that is false at every finitely representable epistemic state. Finally, in Section 6.6 we discuss the implications of these results for **TELL** and **ASK**.

6.1 Equivalent epistemic states

The easiest way to see that there is a difference between the two views of knowledge mentioned above is to consider a simple cardinality argument. If we take knowledge to be characterized by the set of all sentences known, then because there are only $\aleph_0$ sentences in $\mathcal{KL}$, there can be at most $2^{\aleph_0}$ distinct states of knowledge. But if we take knowledge to be characterized by a set of world states, then because there are $2^{\aleph_0}$ world states, there would be $2^{2^{\aleph_0}}$ distinguishable states of knowledge.

It follows from this observation that many epistemic states know exactly the same set of sentences. We call two epistemic states e and e' <u>equivalent</u> (which we write $e \approx e'$) iff for every $\alpha \in \mathcal{KL}$, $e \models \boldsymbol{K}\alpha$ iff $e' \models \boldsymbol{K}\alpha$. This clearly defines an equivalence relation over epistemic states.

For example, consider e_0, the set of all world states. We will show that e_0 and the epistemic state formed by removing a single world state from e_0 are equivalent. Observe that for any α, w, and e, if $e, w \models \alpha$, and if w' is the same as w except for the truth value it gives to some primitive sentence whose predicate symbol does not appear in α, then $e, w' \models \alpha$ (by induction on α). Thus, if α is true for *all* elements of $(e_0 - w)$,[1] it will be true for w too. Consequently, $e_0 \models K\alpha$ iff $(e_0 - w) \models K\alpha$. In other words removing a single world state from e_0 does not affect the sentences believed, and so results in an equivalent state.

Because we want to think of knowledge functionally, in terms of the operations a knowledge-based system can perform, and because these operations are mediated by linguistic arguments, the difference between e_0 and $e_0 - w$ is not something we really care about. The **ASK** operation would not be able to tell the difference between them, which can be thought of as an artifact of the modeling process. This suggests that we should restrict our attention to *equivalence classes* of epistemic states. We can do this by finding suitable representatives for each equivalence class.

To do so, first observe that a world state can be added to an epistemic state if it satisfies everything that is known:

Theorem 6.1.1: *For any e and w,*

$$e \approx (e + w) \text{ iff for every } \alpha \text{ such that } e \models K\alpha, e, w \models \alpha.$$

Proof: Let e' be $e + w$. First we assume that $e \approx e'$ and show that if $e \models K\alpha$, then $e, w \models \alpha$. Observe that in general, for any β and any w', $e, w' \models \beta$ iff $e', w' \models \beta$, by a simple induction argument. So suppose $e \models K\alpha$. Then $e' \models K\alpha$ since $e \approx e'$, in which case, $e', w \models \alpha$, and so $e, w \models \alpha$.

For the converse, assume that for every α such that $e \models K\alpha$, $e, w \models \alpha$. We will show $e \approx e'$ by showing that for any β and any w', $e, w' \models \beta$ iff $e', w' \models \beta$. The proof is by induction. It clearly holds for atomic sentences and equalities, and by induction for negations, conjunctions, and quantifications. Also if $e' \models K\beta$, then $e \models K\beta$, since e' is a superset of e. So finally, suppose that $e' \models \neg K\beta$. Then, for some $w' \in e'$, we have $e', w' \models \neg \beta$, and so $e, w' \models \neg \beta$ by induction. Now there are two cases: if $w' \in e$, then $e \models \neg K\beta$ directly; if $w' = w$, then again $e \models \neg K\beta$, by assumption about w. This completes the proof. ∎

So under certain conditions, we can add world states to an epistemic state and preserve equivalence. Let us say that an epistemic state e is <u>maximal</u> iff for every w, if $e \approx e + w$ then $w \in e$. So a maximal epistemic state is one where the addition of any world state would involve a change in belief for some sentence. Clearly e_0 is maximal and $e_0 - w$ is

---
1 When X is a set, we use the notation $X - x$ to mean $X \setminus \{x\}$. Similarly, $X + x$ means $X \cup \{x\}$.

Knowledge Bases as Representations of Epistemic States

not. We can show that every equivalence class has a *unique* maximal element, which can serve as the representative for the class:

Theorem 6.1.2: *For any epistemic state e, there is a unique maximal state e^+ that is equivalent to it.*

Proof: For any e, let $e^+ = \{w \mid e \approx e + w\}$. We will first show that $e \approx e^+$ by showing that for any w and α, $e, w \models \alpha$ iff $e^+, w \models \alpha$, again by induction. As in the previous theorem, the only difficult case is when $e^+ \models \neg K\alpha$. Then, for some $w \in e^+$, we have $e^+, w \models \neg\alpha$, and so $e, w \models \neg\alpha$ by the induction hypothesis. But since $w \in e^+$, we know that $e \approx e + w$. Therefore, $(e + w), w \models \neg\alpha$ also, and so $(e + w) \models \neg K\alpha$ and consequently, $e \models \neg K\alpha$. This establishes that $e \approx e^+$.

To show that e^+ is the unique maximal set, we show that for any e', if $e \approx e'$ then $e' \subseteq e^+$. That is, we need to show that if $e \approx e'$ then for any $w \in e'$, $e \approx e + w$, and so $w \in e^+$. To show this, we simply observe that if $w \in e'$, then w satisfies everything that is known in e as well as in e', and so by the previous theorem, we have that $e \approx e + w$. ∎

Thus maximal states can be used as representatives of the equivalence classes. Moreover, by Theorem 6.1.1 we have that

Corollary 6.1.3: *A state e is maximal iff there is a set Γ such that*
$$e = \{w \mid e, w \models \alpha, \text{ for every } \alpha \in \Gamma\}.$$

In much of what follows, we will restrict our attention to maximal epistemic states, since these cover all of the possibilities admitted by the logic $\mathcal{KL}$:

Corollary 6.1.4: *For any (possibly infinite) set of sentences Γ, if Γ is satisfiable, then it is satisfied by a maximal epistemic state.*

So in terms of the logic $\mathcal{KL}$, maximal sets are fully *sufficient*; in Section 6.3, we will consider the converse question: are *all* maximal epistemic states required, or can we get by with a subset of them?

6.2 Representing knowledge symbolically

As Corollary 6.1.3 shows, maximal epistemic states can be completely characterized by the sentences that are known. Let us call a set of sentences Γ a *belief set* iff there is an epistemic state e such that $\Gamma = \{\alpha \mid e \models K\alpha\}$. So Γ is a belief set for e if it is everything believed in e. Then we have the following:

Theorem 6.2.1: *There is a bijection between belief sets and maximal epistemic states.*

It follows then that there are only as many maximal states as there are belief sets.

On the other hand, when we think of knowledge linguistically, at least informally, we usually do not think in terms of *all* sentences known. Rather we think in terms of a symbolic representation of what is known, that is, a collection of sentences (or perhaps some other symbolic data structures), typically finite, that is sufficient to characterize the complete belief set.

For example, if we use a sentence α of $\mathcal{KL}$ as part of our representation, then not only is α known, but so are all of its logical consequences. Moreover, by introspection, $K\alpha$ and $K\neg K\neg\alpha$ are also known. In fact, any β such that $(K\alpha \supset K\beta)$ is valid should also be known.

But this is still not the full belief set. To see why, suppose that ϕ is objective, and the epistemic state we are trying to represent is $e = \{w \mid w \models \phi\}$. The belief set associated with e clearly starts with a belief in ϕ, and contains all sentences β as above. But now, assuming that ϕ is satisfiable, let ψ be any objective sentence such that $(\phi \supset \psi)$ is *not* valid. Then, there is a $w \in e$ such that $w \models \neg\psi$, and so $e \models \neg K\psi$, and thus, $e \models K\neg K\psi$. Therefore, $\neg K\psi$ is part of the belief set too. Notice that $(K\phi \supset K\neg K\psi)$ is *not* valid: just because you believe ϕ, it does not *follow* that you do not believe ψ.

So if we are to extract a belief set (and an epistemic state) from a representation like ϕ, we have to consider not only what follows from believing ϕ, but also what follows from *not* believing other sentences. In other words, a representation of an epistemic state must capture what is believed and the fact that what is represented is *all* that is believed. This is of course how we understand a knowledge base: it represents what is known and all that is known.

Can we use any collection of sentences of $\mathcal{KL}$ to represent an epistemic state? This is a difficult question in general, and we will defer it to Chapter 8. But one special case is much simpler: the objective sentences. Clearly our emphasis has been on objective knowledge about the world, and it ought to be possible to use objective sentences to represent knowledge.

So imagine that we start with KB, an arbitrary set (not necessarily finite) of objective sentences of $\mathcal{KL}$. The *epistemic state represented by* KB, which we write $\Re[\![KB]\!]$, is the one where all the sentences in the KB are known, and nothing else. Formally,

$$\Re[\![KB]\!] = \{w \mid w \models \phi, \text{ for every } \phi \in KB\}.$$

Note that according to this definition, $\Re[\![KB]\!]$ is always maximal. When the KB in question is a finite set $\{\phi_1, \ldots, \phi_k\}$, we have that $\Re[\![KB]\!]$ is the same as

TELL$[(\phi_1 \wedge \ldots \wedge \phi_k), e_0]$,

so that the epistemic state represented by KB is the state that results from being told in the initial state that everything in the KB is true. Let us call an epistemic state e *representable* if for some set KB of objective sentences, $e = \Re[\![KB]\!]$, and we call it *finitely representable* if there is a finite KB such that $e = \Re[\![KB]\!]$.

We can now distinguish (conceptually, at least) among three varieties of epistemic states. Given an epistemic state e, if there is a set Γ such that

$$e = \{w \mid \text{for every } \alpha \in \Gamma, e, w \models \alpha\},$$

then e is maximal; if there is an *objective* Γ satisfying the above, then e is representable; finally, if there is a Γ that is both *finite* and *objective*, then e is finitely representable.

One very important and immediate property of representable epistemic states is that they are completely determined by the objective knowledge they contain.

Theorem 6.2.2: *If e_1 and e_2 are representable states, then*

$$e_1 = e_2 \text{ iff for every objective } \phi, e_1 \models K\phi \text{ iff } e_2 \models K\phi.$$

So any two representable states that agree on the objective facts agree on everything else. This is not true in general for maximal epistemic states, as can be seen from the two states e_1 and e_2 used in Theorem 4.6.2 of Chapter 4.

Clearly every finitely representable state is representable, and every representable state is maximal. But what about the converses? In the next sections, we will examine these questions in detail.

6.3 Some epistemic states are not representable

In the previous section we showed that we could limit our attention to maximal epistemic states without any loss of generality whatsoever. For representable states, however, this is not the case:

Theorem 6.3.1: *There is an infinite satisfiable set of sentences of $\mathcal{KL}$ that is not satisfied by any representable epistemic state.*

The proof uses details from the proof of Theorem 4.6.2 from Chapter 4. To recap, let Ω be the set $\{^{\#}1, ^{\#}3, ^{\#}5, \ldots\}$, let Φ be the set of objective sentences consisting of $\{(t = {^{\#}}1)\}$ for every primitive term t, $\{\neg\phi\}$ for every primitive sentence ϕ whose predicate letter is not P, and finally $\{P(n)\}$ for every $n \in \Omega$. Let e_1 be $\{w \mid w \models \Phi\}$. Let $\overline{w}$ be the (unique) element of e_1 such that for every $n \notin \Omega$, $w \models \neg P(n)$, and let e_2 be $e_1 - \overline{w}$. Finally, define Γ_1 and Γ_2 by

$$\Gamma_1 = \{\boldsymbol{K}\phi \mid e_2 \models \boldsymbol{K}\phi\} \cup \{\neg \boldsymbol{K}\phi \mid e_2 \models \neg \boldsymbol{K}\phi\}$$
$$\Gamma_2 = \{\sigma \mid e_2 \models \sigma\}.$$

The set Γ_2 is the one that we will show is not satisfied by any representable state.

First observe that Γ_2 is indeed satisfiable, since e_2 satisfies it. Also, Γ_1 is a subset of Γ_2, so that anything claiming to satisfy the latter must also satisfy the former. Next, e_1 is a representable state, represented by Φ. Finally note that from the proof of Theorem 4.6.2, although e_1 satisfies Γ_1, it does not satisfy Γ_2 since it fails to satisfy the sentence $\boldsymbol{K}\exists x[P(x) \wedge \neg \boldsymbol{K}P(x)]$.

So to prove the theorem: let e be any representable state that satisfies Γ_1. By Theorem 6.2.2, $e = e_1$, and so e does not satisfy Γ_2. Thus, no representable state satisfies Γ_2.

6.4 Representable states are sufficient

In the previous section, we showed that in terms of satisfiability in $\mathcal{KL}$, we needed to allow for epistemic states that were not representable. To prove this, we had to consider an infinite set of sentences in $\mathcal{KL}$. In any realistic application, however, we will only be using *finite* sets of sentences. The question we now ask is whether the above theorem would continue to hold.

Fortunately, the answer here is no:

Theorem 6.4.1: *Any sentence (or finite set of sentences) of $\mathcal{KL}$ that is satisfiable is satisfied by some representable epistemic state. Equivalently, a sentence of $\mathcal{KL}$ is valid iff it is true at all world states and all representable epistemic states.*

This is an important result since it shows that as far as the *logic* of $\mathcal{KL}$ is concerned, representable states are sufficient: validity in the logic is exactly the same as truth in all representable states.

The proof, however, is not trivial, and proceeds as follows: starting with some satisfiable sentence γ, we will construct an infinite satisfiable set Γ that includes γ and that is satisfied by a representable state. The construction is similar to that used in the completeness proof of Chapter 4, in that we construct the set iteratively, adding sentences while preserving the set's satisfiability. To allow the set to be satisfied by a representable state, we must ensure that all required knowledge in that state can be reduced to objective terms. To do so, we will use new predicates not appearing in the set to capture any non-objective

Knowledge Bases as Representations of Epistemic States

knowledge required by γ. That is, we use new predicate letters to convert non-objective knowledge into objective knowledge.

First some notation: suppose $\phi[x_1, \ldots, x_k]$ is an objective formula with free variables $x_1, \ldots, x_k$, and P is a predicate letter. We will let $P \bullet \phi$ be the sentence

$$\forall x_1 \ldots \forall x_k \mathbf{K}[P(x_1, \ldots, x_k) \equiv \mathbf{K}\phi].$$

Thus the subjective sentence $P \bullet \phi$ expresses the property that it is known that instances of P correspond exactly to the *known* instances of ϕ.

Now we need the following lemma:

Lemma 6.4.2: *Suppose predicate P does not appear in formula ϕ or sentence α, and that $(P \bullet \phi \supset \alpha)$ is valid. Then so is α.*

Proof: Assume to the contrary that the given α is not valid, and so $e, w \models \neg\alpha$. For any w, let $w^\bullet$ be exactly like w except that $w^\bullet[P(n_1, \ldots, n_k)] = 1$ iff $e \models \mathbf{K}\phi[n_1, \ldots, n_k]$, and let $e^\bullet$ be the set of $w^\bullet$ for all $w \in e$. Then we have that $e \models \mathbf{K}\phi[n_1, \ldots, n_k]$ iff $e^\bullet \models \mathbf{K}\phi[n_1, \ldots, n_k]$, since ϕ does not use P, and so, $e^\bullet \models P \bullet \phi$. However, because $e, w \models \neg\alpha$, we have that $e^\bullet, w \models \neg\alpha$, since α does not use P either. This contradicts the assumption that $(P \bullet \phi \supset \alpha)$ is valid. ∎

From this, we get:

Corollary 6.4.3: *If Γ is finite and satisfiable, then for any objective formula ϕ, there is a predicate P such that $\Gamma \cup \{P \bullet \phi\}$ is satisfiable.*

and then:

Corollary 6.4.4: *If γ is satisfiable, then there is a satisfiable set Γ containing γ and such that for every objective formula ϕ, there is a predicate P such that $P \bullet \phi$ is in Γ.*

The claim here is that this set Γ is not only satisfiable, but is satisfied by a representable state. Let e be a maximal state that satisfies Γ, and let Φ be the set of all objective sentences ϕ such that $e \models \mathbf{K}\phi$. We will show that Φ represents e.

First some terminology: for any formula α, define $\alpha^\bullet$ to be the following formula:

if α is objective, then $\alpha^\bullet = \alpha$;
$(\neg\alpha)^\bullet = \neg\alpha^\bullet$;
$(\alpha \wedge \beta)^\bullet = (\alpha^\bullet \wedge \beta^\bullet)$;

$(\forall x \alpha)^{\bullet} = \forall x \alpha^{\bullet}$;
$(K\alpha)^{\bullet} = P(x_1, \ldots, x_k)$, where $x_1, \ldots, x_k$ are the free variables in α and $P \bullet \alpha^{\bullet} \in \Gamma$.

Note that $\alpha^{\bullet}$ is always objective. Then we have the following:

Lemma 6.4.5: *For the given e above, for any w such that $w \models \Phi$, for every formula α, and for all names $n_1, \ldots, n_k$*

$$e, w \models \alpha[n_1, \ldots, n_k] \text{ iff } w \models \alpha^{\bullet}[n_1, \ldots, n_k].$$

Proof: The proof is by induction on α. The lemma clearly holds for atomic sentences, equalities, and by induction, for negations, conjunctions, and quantifications (given that α can be an open formula with any number of free variables).

Finally consider $K\alpha[n_1, \ldots, n_k]$. First observe that $e \models K\alpha[n_1, \ldots, n_k]$ iff $e \models K\alpha^{\bullet}[n_1, \ldots, n_k]$: the former holds iff for every $w' \in e$, $e, w' \models \alpha[n_1, \ldots, n_k]$ iff for every $w' \in e$, $w' \models \alpha^{\bullet}[n_1, \ldots, n_k]$ (by the induction hypothesis, since $w' \models \Phi$ when $w' \in e$) iff the latter holds.

Now suppose that $P \bullet \alpha^{\bullet}$ is the element of Γ guaranteed for $\alpha^{\bullet}$ in the construction. So $e \models P \bullet \alpha^{\bullet}$. There are two cases to consider: if $e \models K\alpha[n_1, \ldots, n_k]$, then by the above, $e \models K\alpha^{\bullet}[n_1, \ldots, n_k]$, and therefore $e \models KP(n_1, \ldots, n_k)$, which means that $P(n_1, \ldots, n_k) \in \Phi$, and thus, $w \models P(n_1, \ldots, n_k)$; similarly, if $e \models \neg K\alpha[n_1, \ldots, n_k]$, then $e \models \neg K\alpha^{\bullet}[n_1, \ldots, n_k]$, and so $e \models K\neg P(n_1, \ldots, n_k)$ and $w \models \neg P(n_1, \ldots, n_k)$, by the same argument. Either way, $e \models K\alpha[n_1, \ldots, n_k]$ iff $w \models (K\alpha)^{\bullet}[n_1, \ldots, n_k]$, which completes the proof. ∎

When α has no free variables, we get as an obvious corollary:

Corollary 6.4.6: *Let $w \in e$. Then $e, w \models \alpha$ iff $w \models \alpha^{\bullet}$, and so $e \models K\alpha$ iff $e \models K\alpha^{\bullet}$.*

Finally, to show that e is represented by Φ, we need to show that $e = \{w \mid w \models \Phi\}$. Clearly, if $w \in e$, then $w \models \Phi$. For the converse, we assume that $w \models \Phi$, and show that $w \in e$. Since e is maximal, by Theorem 6.1.1, we need only show that w satisfies everything known in e, that is, that for any α such that $e \models K\alpha$, we have that $e, w \models \alpha$. So suppose that $e \models K\alpha$. By the above corollary, we have that $e \models K\alpha^{\bullet}$, and so $w \models \alpha^{\bullet}$, since $\alpha^{\bullet} \in \Phi$. Then by the lemma above, we get that $e, w \models \alpha$, which completes the proof of the theorem.

So what this theorem shows is that when we are talking about a state of knowledge using only a finite set of sentences of $\mathcal{KL}$, we are justified in interpreting this as pertaining to a representable state of knowledge. We saw in the previous section that it is possible to

Knowledge Bases as Representations of Epistemic States

---

1. $\forall xyz[R(x, y) \land R(y, z) \supset R(x, z)]$
   R is transitive.
2. $\forall x \neg R(x, x)$
   R is irreflexive.
3. $\forall x[\mathbf{K}P(x) \supset \exists y. R(x, y) \land \mathbf{K}P(y)]$
   For every known instance of P, there is another one that is R related to it.
4. $\forall x[\mathbf{K}\neg P(x) \supset \exists y. R(x, y) \land \mathbf{K}\neg P(y)]$
   For every known non-instance of P, there is another one that is R related to it.
5. $\exists x \mathbf{K}P(x) \land \exists x \mathbf{K}\neg P(x)$
   There is at least one known instance and known non-instance of P.
6. $\exists x \neg \mathbf{K}P(x)$
   There is something that is not known to be an instance of P.

Figure 6.1: A sentence unsatisfiable in finite states

force the state of knowledge to be non-representable, but to do so requires an infinite set of sentences.

6.5 Finite representations are not sufficient

When we think of a representation of knowledge, we usually have in mind a finite one, that is, a finite collection of symbolic structures that can be stored and manipulated computationally. It would be nice if the characterization of knowledge offered by $\mathcal{KL}$ conformed to this view. So the question here is whether the theorem of the previous section can be strengthened: is validity in $\mathcal{KL}$ the same as truth in all finitely representable epistemic states?

Unfortunately, this is not the case. As we will show, if we were to limit ourselves to finitely representable epistemic states, we would have to change the logic $\mathcal{KL}$, in the sense that new sentences would be valid. We will discuss the implications of this later in Section 6.6. First, we state the fact formally:

Theorem 6.5.1: *There is a satisfiable sentence π such that π is false at every finitely representable epistemic state. Equivalently, there is a sentence that is not valid in $\mathcal{KL}$, but that is true at every finitely representable epistemic state.*

The proof involves a sentence π that states that there is an infinite set of known instances of a predicate P and an infinite set of known non-instances of P. We then show that although π is satisfiable, it cannot be satisfied by any finitely representable state.

The π in question is the conjunction of the sentences appearing in Figure 6.1. First, observe that π is satisfiable. Choose an ordering of the standard names, n_1, and n_2, and so on. Let w be any world state such that $R(m, n)$ is true exactly when m appears earlier than n in the ordering. Let e be the set of all world states that satisfy all of the following objective sentences:

$$\{P(n_1), \neg P(n_2), P(n_3), \neg P(n_4), \ldots.\}$$

Then it is easy to verify that $e, w \models \pi$.

Notice what π is doing here: by making R be transitive and irreflexive, we are forcing it to behave like *less than*, and so we are forcing every known instance of P to have a "greater" one, and similarly, for the known non-instances. The second to last conjunct makes sure that these sets are not empty, so that they must be infinite, That is, we must have an infinite chain of known instances, and an infinite one of known non-instances. The very last conjunct makes sure that the epistemic state is consistent, so that not every sentence is known.

To complete the proof, we first need the following easy to prove property of representable states:

Theorem 6.5.2: *Suppose* KB *is objective and* $e = \Re[\![\text{KB}]\!]$. *Then for any objective sentence* ϕ, $e \models K\phi$ *iff* KB $\cup \{\neg\phi\}$ *is unsatisfiable.*

We also need the following property of objective sentences:

Lemma 6.5.3: *Suppose ϕ is a satisfiable objective sentence. Let*

$$A = \{n \mid (\phi \supset P(n)) \text{ is valid}\} \quad B = \{n \mid (\phi \supset \neg P(n)) \text{ is valid}\}.$$

Then either A or B is finite.

Proof: Consider all the names appearing in ϕ. If A only contains these names, then clearly A is finite, and we are done. Otherwise, there must be an n such that $(\phi \supset P(n))$ is valid, but such that n does not appear in ϕ. Let m be any other name that does not appear in ϕ. By Theorem 4.4.2 of Chapter 4, $(\phi \supset P(m))$ must be valid too. Consequently, A contains all the names that do not appear in ϕ. Since ϕ is satisfiable, A and B must be disjoint, and so B can only contain names that do appear in ϕ, and so must be finite. ∎

Now suppose to the contrary that π is satisfied by some finitely representable state. That is, for some e and w, we have that $e, w \models \pi$, where $e = \Re[\![\phi]\!]$. Because of the last conjunct in π, e must be non-empty, and so ϕ must be satisfiable. By the theorem above, the known instances of P will be the set A in the above lemma, and the known non-instances of P will be the set B. Moreover, by the lemma, one of A or B must be finite. Thus, if e is finitely representable and consistent, it cannot have an infinite set of known instances and known non-instances of P, which contradicts π. This completes the proof of the theorem.

What we have shown is that it is possible using a single sentence of $\mathcal{KL}$ to assert that there is knowledge that cannot be represented finitely. If we were to restrict our attention to finite states alone, we would have to arrange the semantics so that the negations of sentences like π somehow came out valid, an unlikely prospect.

6.6 Representability and TELL

After such a tortuous route, we should summarize where we stand in terms of the types of epistemic states we have considered:

1. *maximal states* are fully general in that any satisfiable set of sentences is satisfied by a maximal state.

2. *representable states*, that is, those that can be represented by a set of objective sentences, are also sufficient in that a sentence of $\mathcal{KL}$ is satisfiable iff it is true at some representable state. However, this generality does not extend to infinite sets of sentences, as it did with maximal states.

3. *finitely representable states*, that is, those that can be represented by a finite set of objective sentences, are *not* adequate for the semantics of $\mathcal{KL}$ in that there are satisfiable sentences that are false at every finitely representable state.

Thus the three categories of epistemic states are semantically distinct.

So where does this leave us in terms of providing a specification for knowledge-based systems via the **TELL** and **ASK** operations. Intuitively, it would be nice to say that finitely representable epistemic states (that is, those resulting from finite symbolic KBs) are our only concern. But the results above show that these are overly restrictive as far as $\mathcal{KL}$ is concerned. We need to allow for all representable states, including those resulting from an infinite KB, although it is far from clear how we are supposed to "implement" them.

But there is another problem. Consider the definition of the **TELL** operation from Chapter 5, and the states e_1 and e_2 used in the proof in Section 6.3. As we showed, e_1 is representable, but e_2, defined as $e_1 - \overline{w}$, is not. The problem is that e_2 is the result of a **TELL** operation:

$$e_2 = \textbf{TELL}[\exists x(P(x) \wedge \neg KP(x)), e_1].$$

Thus we have the following unfortunate result:

Theorem 6.6.1: *Representable states are not closed under* **TELL**.

In other words, given a representable state as argument, the result of **TELL** need not be representable. This is a problem since in allowing for infinitely representable states, we can move to a non-representable one by applying a **TELL** operation.

How can we resolve these difficulties? As it turns out, there is a simple and elegant answer. First of all, we need a certain property of $\mathcal{KL}$ which will be demonstrated in the next chapter: although representable states are not closed under **TELL**, *finitely representable* states are. That is, the main result of the next chapter is that given a finitely represented epistemic state as argument, the result of a **TELL** operation can always be represented finitely. This is encouraging since we are clearly more interested in finite representations of knowledge than in the (non-physically realizable) infinite ones.

On the other hand, we have already cautioned against limiting our attention to finitely representable states. As we showed, the logic of $\mathcal{KL}$ must allow for the infinite ones. The solution is that, while we have to allow for these non-finite states, we do not need to ever implement them.

Imagine, epistemic states coming to exist as a result of a sequence of **TELL** operations. We start with e_0, the least informed state, and as a result of being told sentences α_1, α_2, and so on, we move through states e_1, e_2, and so on. Because e_0 is finitely representable, by the theorem of the next chapter, each of the e_i will also be finitely representable. The infinitely represented states, then can be thought of as *limit points* in this process, where an infinite number of sentences have been asserted. This means that we never actually arrive at an infinite state, but that we can get arbitrarily close. And since we never get there, we never get to go *beyond* them either with additional **TELL** operations. Thus we never have to consider a non-finitely representable state as an argument to **TELL**. This picture fully resolves the problem since the full range of states to consider are all those that result from a sequence of **TELL** operations *including the limit points* even though the limit points are themselves never further arguments to **TELL**.

In the next chapter, we supply the remaining piece, showing that finitely representable states are indeed closed under **TELL** operations, and hence that a specification of knowledge-based systems can be meaningfully limited to representable epistemic states.

6.7 Bibliographic notes

This chapter deals with the relationship between representations of knowledge and abstract epistemic states. Somewhat surprisingly, although the idea of representing knowledge symbolically is a familiar one within AI, to our knowledge, very little has been written about the correspondence between representations and states of knowledge. One reason is that there appear to be two somewhat separate communities involved: the knowledge representation community, as seen, for example in the Conference on Principles of Knowledge Representation and Reasoning [19], and the logics of knowledge community, as seen, for example, in the Conference on Theoretical Aspects of Rationality and Knowledge [133]. Early version of the results reported here appeared in [80] and [82]. These results depend on our view of a KB as encoding objective knowledge about the world. The idea that facts about what is or is not known could also be part of a KB and thus contribute to what is known is a much more complex notion, and the basis of autoepistemic logic, discussed in Chapter 9.

6.8 Exercises

1. Show that maximal states satisfy $e = \{w \mid e, w \models K\alpha \supset \alpha\}$.
2. Prove Theorem 6.2.1.
3. Prove Theorem 6.2.2.
4. Prove Theorem 6.5.2.
5. Show that Theorem 6.2.2 can be strengthened for the quantifier-free subset of $\mathcal{KL}$: any two epistemic states that have the same objective knowledge are equivalent.
6. Show that Theorem 6.5.1 is false for the quantifier-free subset of $\mathcal{KL}$: any quantifier-free sentence that is true at all finitely representable states is valid.

7 The Representation Theorem

In our analysis of what it would mean for a system to have knowledge, we started with an informal picture of a knowledge-based system, that is, one containing a *knowledge base*: a collection of symbolic structures representing what is known to be true about the world. As we developed the logic $\mathcal{KL}$, we gradually replaced this symbolic understanding of knowledge with a more abstract one where we talked about an *epistemic state*: a set of world states any of which could be, according to what is known, the correct specification of what is true in the world.

In Chapter 5, we showed how we could define **TELL** and **ASK** operations in terms of these abstract epistemic states without appealing to any notion of symbolic representation, except as part of the interface language for assertions and questions. In the previous chapter, we showed that even though there were far more epistemic states than possible symbolic knowledge bases, as far as the logic $\mathcal{KL}$ was concerned, we could restrict our attention to those epistemic states that were representable by knowledge bases.

The question to be addressed in this chapter is the impact of this representational view of epistemic states on the **TELL** and **ASK** operations. What we will show here is how **TELL** and **ASK** can be realized or implemented using ordinary first-order knowledge bases. In particular, we will show that is possible to reduce the use of $\mathcal{KL}$ in these operations to that of $\mathcal{L}$ in the following way:

- Given a finite KB representing an epistemic state and any sentence of $\mathcal{KL}$ used as an argument to **ASK**, we can eliminate the *K* operators from the question, reducing the question to an objective one, and answer it using ordinary (first-order) theorem-proving operations.
- Given a finite KB representing an epistemic state and any sentence of $\mathcal{KL}$ used as an argument to **TELL**, we can eliminate the *K* operators from the assertion, reducing the assertion to an objective one, and represent the new epistemic state by conjoining this objective assertion and the original KB.

So starting with a finitely representable epistemic state, it will always be possible to find a finite representation of the result of a **TELL** operation, even if the assertion uses *K* operators. Moreover, the operation is *monotonic* in the sense that it involves adding a new objective sentence conjunctively to the previous representation. This result also establishes the fact that finitely representable states are closed under **TELL**. Perhaps more importantly, it provides a clear link between the abstract knowledge level view of knowledge and the more concrete symbol level view where symbolic representations are manipulated.

Since this representation theorem involves several steps that are interesting in their own right, before presenting the proof, we begin by discussing the argument informally.

7.1 The method

To see how the representation theorem will work, it is useful to consider a very simple example. Suppose that we have a state e represented by a KB consisting of two sentences: $P(^\#1)$, and $P(^\#2)$. What we want to consider is how to represent the result of asserting a sentence containing K operators. For example, consider the result of

$$\textbf{TELL}[\exists x(P(x) \land \neg KP(x)), \Re[\![KB]\!]].$$

The intent of this assertion is to say that there is an instance of P apart from the known ones. Since in this case the known ones are $^\#1$ and $^\#2$, the resulting state can be represented by

$$\{P(^\#1),\ P(^\#2),\ \exists x[P(x) \land \neg((x = {}^\#1) \lor (x = {}^\#2))]\}.$$

So we add the assertion to the KB except that the subwff $KP(x)$ is replaced by the wff

$$((x = {}^\#1) \lor (x = {}^\#2)).$$

In general this is the tactic we will follow: replace subwffs containing a K by objective wffs that carry the same information, and then conjoin the result with the original KB.

But what does it mean to carry the same information? It is certainly not true that the two wffs above are logically equivalent. What we do have, however, is that for the initial state, $e = \Re[\![KB]\!]$, the known instances of P are precisely $^\#1$ and $^\#2$. Thus, we get for any n that

$$e \models KP(n) \quad \text{iff} \quad n \in \{^\#1, {}^\#2\} \quad \text{iff} \quad \models ((n = {}^\#1) \lor (n = {}^\#2)).$$

If our initial KB had also contained $P(^\#3)$, we would have wanted the formula

$$((x = {}^\#1) \lor (x = {}^\#2) \lor (x = {}^\#3)).$$

Clearly the objective formula we need to replace $KP(x)$ is a function of the initial epistemic state, $\Re[\![KB]\!]$.

Consider a more difficult case. Suppose KB had been the sentence

$$\forall x[(x \neq {}^\#3) \supset P(x)].$$

In this case, there are an *infinite* number of known instances of P so we cannot disjoin them as above. However, we can still represent the set finitely using the wff $(x \neq {}^\#3)$, since

$$e \models KP(n) \quad \text{iff} \quad n \notin \{^\#3\} \quad \text{iff} \quad \models (n \neq {}^\#3).$$

The result of the same assertion can be represented in this case by

$$\{\forall x[(x \neq {}^\#3) \supset P(x)],\ \exists x[P(x) \land \neg(x \neq {}^\#3)]\},$$

which is logically equivalent to

$$\{\forall x. P(x)\}.$$

The Representation Theorem

In other words, if we start with all but $^\#3$ as the known instances of P, and then we are told that there is another P apart from the known ones, we end up knowing that everything is an instance of P.

So the procedure we will follow in general is this: Given a KB and a subwff $K\phi$ appearing in an assertion, we find an *objective* formula with the same free variables as ϕ, which we call RES⟦ϕ, KB⟧, and use it to replace $K\phi$. Once all such subwffs have been replaced, the resulting objective sentence is added to the KB. For this to work, we need RES⟦ϕ, KB⟧ to satisfy the property that for any n

$$e \models K\phi_n^x \quad \text{iff} \quad \models \text{RES}⟦\phi, e⟧_n^x,$$

for $e = \Re⟦\text{KB}⟧$, (and suitably generalized for additional free variables). In other words, we need the formula RES⟦ϕ, KB⟧ to correctly capture the known instances of ϕ for the epistemic state $\Re⟦\text{KB}⟧$.

7.2 Representing the known instances of a formula

The definition of RES is a recursive one, based on the number of free variables in the wff ϕ. For the base case, we need to consider what to do if ϕ has no free variables. For example, if we were to assert

$$P(^\#1) \wedge (P(^\#2) \vee KP(^\#3)),$$

then since the subwff $KP(^\#3)$ is subjective, it is either known to be true or known to be false. In the former case, the assertion overall should reduce to

$$P(^\#1),$$

and in the latter case, to

$$P(^\#1) \wedge P(^\#2).$$

The simplest way of achieving this is to let RES return an always-true objective sentence in the former case and an always-false objective sentence in the latter. It will be important not to introduce any new standard names in the process so we will use $\forall z(z = z)$ and its negation as the two sentences. We will call the former TRUE and the latter FALSE.

So for example, if KB is $\{P(^\#1), P(^\#2)\}$, then we have that

RES⟦$P(^\#1)$, KB⟧ = TRUE,
RES⟦$(P(^\#2) \vee P(^\#6))$, KB⟧ = TRUE,
RES⟦$(P(^\#3) \vee P(^\#6))$, KB⟧ = FALSE.

The decision to return the true or the false sentence is based on whether the ϕ in question is known. Because ϕ is objective, by Theorem 6.5.2, this is the same as whether or not the objective sentence (KB $\supset \phi$) is valid.

Now consider the case of $\phi(x)$ containing a single free variable x. The idea here is to construct a wff that carries the same information as an infinite disjunction that runs through all the standard names, and for each name n, considers whether or not $\phi(n)$ is known. If it is, we keep $(x = n)$ as part of the disjunction; if it is not, we discard it. For example, for the above KB, we want something like

$$\text{RES}[\![P(x), \text{KB}]\!] = ((x = {}^{\#}1) \lor (x = {}^{\#}2)),$$

since for every other n, we have that $P(n)$ is not known to be true. The test for $\phi(n)$ being known is actually a recursive call to RES with one fewer free variable, so we will really get something more like this:

$$((x = {}^{\#}1) \land \text{RES}[\![P({}^{\#}1), \text{KB}]\!]) \lor$$
$$((x = {}^{\#}2) \land \text{RES}[\![P({}^{\#}2), \text{KB}]\!]) \lor$$
$$((x = {}^{\#}3) \land \text{RES}[\![P({}^{\#}3), \text{KB}]\!]) \lor$$
$$((x = {}^{\#}4) \land \text{RES}[\![P({}^{\#}4), \text{KB}]\!]) \lor \ldots,$$

which simplifies to the same thing, since only the first two return TRUE.

The only thing left to do is to convert this infinite disjunction to a finite formula. To do so, we focus on the names appearing in either the KB or ϕ. Since the KB is assumed to be finite, there are only a finite number of these. Assuming for the moment that $\phi(x)$ only uses the name ${}^{\#}3$, for the above KB, this gives us the first three terms of the disjunction:

$$((x = {}^{\#}1) \land \text{RES}[\![\phi({}^{\#}1), \text{KB}]\!]),$$
$$((x = {}^{\#}2) \land \text{RES}[\![\phi({}^{\#}2), \text{KB}]\!]),$$
$$((x = {}^{\#}3) \land \text{RES}[\![\phi({}^{\#}3), \text{KB}]\!]).$$

For all the remaining n, that is, for the infinite set of names not appearing in either KB or ϕ, we use Theorem 2.8.5 and its corollaries to establish that we need only consider a single name, since all such names will behave the same. In other words, instead of asking if $\phi(n)$ is known for an infinite set of names, we choose a single new name n', and ask if $\phi(n')$ is known. However, we cannot simply use the disjunct

$$((x = n') \land \text{RES}[\![\phi(n'), \text{KB}]\!]),$$

since this is what we do when n' appears normally in the KB or ϕ (like ${}^{\#}1$). Rather, we construct the final disjunct so that it does not mention n' directly: instead of $(x = n')$ we use

$$((x \neq {}^{\#}1) \land (x \neq {}^{\#}2) \land (x \neq {}^{\#}3)),$$

and instead of $\text{RES}[\![\phi(n'), \text{KB}]\!]$, which might end up containing n', we use

$$\text{RES}[\![\phi(n'), \text{KB}]\!]_x^{n'}$$

which (abusing notation somewhat) means replacing any n' that occurs by the free variable x. Putting all this together, then, for a KB that uses just ${}^{\#}1$ and ${}^{\#}2$ and a ϕ that uses just ${}^{\#}3$, we get

$$\text{RES}[\![\phi(x), \text{KB}]\!] =$$
$$((x = {}^{\#}1) \wedge \text{RES}[\![\phi({}^{\#}1), \text{KB}]\!]) \vee$$
$$((x = {}^{\#}2) \wedge \text{RES}[\![\phi({}^{\#}2), \text{KB}]\!]) \vee$$
$$((x = {}^{\#}3) \wedge \text{RES}[\![\phi({}^{\#}3), \text{KB}]\!]) \vee$$
$$((x \neq {}^{\#}1) \wedge (x \neq {}^{\#}2) \wedge (x \neq {}^{\#}3) \wedge \text{RES}[\![\phi(n'), \text{KB}]\!]_x^{n'}),$$

where n' is some name other than ${}^{\#}1$, ${}^{\#}2$, or ${}^{\#}3$. For example, for the above KB, we have that

$$\text{RES}[\![P(x), \text{KB}]\!] =$$
$$((x = {}^{\#}1) \wedge \text{TRUE}) \vee$$
$$((x = {}^{\#}2) \wedge \text{TRUE}) \vee$$
$$((x \neq {}^{\#}1) \wedge (x \neq {}^{\#}2) \wedge \text{FALSE}),$$

which correctly simplifies to $((x = {}^{\#}1) \vee (x = {}^{\#}2))$. If instead the KB had been of the form $\forall x[(x \neq {}^{\#}3) \supset P(x)]$, then we would have had

$$\text{RES}[\![P(x), \text{KB}]\!] =$$
$$((x = {}^{\#}3) \wedge \text{FALSE}) \vee ((x \neq {}^{\#}3) \wedge \text{TRUE}),$$

which simplifies to $(x \neq {}^{\#}3)$, as desired. Further examples are in the exercises.

We now provide the definition of RES in its full generality:

Definition 7.2.1: Let ϕ be an objective formula and KB be a finite set of objective sentences. Suppose that $n_1, \ldots, n_k$, are all the names in ϕ or in KB, and that n' is some name that does not appear in ϕ or in KB. Then $\text{RES}[\![\phi, \text{KB}]\!]$ is defined by:

1. If ϕ has no free variables, then $\text{RES}[\![\phi, \text{KB}]\!]$ is
   TRUE, if $\text{KB} \models \phi$, and FALSE, otherwise.
2. If x is a free variable in ϕ, then $\text{RES}[\![\phi, \text{KB}]\!]$ is
   $((x = n_1) \wedge \text{RES}[\![\phi_{n_1}^x, \text{KB}]\!]) \vee \ldots$
   $((x = n_k) \wedge \text{RES}[\![\phi_{n_k}^x, \text{KB}]\!]) \vee$
   $((x \neq n_1) \wedge \ldots \wedge (x \neq n_k) \wedge \text{RES}[\![\phi_{n'}^x, \text{KB}]\!]_x^{n'}).$

To make this definition completely determinate, we can choose n' and x to be the first (in lexicographic order) standard name and variable that satisfy their respective criterion.

The main property of this definition that we require is:

Lemma 7.2.2: *For any finite KB with $e = \Re[\![\text{KB}]\!]$, any objective formula ϕ with free variables $x_1, \ldots, x_k$, and any standard names $n_1, \ldots, n_k$,*

$$e \models \mathbf{K}\phi_{n_1 \cdots n_k}^{x_1 \cdots x_k} \quad \textit{iff} \quad \models \text{RES}[\![\phi, \text{KB}]\!]_{n_1 \cdots n_k}^{x_1 \cdots x_k}.$$

Proof: Since $e \models K\phi_{n_1}^{x_1}\cdots_{n_k}^{x_k}$ iff $\models (KB \supset \phi_{n_1}^{x_1}\cdots_{n_k}^{x_k})$ by Theorem 6.5.2, it suffices to show that

$$\models \text{RES}[\![\phi, KB]\!]_{n_1}^{x_1}\cdots_{n_k}^{x_k} \quad \text{iff} \quad \models (KB \supset \phi)_{n_1}^{x_1}\cdots_{n_k}^{x_k}.$$

The proof is by induction on the number of free variables in ϕ.

If ϕ has no free variables, the lemma clearly holds since $\text{RES}[\![\phi, KB]\!]$ will be valid iff it is equal to TRUE, which happens iff $(KB \supset \phi)$ is valid.

Now suppose that ϕ has k free variables, and that by induction, for any n we have that $\phi_n^{x_1}$ satisfies the lemma. Now consider $\text{RES}[\![\phi, KB]\!]_{n_1 n_2}^{x_1 x_2}\cdots_{n_k}^{x_k}$, and call this sentence ψ. Looking at the name n_1, there are two cases to consider, depending on whether or not n_1 appears in KB or ϕ. If it does appear, all disjuncts in ψ but the one naming n_1 simplify to false, and so $\models \psi$ iff $\models \text{RES}[\![\phi_{n_1}^{x_1}, KB]\!]_{n_2}^{x_2}\cdots_{n_k}^{x_k}$, which by induction happens iff $\models (KB \supset \phi_{n_1}^{x_1})_{n_2}^{x_2}\cdots_{n_k}^{x_k}$, and so the lemma is satisfied.

If on the other hand, n_1 does not appear in KB or ϕ, then all but the last disjunct in ψ simplifies to false, and so $\models \psi$ iff $\models \text{RES}[\![\phi_{n'}^{x_1}, KB]\!]_{x_1 n_1 n_2}^{n' x_1 x_2}\cdots_{n_k}^{x_k}$, where n' is some name that also does not appear in either KB or ϕ. Now consider the formula $\text{RES}[\![\phi_{n'}^{x_1}, KB]\!]_{x_1}^{n'}$. A trivial induction argument shows that this objective formula does not contain either n_1 or n', since RES does not introduce any new names in its result. Now we will apply Corollary 2.8.6 using a bijection $*$ that swaps the names n_1 and n' but leaves all other names unchanged. We get that

$$\models \text{RES}[\![\phi_{n'}^{x_1}, KB]\!]_{x_1 n_1 n_2}^{n' x_1 x_2}\cdots_{n_k}^{x_k} \quad \text{iff} \quad \models \text{RES}[\![\phi_{n'}^{x_1}, KB]\!]_{x_1 n' n_2*}^{n' x_1 x_2}\cdots_{n_k*}^{x_k}.$$

But the formula $\text{RES}[\![\phi_{n'}^{x_1}, KB]\!]_{x_1 n'}^{n' x_1}$ is just $\text{RES}[\![\phi_{n'}^{x_1}, KB]\!]$, and by induction,

$$\models \text{RES}[\![\phi_{n'}^{x_1}, KB]\!]_{n_2*}^{x_2}\cdots_{n_k*}^{x_k} \quad \text{iff} \quad \models (KB \supset \phi)_{n' n_2*}^{x_1 x_2}\cdots_{n_k*}^{x_k}.$$

Again applying Corollary 2.8.6 we have that

$$\models (KB \supset \phi)_{n' n_2*}^{x_1 x_2}\cdots_{n_k*}^{x_k} \quad \text{iff} \quad \models (KB \supset \phi)_{n_1 n_2}^{x_1 x_2}\cdots_{n_k}^{x_k}$$

as desired, which completes the proof. ∎

This lemma shows that RES properly captures the known instances of ϕ. As a corollary to this, we have

Corollary 7.2.3: *Let KB, e, ϕ and the names n_i be as in the lemma. Let w be an arbitrary world state. Then $e \models K\phi_{n_1}^{x_1}\cdots_{n_k}^{x_k}$ iff $w \models \text{RES}[\![\phi, KB]\!]_{n_1}^{x_1}\cdots_{n_k}^{x_k}$.*

The proof is that because the result of RES is a wff that does not use predicate symbols, function symbols, or K operators, its instances are either valid or their negations are valid.

The Representation Theorem

7.3 Reducing arbitrary sentences to objective terms

Now that we have seen how we can replace $K\phi$ by an objective wff that correctly represents its instances, we need to reconsider more precisely the idea discussed earlier of replacing all such subwffs in a sentence, so that the result can then be conjoined with a KB. We call this operation *reducing* a formula to objective terms, and for any wff α of $\mathcal{KL}$, we use the notation $\|\alpha\|_{\text{KB}}$ to mean the objective reduction α with respect to KB. Formally the definition is as follows:

Definition 7.3.1: Given a finite KB and α an arbitrary wff of $\mathcal{KL}$, $\|\alpha\|_{\text{KB}}$ is the objective wff defined by

$\|\alpha\|_{\text{KB}} = \alpha$, when α is objective;

$\|\neg\alpha\|_{\text{KB}} = \neg\|\alpha\|_{\text{KB}}$;

$\|(\alpha \wedge \beta)\|_{\text{KB}} = (\|\alpha\|_{\text{KB}} \wedge \|\beta\|_{\text{KB}})$;

$\|\forall x \alpha\|_{\text{KB}} = \forall x \|\alpha\|_{\text{KB}}$;

$\|\boldsymbol{K}\alpha\|_{\text{KB}} = \text{RES}[\![\|\alpha\|_{\text{KB}}, \text{KB}]\!]$.

Note that this recursive definition works from the "inside out" in that we first reduce the argument to K to objective terms before applying RES. If the argument to K happens to be objective, $\|\cdot\|$ will not change it, and RES is called directly. Otherwise, that is, when we have nested K operators, the call to $\|\cdot\|$ produces an objective wff which can then be passed to RES.[1]

For example, if KB is $\{P(^\#1), P(^\#2)\}$ as we had before, then

$\|\exists x[P(x) \wedge \neg \boldsymbol{K}P(x)]\|_{\text{KB}} =$
$\exists x[P(x) \wedge \neg((x = {}^\#1) \vee (x = {}^\#2))]$.

Note that the only part of the sentence that gets changed by $\|\cdot\|$ is the part involving $\boldsymbol{K}$. Unlike RES, the goal of $\|\cdot\|$ is not to produce the known instances of a wff; rather, it is to take an arbitrary wff and produce an objective version that is true in exactly the same world states by encoding what is needed from the epistemic state. More precisely, we have the following:

Lemma 7.3.2: *For any finite KB with $e = \Re[\![\text{KB}]\!]$, any world state w, any formula α of*

---

[1] Strictly speaking, the well-formedness of this definition should not simply be assumed; we need to prove (by induction) that $\|\cdot\|$ always returns an objective formula.

$\mathcal{KL}$ *with free variables* $x_1, \ldots, x_k$, *and any standard names* $n_1, \ldots, n_k$,
$$e, w \models \alpha_{n_1\cdots n_k}^{x_1\cdots x_k} \quad \textit{iff} \quad w \models \|\alpha\|_{\text{KB}}{}_{n_1\cdots n_k}^{x_1\cdots x_k}.$$

Proof: The proof is by induction on the structure of α. If α is atomic or an equality, the lemma clearly holds since α is then objective. The lemma also holds by induction for negations, conjunctions, and quantifications. Now consider the formula $\boldsymbol{K}\alpha$. We have that
$$e \models \boldsymbol{K}\alpha_{n_1\cdots n_k}^{x_1\cdots x_k}$$
iff for every $w' \in e$, we have
$$e, w' \models \alpha_{n_1\cdots n_k}^{x_1\cdots x_k}$$
iff (by induction) we have for every $w' \in e$,
$$w' \models \|\alpha\|_{\text{KB}}{}_{n_1\cdots n_k}^{x_1\cdots x_k}$$
iff we have
$$e \models \boldsymbol{K}\|\alpha\|_{\text{KB}}{}_{n_1\cdots n_k}^{x_1\cdots x_k}.$$
Since the formula within the $\boldsymbol{K}$ operator here is now objective, by Corollary 7.2.3, this holds iff
$$w \models \text{RES}[\![\|\alpha\|_{\text{KB}}, \text{KB}]\!]_{n_1\cdots n_k}^{x_1\cdots x_k}$$
which, by definition of $\|\cdot\|$, is the same as
$$w \models \|\boldsymbol{K}\alpha\|_{\text{KB}}{}_{n_1\cdots n_k}^{x_1\cdots x_k}. \blacksquare$$

So this lemma shows that $\|\cdot\|$ preserves the truth value of α with respect to the world state, but removes the dependency of α on the epistemic state. This is just what we need to represent a new epistemic state in objective terms.

7.4 TELL and ASK at the symbol level

With the lemma of the previous section, we saw how we could take an arbitrary formula of $\mathcal{KL}$, and reduce it to objective form in a way that captures its dependency on the epistemic state. This will allow us to deal with **TELL** and **ASK** completely in objective terms, provided that we start with a finitely representable objective state.

Theorem 7.4.1: [The Representation Theorem] *Let KB be any finite set of objective sentences and α be any sentence of $\mathcal{KL}$. Then:*
1. $\text{TELL}[\alpha, \mathfrak{R}[\![\text{KB}]\!]] = \mathfrak{R}[\![(\text{KB} \wedge \|\alpha\|_{\text{KB}})]\!]$.
2. $\text{ASK}[\alpha, \mathfrak{R}[\![\text{KB}]\!]] = yes \quad \textit{iff} \quad \text{KB} \models \|\alpha\|_{\text{KB}}$.

The proof is immediate from the fact that for any w, and for $e = \mathfrak{R}[\![\text{KB}]\!]$,
$$e, w \models \alpha \quad \text{iff} \quad w \models \|\alpha\|_{\text{KB}},$$

which is just Lemma 7.3.2 when α has no free variables. We can also state a variant for **ASK**:

Corollary 7.4.2: *Under the same conditions as the Theorem,*
$$\text{ASK}[\alpha, \Re[\![\text{KB}]\!]] = yes \quad iff \quad \|K\alpha\|_{\text{KB}} \text{ is TRUE}.$$

This important theorem tells us that the result of a **TELL** can always be represented by conjoining an objective sentence to the KB, and that an **ASK** can always be calculated in terms of the (objective) logical implications of the KB. Moreover, as can be seen by examining the definition of $\|\cdot\|$ and RES, the reduction to objective terms itself can be done using only the (objective) logical implications of the KB. The conclusion: we can calculate the answers to **TELL** and to **ASK** for arbitrary sentences of $\mathcal{KL}$ using ordinary first-order theorem proving.

Of course, this does not make it easy to perform these operations, since in general, it is *impossible* to calculate the objective logical implication of a KB. Moreover, the way RES was defined was not particularly realistic, since it involved constructing a formula that could be as large as twice the total number of constants in the KB. But the representation theorem at least shows that the operation is definable in terms of these ordinary first-order operations, which opens the door to possible optimizations in special cases.

Another way of looking at this theorem is to consider a symbolic "implementation" of **TELL** and **ASK** which works directly on representations of the epistemic states. Call these procedures TELL′ and ASK′ respectively, where

- ASK′[α, KB] is defined as
    1. Calculate $\|\alpha\|_{\text{KB}}$ using the recursive definition. Call this ϕ.
    2. Test if (KB $\models \phi$); if it does, return *yes*; otherwise, *no*.
- TELL′[α, KB] is defined as
    1. Calculate $\|\alpha\|_{\text{KB}}$ using the recursive definition. Call this ϕ.
    2. Return (KB $\wedge \phi$).

The representation theorem can thought of as a proof of *correctness* for these symbol level procedures.

7.5 The example KB reconsidered

Let us now return to the example KB introduced in Section 5.7 to examine the workings of the representation theorem. We begin by looking at an example of **ASK**, in particular, Question 8:

8. $\exists x[Teach(x, sue) \land \neg \mathbf{K}Teach(x, sue)]$ TRUE

We have already considered why the answer should be *yes* on semantic grounds. In terms of the representation theorem, we need to reduce the question to objective terms. To do so, we need to calculate the known instances of $Teach(x, sue)$. If we apply the definition of RES, we get that

$\text{RES}[\![Teach(x, sue), KB]\!] =$

    $((x = tom) \land \text{RES}[\![Teach(tom, sue), KB]\!]) \lor$
    $((x = sam) \land \text{RES}[\![Teach(sam, sue), KB]\!]) \lor$
    $((x = tina) \land \text{RES}[\![Teach(tina, sue), KB]\!]) \lor$
    $((x = tara) \land \text{RES}[\![Teach(tara, sue), KB]\!]) \lor$
    $((x = sue) \land \text{RES}[\![Teach(sue, sue), KB]\!]) \lor$
    $((x = ted) \land \text{RES}[\![Teach(ted, sue), KB]\!]) \lor$
    $((x = sara) \land \text{RES}[\![Teach(sara, sue), KB]\!]) \lor$
    $((x = sandy) \land \text{RES}[\![Teach(sandy, sue), KB]\!]) \lor$
    $((x \neq tom) \land (x \neq sam) \land (x \neq tina) \land (x \neq tara) \land$
        $(x \neq sue) \land (x \neq ted) \land (x \neq sara) \land (x \neq sandy) \land$
        $\text{RES}[\![Teach(tania, sue), KB]\!]_x^{tania}),$

where *tania* is the chosen new name not appearing in the KB. Here all the recursive calls are of the form $Teach(n, sue)$ and have no free variables. So they will either return TRUE or FALSE, depending on whether $KB \models Teach(n, sue)$. For all but the standard name *tina*, the answer will be FALSE. So simplifying, we have

$\text{RES}[\![Teach(x, sue), KB]\!] = (x = tina),$

meaning that Tina is the only *known* teacher of Sue. Then we get that

$\|\exists x Teach(x, sue) \land \neg \mathbf{K}Teach(x, sue)\|_{KB} =$
    $\exists x[Teach(x, sue) \land \neg(x = tina)].$

So the original question reduces to whether or not Sue has a teacher apart from Tina. The answer depends on whether or not

$KB \models \exists x[Teach(x, sue) \land \neg(x = tina)].$

Since the entailment holds (because of what is known about Tom and Ted), the answer is *yes*, as desired.

Notice the two step operation: we first reduce the question to objective terms (using the implications of the KB) and then we determine if the result is implied by the KB.

The Representation Theorem

Alternatively, we could have used the corollary to the representation theorem, and simply calculated

$$\|K\exists x[\mathit{Teach}(x, \mathit{sue}) \land \neg K\mathit{Teach}(x, \mathit{sue})]\|_{\mathrm{KB}}.$$

First the argument to the outermost K must be reduced, which produces

$$\exists x[\mathit{Teach}(x, \mathit{sue}) \land \neg(x = \mathit{tina})],$$

as before because of RES⟦$\mathit{Teach}(x, \mathit{sue})$, KB⟧. Then we need to apply RES to this sentence which, because it is implied by the KB, gives us TRUE, and so once again the answer is *yes*.

As a second example, consider Question 11:

11. $\exists y K \forall x [\mathit{Teach}(x, y) \supset K\mathit{Teach}(x, y)]$ TRUE

We need to apply RES to the innermost formula dominated by a K, $\mathit{Teach}(x, y)$. This has two free variables, and so the depth of recursion will be two. Assuming the y is used first, say, we get

$$\text{RES}⟦\mathit{Teach}(x, y), \mathrm{KB}⟧ =$$
$$((y = \mathit{tom}) \land \text{RES}⟦\mathit{Teach}(x, \mathit{tom}), \mathrm{KB}⟧) \lor$$
$$((y = \mathit{sam}) \land \text{RES}⟦\mathit{Teach}(x, \mathit{sam}), \mathrm{KB}⟧) \lor$$
$$((y = \mathit{tina}) \land \text{RES}⟦\mathit{Teach}(x, \mathit{tina}), \mathrm{KB}⟧) \lor$$
$$((y = \mathit{tara}) \land \text{RES}⟦\mathit{Teach}(x, \mathit{tara}), \mathrm{KB}⟧) \lor$$
$$((y = \mathit{sue}) \land \text{RES}⟦\mathit{Teach}(x, \mathit{sue}), \mathrm{KB}⟧) \lor$$
$$((y = \mathit{ted}) \land \text{RES}⟦\mathit{Teach}(x, \mathit{ted}), \mathrm{KB}⟧) \lor$$
$$((y = \mathit{sara}) \land \text{RES}⟦\mathit{Teach}(x, \mathit{sara}), \mathrm{KB}⟧) \lor$$
$$((y = \mathit{sandy}) \land \text{RES}⟦\mathit{Teach}(x, \mathit{sandy}), \mathrm{KB}⟧) \lor$$
$$((y \neq \mathit{tom}) \land (y \neq \mathit{sam}) \land (y \neq \mathit{tina}) \land (y \neq \mathit{tara}) \land$$
$$(y \neq \mathit{sue}) \land (y \neq \mathit{ted}) \land (y \neq \mathit{sara}) \land (y \neq \mathit{sandy}) \land$$
$$\text{RES}⟦\mathit{Teach}(x, \mathit{sally}), \mathrm{KB}⟧_x^{\mathit{sally}}),$$

where *sally* is the chosen new name that does not appear in the KB. For each name n, we need to calculate RES⟦$\mathit{Teach}(x, n)$, KB⟧ which, as in the previous example, gives us the known teachers of n. For example, if n has no known teachers, such as for *sally* or *tom*, we get the following:

RES⟦$Teach(x, sally)$, KB⟧ =
    (($x = tom$) ∧ RES⟦$Teach(tom, sally)$, KB⟧) ∨
    (($x = sam$) ∧ RES⟦$Teach(sam, sally)$, KB⟧) ∨
    (($x = tina$) ∧ RES⟦$Teach(tina, sally)$, KB⟧) ∨
    (($x = tara$) ∧ RES⟦$Teach(tara, sally)$, KB⟧) ∨
    (($x = sue$) ∧ RES⟦$Teach(sue, sally)$, KB⟧) ∨
    (($x = ted$) ∧ RES⟦$Teach(ted, sally)$, KB⟧) ∨
    (($x = sara$) ∧ RES⟦$Teach(sara, sally)$, KB⟧) ∨
    (($x = sandy$) ∧ RES⟦$Teach(sandy, sally)$, KB⟧) ∨
    (($x = sally$) ∧ RES⟦$Teach(sally, sally)$, KB⟧) ∨
    (($x \neq tom$) ∧ ($x \neq sam$) ∧ ($x \neq tina$) ∧ ($x \neq tara$) ∧ ($x \neq sue$) ∧
      ($x \neq ted$) ∧ ($x \neq sara$) ∧ ($x \neq sandy$) ∧ ($x \neq sally$) ∧
      RES⟦$Teach(tony, sara)$, KB⟧$_x^{tony}$),

where *tony* is the new name. Note that in this case, *tony* must be distinct from all the names in the KB and from the name of *sally* as well, which was introduced earlier in the recursion. Here each of the recursive calls returns FALSE, and so RES⟦$Teach(x, sally)$⟧ simplifies to FALSE. So overall, after simplification, we get

RES⟦$Teach(x, y)$, KB⟧ =
    (($y = sam$) ∧ ($x = tom$)) ∨
    (($y = sue$) ∧ ($x = tina$)) ∨
    (($y = sandy$) ∧ ($x = ted$)),

which captures all the known instances of the *Teach* predicate. Given this, we can reduce the formula

∀$x[Teach(x, y) \supset \mathbf{K}Teach(x, y)]$,

which says that all of the teachers of *y* are known, to

∀$x[Teach(x, y) \supset$ (($y = sam$) ∧ ($x = tom$)) ∨
    (($y = sue$) ∧ ($x = tina$)) ∨ (($y = sandy$) ∧ ($x = ted$))].

This formula, call it ψ, will now become the argument to RES for the outermost **K** operator. It has a single free variable *y*, and we wish to find names *n* for which $\psi(n)$ is known to be true. As usual, we need to consider all the names in the KB:

$\text{RES}[\![\psi(y), \text{KB}]\!] =$

$((y = tom) \land \text{RES}[\![\psi(tom), \text{KB}]\!]) \lor$
$((y = sam) \land \text{RES}[\![\psi(sam), \text{KB}]\!]) \lor$
$((y = tina) \land \text{RES}[\![\psi(tina), \text{KB}]\!]) \lor$
$((y = tara) \land \text{RES}[\![\psi(tara), \text{KB}]\!]) \lor$
$((y = sue) \land \text{RES}[\![\psi(sue), \text{KB}]\!]) \lor$
$((y = ted) \land \text{RES}[\![\psi(ted), \text{KB}]\!]) \lor$
$((y = sara) \land \text{RES}[\![\psi(sara), \text{KB}]\!]) \lor$
$((y = sandy) \land \text{RES}[\![\psi(sandy), \text{KB}]\!]) \lor$
$((y \neq tom) \land (y \neq sam) \land (y \neq tina) \land (y \neq tina) \land$
$\quad (y \neq sue) \land (y \neq ted) \land (y \neq sara) \land (y \neq sandy) \land$
$\quad \text{RES}[\![\phi(steve), \text{KB}]\!]_y^{steve}).$

For each name n, the sentence $\psi(n)$ is true if either n has no teachers or n is *sam* and his only teacher is *tom*, or n is *sue* and her only teacher is *tina*, or n is *sandy* and her only teacher is *ted*. The values of n for which this is implied by the KB are *tom*, *tina*, *tara*, *ted* and *steve*, (for which it is known that they have no teachers), and *sandy* (for which it is known that *ted* is her only teacher). For these values of n, $\text{RES}[\![\psi(n), \text{KB}]\!]$ will return TRUE, and for the others, FALSE. Thus, we get that

$\text{RES}[\![\psi(y), \text{KB}]\!] =$

$(y = ted) \lor (y = tom) \lor (y = tina) \lor (y = tara) \lor (y = sandy) \lor$
$[(y \neq tom) \land (y \neq sam) \land (y \neq tina) \land (y \neq tara) \land (y \neq sue) \land$
$\quad (y \neq ted) \land (y \neq sara) \land (y \neq sandy)]$

which simplifies to

$[(y \neq sam) \land (y \neq sue) \land (y \neq sara)].$

Thus for anyone but *sam*, *sue*, or *sara*, it is known that all of the teachers are known (for all but *sandy*, this is because they are known to not have any teachers). So ψ is an example of a non-trivial formula with infinitely many known instances, captured by the last disjunct of RES using inequalities.

Finally, we can reduce the original question

$\exists y \mathbf{K} \forall x [\textit{Teach}(x, y) \supset \mathbf{K}\textit{Teach}(x, y)]$

to

$\exists y.(y \neq sam) \land (y \neq sue) \land (y \neq sara).$

This will be answered *yes*, since it is implied by the KB, because of what is known about Sandy. She is the only individual that has a teacher and for which it is known that all of her teachers are known.

As a final example in this chapter, we will consider the effect of the Assertion 3 in terms of the representation theorem:

3. $\forall x[\textit{Teach}(x, \textit{sara}) \supset \exists y \textbf{\textit{K}} \textit{Teach}(x, y)]$ $\hspace{2em} [e_2 \rightarrow e_3]$

Earlier, we considered the result of this assertion after two preliminary assertions. To simplify here, we assume that this is being asserted starting in the original KB.

What this sentence says is that anyone who teaches Sara must have someone they are known to teach. To reduce this sentence, we must calculate $\text{RES}[\![\textit{Teach}(x, y), \text{KB}]\!]$, which we did in the previous example. Using that result, the assertion reduces to

$\forall x (\textit{Teach}(x, \textit{sara}) \supset \exists y$
$\hspace{2em} [((y = \textit{sam}) \land (x = \textit{tom})) \lor$
$\hspace{2em} ((y = \textit{sue}) \land (x = \textit{tina})) \lor$
$\hspace{2em} ((y = \textit{sandy}) \land (x = \textit{ted}))]),$

which simplifies to

$\forall x (\textit{Teach}(x, \textit{sara}) \supset [(x = \textit{tom}) \lor (x = \textit{tina}) \lor (x = \textit{ted})]).$

This sentence is objective and can be conjoined to the KB. So the effect of the **TELL** is to assert that Sara's teachers must be among Tom, Tina, or Ted, the only individuals with a known student.

7.6 Wh-questions at the symbol level

In Chapter 5, we introduced a new interaction operation **WH-ASK** which returned the known instances of a formula, defined by

$\textbf{WH-ASK}[\alpha[\vec{x}], e] = \{\vec{n} \mid e \models \textbf{\textit{K}}\alpha[\vec{n}]\}.$

We also mentioned that this set of standard names could be infinite, which presents a problem from an implementation standpoint. But $\|\cdot\|$, which can be used to represent the known instances of a formula, provides a perfect solution to this problem:

$\textbf{WH-ASK}[\alpha, \Re[\![\text{KB}]\!]] = \|\alpha\|_{\text{KB}}.$

Instead of returning a possibly infinite set of known instances of α, we return instead a finite formula from which as many standard names as desired can easily be extracted. From Lemma 7.3.2, it follows that these standard names are precisely the known instances of α.

We also considered in Chapter 5, an interaction operation **DESCRIBE** which returned terms known to be co-referential with a given standard name, defined by

$\textbf{DESCRIBE}[n, e] = \{t \mid e \models \textbf{\textit{K}}(t = n)\}.$

Again, this presents an implementation problem since for a KB like $\{\forall x.x = f(x)\}$, the standard name $^\#1$ is known to be co-referential with an infinite number of other terms:

$$\{^\#1,\ f(^\#1),\ f(f(^\#1)),\ f(f(f(^\#1))),\ \ldots\}.$$

A suggestion here is that instead of returning all co-referential terms, we only return co-referential *primitive* terms, that is, containing exactly one function or constant symbol:

DESCRIBE$[n, e] = \{t \mid t \text{ is primitive and } e \models \boldsymbol{K}(t = n)\}.$

For example, asked to describe a standard name like $^\#23$, we might get the set

$\{jake,\ jack,\ jackie,\ best\_friend(^\#13),\ first\_child(^\#5, ^\#79)\}$

of primitive co-referring terms. We are then free to further elaborate on this set by describing any of the standard names mentioned, and so on to any depth. We leave it as an exercise to show that **DESCRIBE** as redefined above always returns a finite set.

7.7 Bibliographic notes

As discussed in Chapter 4, knowledge has many of the closure properties of entailment, validity, or provability. Further, as seen in Chapter 6, what is known objectively in the epistemic state represented by some KB is precisely the logical entailments of that KB. At its simplest, the Representation Theorem of the current chapter is based on the idea of going through a formula and replacing knowledge of an objective sentence by either TRUE or FALSE according to whether the sentence is entailed by the given KB. This idea is then generalized to non-objective knowledge by working recursively on formulas from the inside out. Finally, we use standard names and equality to deal with formulas with free variables and quantifying in. An early version of these ideas appeared in [80] and [82]. The Representation Theorem was subsequently used to describe integrity constraints on databases in [121].

7.8 Exercises

1. Show that for any KB, RES$[\![(x = ^\#1), \text{KB}]\!]$ is equivalent to $(x = ^\#1)$.
2. Consider the KB that is the conjunction of

$\forall y.R(^\#1, y) \equiv (y = ^\#1) \vee (y = ^\#2)$

$\forall y.R(^\#2, y) \equiv (y \neq ^\#2) \wedge (y \neq ^\#3)$

$\forall y.\neg R(^\#3, y)$

$\forall x, \forall y.((x \neq ^\#1) \wedge (x \neq ^\#2) \wedge (x \neq ^\#3)) \supset (R(x, y) \equiv (x = y))$

Calculate each of the following: RES⟦$R(^\#1, y)$, KB⟧; RES⟦$R(^\#5, y)$, KB⟧; RES⟦$R(x, ^\#2)$, KB⟧; and RES⟦$R(x, ^\#5)$, KB⟧.

3. Show using the representation theorem, why the answer to the question

    $\exists x.Teach(x, sam) \land \neg \mathbf{K}Teach(x, sam)$

    for the example KB in Section 5.7 is UNKNOWN.

4. The definition of RES requires constructing a formula using every standard name mentioned in the KB. Describe a more practical class of KB's and queries where it would not be necessary to enumerate all the standard names in the KB.

5. Call an epistemic state *quasi-finitely representable* if it can be represented by a KB (finite or infinite) that uses only finitely many standard names.

    (a) Prove that the representation theorem works for quasi-finite epistemic states, and hence that these are closed under **TELL**.

    (b) Prove that the sentence π of Theorem 6.5.1 is not satisfied by a quasi-finite epistemic state, and hence that the logic of $\mathcal{KL}$ requires epistemic states that are not quasi-finite.

6. Give an example where RES would return an incorrect value if the standard names used in the definition ranged only over those in the KB, but not over those in the first argument.

7. Prove that for a finite KB, **DESCRIBE**$[n, \Re⟦KB⟧]$ is always finite.

8 Only-Knowing

In previous chapters, we covered in detail the language $\mathcal{KL}$ and how it could be used as the interface language for **TELL** and **ASK** operations. We also saw how its objective fragment, $\mathcal{L}$, could always be used to represent what was known. In this chapter, we begin the examination of a third use for a logical language: as a specification of the behaviour of a knowledge base under the **TELL** and **ASK** operations.

Since we already have a semantic definition of these operations and, as a result of the Representation Theorem of the previous chapter, an equivalent symbolic characterization, why do we need yet another specification? The answer is simply that this logical specification will allow us to generalize very nicely the **TELL** and **ASK** operations in a way that will make a close connection to some of the work in nonmonotonic reasoning, explored in the next chapter.

8.1 The logic of answers

Suppose we start with an epistemic state e represented by $P(^\#1)$. In this state, we have, for example, that

$\quad$ **ASK**$[\exists x P(x), e] = yes$.

One question we can ask about this answer is this:

$\quad$ What property of the logic of $\mathcal{KL}$ tells us that this answer is correct?

By looking at the definition of **ASK**, we can see that all the world states in e satisfy $\exists x P(x)$. In other words, we answer *yes* because any w that satisfies the KB must also satisfy $\exists x P(x)$. This is just another way of saying that we will answer *yes* for any α for which

$\quad \models (\text{KB} \supset \alpha)$,

as expected.

But there is clearly more to the story of **ASK**. For example, we also have that

$\quad$ **ASK**$[K \exists x P(x), e] = yes$

and even

$\quad$ **ASK**$[\exists x K P(x), e] = yes$,

where the α here is not implied by the KB. In this case, the answer arises due to introspection: *knowing* the α is implied by knowing the KB. Thus, we will answer *yes* for any α for which

$\quad \models (K\text{KB} \supset K\alpha)$.

The reason is that since $e \models K\text{KB}$, we get that $e \models K\alpha$ and thus **ASK** must return *yes*. This also subsumes the previous case since if KB implies α, then KKB implies $K\alpha$.

Although this explanation handles positive introspection properly, it does not work for negative introspection. For example, we also have that

$\textbf{ASK}[\neg KP(^\#2), e] = yes.$

What property of the logic of $\mathcal{KL}$ explains this? In this case, knowing KB does not imply knowing α. In fact there is nothing in KB that suggests anything one way or another about $P(^\#2)$. Just because $P(^\#1)$ is known, $P(^\#2)$ may or may not be known. If $P(^\#2)$ is in fact known, then so will be $KP(^\#2)$, by positive introspection; if it is not known, then $\neg KP(^\#2)$ will be known by negative introspection.

How then did **ASK** come to settle on the second case? Informally, the answer is that because the negation of $P(^\#2)$ is consistent with what is known, that is, because

$\not\models (P(^\#1) \supset P(^\#2)),$

$P(^\#2)$ is not known. In other words, although there is nothing about $P(^\#2)$ implied by knowing $P(^\#1)$, if this is *all* that is known, then we can say something about $P(^\#2)$, namely that it is not known.

To make this distinction, we need to clearly separate the difference between saying that α is known and α is all that is known. Of course, we never mean that α is the unique single sentence known to be true, since at the very least we will know the logical consequences of α and other formulas by positive introspection. But when we say that a state e is represented by the sentences in KB, we are saying more than just that these sentences are known. We are implicitly saying that these represent all that is known.

The difference between the two readings shows up most clearly with objective sentences. If KB and ϕ are objective, and KB does not imply ϕ, then if KB is known, ϕ may or may not be known; but if KB is all that is known, ϕ is not known, and so $\neg K\phi$ will be known by negative introspection, as above.

But characterizing the answer to **ASK** for non-objective sentences involving negative introspection is somewhat more complex. Rather than try to devise a complicated strategy using satisfiability (or consistency) instead of validity, we will take a very different approach: we will extend the language $\mathcal{KL}$ so that we can distinguish between saying "α is known" and "α is all that is known." As always, we will write the former as $K\alpha$. For the latter, we will introduce a new modal operator O so that $O\alpha$ is read as α is all that is known, or that only α is known. We will also sometimes use the expression "*only-knowing* α." What we will end up establishing is that **ASK** returns *yes* for question α iff

$\models (O\text{KB} \supset K\alpha),$

that is, only-knowing the KB implies knowing α. This will subsume the previous two cases above since if the KB is all that is known, then the KB is known. Thus we will have characterized the behaviour of **ASK** as applied to finitely representable states completely by the valid sentences of this extended logic. And, as we saw in the last chapter, we can restrict our attention to such states when it comes to arguments for **TELL** and **ASK**.

8.2 The language $\mathcal{OL}$

The language $\mathcal{OL}$ has exactly the same syntactic formation rules as that of $\mathcal{KL}$ but with one addition:

- If α is a wff, then $O\alpha$ is one too.

Note that the argument to O need not be objective or even a sentence of $\mathcal{KL}$. For example,

$$O[P(^\#1) \wedge \neg O(P(^\#2))] \vee KO(P(^\#3))$$

is a proper sentence of $\mathcal{OL}$. It is also considered to be a *subjective* sentence of $\mathcal{OL}$, since all predicate and function symbols are within the scope of a modal operator. A sentence of $\mathcal{OL}$ is called *basic* if it is also a sentence of $\mathcal{KL}$ (that is, contains no O operators.)

Turning now to the semantics of $\mathcal{OL}$, we will have the usual rules of interpretation for all the connectives from $\mathcal{KL}$. All we need to do, then, is to specify when $e \models O\alpha$ holds, after which satisfaction, validity, and implication will be as before.

The idea of only-knowing α means knowing no more than α about the world. So α will be all that is known in e when α is known, but e has as little world knowledge as possible. Since, as we discussed in Chapter 3, more knowledge means fewer world states and less knowledge means more world states, we want e to have as many world states as possible, although it clearly cannot contain any where α comes out false. In other words, α is all that is known in e iff e consists of *exactly* the world states where α is true, no more (since α is known) and no less (since it is all that is known).

More formally, we augment the semantic specification of $\mathcal{KL}$ by a single new rule of interpretation:

- $e, w \models O\alpha$ iff for every w', $w' \in e$ iff $e, w' \models \alpha$.

Note that this is (inductively) well-defined for any α in $\mathcal{OL}$. So whereas the semantics of K requires e to be a subset of the states where α is true, that is,

- $e, w \models K\alpha$ iff for every w', if $w' \in e$ then $e, w' \models \alpha$,

the semantics of O requires equality. That is, an "if" has been augmented to an "iff." Thus $O\alpha$ logically implies $K\alpha$, but not vice versa.

Also worth noting is that because we are insisting that e be the set of *all* states where α is true, we need to be careful when it comes to equivalent epistemic states. For example, imagine that we have two completely equivalent states but that one is a subset of another, as we described in Chapter 6. Since these two states know exactly the same basic sentences, we obviously want them to agree on all sentences of the form $O\alpha$ as well. But by the above definition, they would not. One would only be a subset of the set of states where α was true. Rather than complicate the semantics somehow (using the equivalence relation) to handle this situation, we will simply restrict our attention to *maximal* epistemic states, as we have done in previous chapters. Indeed, with maximal sets it is straightforward to show that the basic beliefs of an epistemic state uniquely determine all beliefs at that state, including those that mention O. Let a <u>basic belief set</u> Γ be defined just like a belief set in $\mathcal{KL}$, that is, $\Gamma = \{\alpha \mid \alpha \text{ is basic and } e \models K\alpha\}$.

Lemma 8.2.1: *If e and e' are maximal sets that have the same basic belief set, then for any subjective sentence σ of $\mathcal{OL}$, $e \models \sigma$ iff $e' \models \sigma$.*

Proof: Since e and e' have the same belief set, they are equivalent. Since they are both maximal, they must be equal by Theorem 6.1.2, and so satisfy the same subjective sentences. ∎

With that we define satisfiability, validity, and logical implication in $\mathcal{OL}$ just like in $\mathcal{KL}$ except that we explicitly restrict ourselves to maximal sets of worlds only.

8.3 Some properties of $\mathcal{OL}$

The simplest and most common case of only-knowing that we will consider is when the argument is an objective sentence ϕ. Saying $O\phi$ is simply saying that what is known can be finitely represented by ϕ. There is exactly one epistemic state where this is true:

Theorem 8.3.1: *For any objective ϕ, there is a unique maximal e such that $e \models O\phi$.*

Proof: Let $e = \Re[\![\phi]\!]$. Clearly, $e \models O\phi$, and no other e' can contain any other world states or fail to contain those in e. ∎

As a trivial corollary, we have

Corollary 8.3.2: *If ϕ is objective and σ is subjective, then either*
$$\models (O\phi \supset \sigma) \quad \text{or} \quad \models (O\phi \supset \neg\sigma).$$

So given that only ϕ is known, everything else about the epistemic state is logically implied. Note that this is not true for K. If ϕ and ψ are distinct atomic sentences, then we have that
$$\models (O\phi \supset \neg K\psi),$$

yet $\not\models (K\phi \supset \neg K\psi)$ and $\not\models (K\phi \supset K\psi)$. In other words, $K\phi$ leaves open whether or not $K\psi$. We also have:

Theorem 8.3.3: *Suppose ϕ and ψ are objective. Then*
$$\models (O\phi \supset K\psi) \quad \textit{iff} \quad \models (\phi \supset \psi).$$

Proof: Suppose that $\models (O\phi \supset K\psi)$, and that e is such that $e \models O\phi$ as guaranteed by the previous theorem, and so $e \models K\psi$. For any w, if $w \models \phi$, then $w \in e$, and so $w \models \psi$. Conversely, assume that $\models (\phi \supset \psi)$, and so, $\models (K\phi \supset K\psi)$. For any e, if $e \models O\phi$, then $e \models K\phi$, and so $e \models K\psi$. ∎

Finally, notice that nothing in the proof of the theorem depends on ϕ being finite. Hence we obtain the following corollary, which, in a sense, generalizes the concept of only-knowing to arbitrary sets of sentences, a subject we will not pursue further in this book.

Corollary 8.3.4: *For any set of objective sentences Φ, there is a unique maximal e such that for any objective ψ, $e \models K\psi$ iff $\Phi \models \psi$.*

These results give us a complete characterization of which objective sentences are believed, given that all that is known is also objective.

Turning now to only-knowing purely subjective sentences, here the situation is somewhat trivial. If we say "all that is known about the world is σ," and σ is subjective and so doesn't say anything about the world, then nothing is known about the world. So the epistemic state must be e_0. The only other possibility is the inconsistent epistemic state: in this case, for certain σ, such as $\neg K\alpha$, we have that σ is known because every sentence is known, and nothing else need be known to arrive at this inconsistent state, since $K\alpha$ is also true. More precisely, we have:

Theorem 8.3.5: *For any e and subjective σ, $e \models O\sigma$ iff either $e \models \sigma$ and $e = e_0$, or $e \models \neg\sigma$ and $e = \{\}$.*

Proof: Suppose that $e \models O\sigma$. In one case, we have $e \models \sigma$, in which case $e = e_0$ since for every world state w, $e, w \models \sigma$. In the other case, we have $e \models \neg\sigma$, and so for any world state w, $e, w \models \neg\sigma$, and thus $e \models K\neg\sigma$. However, because we also have $e \models K\sigma$, we must have $e = \{\}$.
Conversely, suppose that $e \models \sigma$ where $e = e_0$. Then clearly $e \models K\sigma$, and since e_0 contains every w, $e \models O\sigma$. Similarly, assume that $e \models \neg\sigma$ where $e = \{\}$. Then, for no w do we have $e, w \models \sigma$, and since $e \models K\sigma$, we get that $e \models O\sigma$. ∎

So if $O\sigma$ satisfiable at all, it is only satisfied in trivial epistemic states like e_0 or $\{\}$, or both. For example, $O\neg K\psi$ is satisfied by both e_0 and $\{\}$. For a similar reason, we have the following:

Corollary 8.3.6: *Suppose that ψ is atomic. Then $\models \neg OK\psi$.*

Proof: Assume, to the contrary, that $e \models OK\psi$. Then, by the theorem, $e = e_0$ or $e = \{\}$. However, $e_0 \models \neg K\psi$, and $\{\} \models K\psi$, contradicting the theorem. ∎

Thus, just as there are sentences of the form $K\alpha$ that are valid in $\mathcal{KL}$ and $\mathcal{OL}$, there are sentences $\neg O\alpha$ that are valid in $\mathcal{OL}$. In other words, there are sentences that simply cannot be all that is known in any state, from e_0 to the inconsistent one.

The following properties provide us with criteria under which sentences can be conjoined or disjoined to what is only-known without actually changing the epistemic state.

Theorem 8.3.7: $\models (O\alpha \wedge K\beta \supset O[\alpha \wedge \beta])$.

Proof: Suppose that $e \models O\alpha \wedge K\beta$. Then $e \models K\alpha \wedge K\beta$, and so $e \models K[\alpha \wedge \beta]$. Now assume that $e, w \models [\alpha \wedge \beta]$. Then $e, w \models \alpha$, and so $w \in e$, since $e \models O\alpha$. ∎

So in expressing all that is known, we can conjoin anything that happens to be known. The second property is:

Theorem 8.3.8: *For any subjective σ, $\models (O\alpha \wedge \sigma \supset O[\alpha \vee \neg\sigma])$.*

Proof: Suppose that $e \models O\alpha \wedge \sigma$. Then $e \models K\alpha$, and so $e \models K[\alpha \vee \neg\sigma]$. Now assume that $e, w \models [\alpha \vee \neg\sigma]$. Then $e, w \models \alpha$, since $e \models \sigma$, and so $w \in e$. ∎

So in expressing all that is known, we can disjoin any false subjective sentence. This is just another way of saying that when it comes to only-knowing, the true subjective sentences add nothing, and the false ones take nothing away.

We will investigate many other properties of $\mathcal{OL}$ in more detail later, including an attempt at axiomatizing the logic in Chapter 10. At this point, however, we already know enough about the logic to apply it to the specification of **ASK** and **TELL** within $\mathcal{OL}$, a task we now turn to.

8.4 Characterizing ASK and TELL

Recall that **ASK** and **TELL** were specified originally in terms of what the epistemic state $\Re[\![KB]\!]$ of a KB knows. With O this specification can be carried out entirely within $\mathcal{OL}$, that is, in terms of certain valid sentences.

Theorem 8.4.1: *Let KB be an objective sentence and α arbitrary. Then*
$$\mathbf{ASK}[\alpha, \Re[\![KB]\!]] = yes \quad \text{iff} \quad \models (O KB \supset K\alpha).$$

Proof: To prove the only-if direction, assume $e \models O KB$. Then $e = \Re[\![KB]\!]$ because KB is objective. Since the answer is *yes*, we have $e \models K\alpha$ by the definition of **ASK**.
Conversely, assume that $\models (O KB \supset K\alpha)$. Clearly $\Re[\![KB]\!] \models O KB$ and, therefore, $\Re[\![KB]\!] \models K\alpha$. So the answer is *yes*. ∎

Note that the theorem holds for any α, not just basic ones. Moreover, the use of O is essential for the theorem to go through.

The characterization of **TELL** turns out to be not quite as straightforward. One might expect that $\mathbf{TELL}[\alpha, \Re[\![KB]\!]] = \Re[\![KB^*]\!]$ iff $\models O KB^* \equiv O[KB \wedge \alpha]$.

While this is true for objective α, it does *not* hold if α is non-objective. For example, $\mathbf{TELL}[(Kp \vee p), e_0] = \Re[\![p]\!]$, but $\not\models Op \equiv O[Kp \vee p]$. To see why the equivalence fails, recall that **TELL** requires that any occurrence of K within the new sentence α be interpreted with respect to the *old* epistemic state $\Re[\![KB]\!]$ (e_0 in the example). Occurrences of K within $O[KB \wedge \alpha]$, on the other hand, refer to the state(s) which only know $KB \wedge \alpha$. As the example shows, these are in general different from $\Re[\![KB]\!]$.

What does hold, on the other hand, is that adding an objective sentence ϕ to the KB as a result of $\mathbf{TELL}[\alpha, \Re[\![KB]\!]]$ is correct just in case α is known to be equivalent to ϕ before **TELL** is performed. Formally:

Theorem 8.4.2: *Let KB and ϕ be objective, α arbitrary. Then*
$$\mathbf{TELL}[\alpha, \Re[\![KB]\!]] = \Re[\![KB \wedge \phi]\!] \quad \text{iff} \quad \models (O KB \supset K(\alpha \equiv \phi)).$$

Proof: Let $e = \Re[\![KB]\!]$. Recall that $\mathbf{TELL}[\alpha, e] = e \cap \{w \mid e, w \models \alpha\}$. To show the if direction, assume that $(O KB \supset K(\alpha \equiv \phi))$ is valid. Then clearly $e \models K(\alpha \equiv \phi)$, that is, $\{w \in e \mid e, w \models \alpha\} = \{w \in e \mid w \models \phi\}$. Hence $\mathbf{TELL}[\alpha, e] = e \cap \{w \mid w \models \phi\} = \Re[\![KB \wedge \phi]\!]$.
Conversely, assume that $\mathbf{TELL}[\alpha, e] = \Re[\![KB \wedge \phi]\!]$. Then $e \cap \{w \mid e, w \models \alpha\} = e \cap \{w \mid w \models \phi\}$. Thus for any $w \in e$, $e, w \models \alpha$ iff $w \models \phi$, from which $e \models K(\alpha \equiv \phi)$ follows immediately. ∎

Corollary 8.4.3: *For any objective KB and any basic α there is an objective ϕ such that* $\models (O\text{KB} \supset K(\alpha \equiv \phi))$.

Proof: The corollary follows immediately from this theorem and the Representation Theorem (Theorem 7.4.1). ∎

8.5 Determinate sentences

We have seen that objective sentences uniquely determine epistemic states, that is, for any ϕ, there is exactly one e such that $e \models O\phi$. Let us call such sentences *determinate*. Notice that there is nothing in the definition that requires a determinate sentence to be objective and it seems worthwhile to look at the more general case. In this section, we will therefore consider arbitrary determinate sentences. The main result will be that the Representation Theorem, which we obtained in Chapter 7 for objective knowledge bases, carries over nicely to the case of arbitrary determinate knowledge bases. We begin by showing that determinate sentences indeed deserve their name, that is, they leave no doubt about what is and is not believed.

Theorem 8.5.1: *A sentence δ is determinate iff for every basic α, exactly one of $(O\delta \supset K\alpha)$ and $(O\delta \supset \neg K\alpha)$ is valid.*

Proof: First, suppose that δ is determinate, and that e is the unique maximal set of worlds satisfying $O\delta$. Then, as in Corollary 8.3.2 either $(O\delta \supset K\alpha)$ or $(O\delta \supset \neg K\alpha)$ is valid according to whether $e \models K\alpha$ or not.

Now suppose that exactly one of $(O\delta \supset K\alpha)$ or $(O\delta \supset \neg K\alpha)$ is valid for every basic α. $O\delta$ must be satisfiable since otherwise it would imply every sentence. Moreover, for any e that satisfies it, we have that $e \models K\alpha$ iff $(O\delta \supset K\alpha)$ is valid, because either $K\alpha$ or $\neg K\alpha$ is implied by $O\delta$. Thus, if e and e' satisfy $O\delta$, then $e \models K\alpha$ iff $e' \models K\alpha$ for every basic α. So e and e' are equivalent and, by Theorem 6.1.2, $e = e'$. Thus, δ is determinate. ∎

Thus, determinate sentences not only tell us exactly what is and what is not believed, they are also the only sentences to do so. As such, they can be used as *representations of knowledge*, since they implicitly specify a complete epistemic state.

To see that there are interesting determinate sentences beyond the objective ones, let γ be the closed world assumption from Chapter 5, saying that all instances of predicate P are known,

$$\forall x (P(x) \supset KP(x)),$$

and consider $KB_1 = \{P(^\#1), P(^\#2), \gamma\}$. As the following lemma shows, there is a unique epistemic state which only knows KB_1:

Lemma 8.5.2: *Let* $e = \Re[\![\forall x(P(x) \equiv [(x = {}^\#1) \lor (x = {}^\#2)])]\!]$. *Then for any* e^*, $e^* \models OKB_1$ *iff* $e = e^*$.

Proof: We begin by showing that $e \models OKB_1$. Clearly, $e \models KKB_1$. Now let $w \notin e$. There are two cases. If $w \not\models P(^\#1) \land P(^\#2)$, then $e, w \not\models KB_1$. Otherwise, $w \models P(n)$ for some $n \notin \{^\#1, {}^\#2\}$. Then $e, w \models P(n) \land \neg KP(n)$, from which $e, w \not\models KB_1$ follows. Therefore, $e \models OKB_1$.

Now let e^* be any epistemic state such that $e^* \models OKB_1$. Consider any world $w \in e$, that is, $w \models P(n)$ iff $n \in \{^\#1, {}^\#2\}$. Since $e^* \models KP(^\#1) \land KP(^\#2)$, we obtain $e^*, w \models KB_1$ and, hence, $w \in e^*$. Since this is true for all $w \in e$, we obtain $e \subseteq e^*$. Since e itself is maximal, no proper superset of e only-knows KB_1. Hence $e^* = e$. ∎

Note that KB_1 makes the closed world assumption just for P. In particular,

$\models OKB_1 \supset K\neg P(^\#3)$, yet
$\models OKB_1 \supset \neg KQ(^\#1)$
$\models OKB_1 \supset \neg K\neg Q(^\#1)$
$\models OKB_1 \supset \neg K\neg Q(^\#2)$
etc.

In general, we have that

$$\models OKB_1 \equiv O\forall x(P(x) \equiv [(x = {}^\#1) \lor (x = {}^\#2)]).$$

We get the similar behaviour in the case where our knowledge about P is infinite. For example, let $KB_2 = \{\forall x((x \neq 3) \supset P(x)), \gamma\}$. We leave it to the reader to prove that $OKB_2 \equiv O[\forall x(P(x) \equiv (x \neq 3))]$ is valid.

If the knowledge base has incomplete information about P, applying the closed world assumption may not lead to a determinate knowledge base. For example, if $KB_3 = \{P(^\#1) \lor P(^\#2), \gamma\}$, then there are two corresponding epistemic states, one where $^\#1$ is the only P and another one where $^\#2$ is the only P. Formally,

$$OKB_3 \equiv O\forall x[P(x) \equiv (x = {}^\#1)] \lor O\forall x[P(x) \equiv (x = {}^\#2)]$$

is valid. We will not prove this here, but in the next chapter, we will have a lot more to say about such nondeterminate sentences. What we will show here is that the only way something like $(\phi \land \gamma)$ can lead to more than one epistemic state is when it is already known from ϕ that γ is false (as above).

First we need the following lemmas:

Lemma 8.5.3: *If $e_1 \subseteq e_2$ and $e_2, w \models \gamma$, then $e_1, w \models \gamma$.*

Proof: Suppose n is any name and $w \models P(n)$. Since $e_2, w \models \gamma$, we have that $e_2 \models KP(n)$, from which it follows that $e_1 \models KP(n)$, since $e_1 \subseteq e_2$. Consequently, for any n, $e_1, w \models (P(n) \supset KP(n))$. ∎

Lemma 8.5.4: *Suppose ϕ is objective, and $\Re[\![\phi]\!] \models \neg K\neg\gamma$. Further suppose that e is an epistemic state such that $e \models O(\phi \wedge \gamma)$. Then for any w, $e, w \models \gamma$ iff $\Re[\![\phi]\!], w \models \gamma$.*

Proof: As with Lemma 5.6.1, we show that for any n, $e \models KP(n)$ iff $\Re[\![\phi]\!] \models KP(n)$. If $\Re[\![\phi]\!] \models KP(n)$, then $e \models KP(n)$, since $e \subseteq \Re[\![\phi]\!]$. Conversely, if $\Re[\![\phi]\!] \models \neg KP(n)$, then since $\Re[\![\phi]\!] \models \neg K\neg\gamma$, there is a w in $\Re[\![\phi]\!]$ such that $w \models \phi$, $\Re[\![\phi]\!], w \models \gamma$ and thus for which, $w \models \neg P(n)$. However, by the lemma above, we then have that $e, w \models (\phi \wedge \gamma)$, and so $w \in e$. Thus, there is a $w \in e$ such that $w \models \neg P(n)$, and so $e \models \neg KP(n)$. ∎

Theorem 8.5.5: *Suppose ϕ is objective, and $\Re[\![\phi]\!] \models \neg K\neg\gamma$. Then $(\phi \wedge \gamma)$ is determinate and $\textbf{TELL}[\gamma, \Re[\![\phi]\!]]$ is the unique epistemic state satisfying $O(\phi \wedge \gamma)$.*

Proof: Let $e' = \textbf{TELL}[\gamma, \Re[\![\phi]\!]] = \{w \mid \Re[\![\phi]\!], w \models (\phi \wedge \gamma)\}$. We first show that $e' \models O(\phi \wedge \gamma)$. By the definition of **TELL**, $e' \models K\phi$, and by Theorem 5.6.2, $e' \models K\gamma$, and so $e' \models K(\phi \wedge \gamma)$. Now suppose that for some w, $e', w \models (\phi \wedge \gamma)$. By Lemma 5.6.1, $\Re[\![\phi]\!], w \models (\phi \wedge \gamma)$, and so $w \in e'$.

Next we need to show that if $e \models O(\phi \wedge \gamma)$ then $e = e'$. So suppose that $e \models O(\phi \wedge \gamma)$. Then for any w, we have that $w \in e$ iff $e, w \models (\phi \wedge \gamma)$ iff (by the lemma immediately above) $\Re[\![\phi]\!], w \models (\phi \wedge \gamma)$ iff (by Lemma 5.6.1) $e', w \models (\phi \wedge \gamma)$ iff $w \in e'$. ∎

So as long as γ is not already known to be false given ϕ, $(\phi \wedge \gamma)$ will be determinate. Moreover, from the point of view of **TELL**, we can see that if we start with an objective KB and assert that γ is true, we not only end up in a state where γ is known (as already established in Theorem 5.6.2), we also have that (KB $\wedge \gamma$) is *all* that is known.

The previous examples of determinate knowledge bases have in common that they can always be converted into equivalent objective knowledge bases. The main result of this section is that this is true in general, that is, it is always possible to represent determinate knowledge in objective terms. Although *believing* does not reduce to believing objective sentences (Theorem 4.6.2), *only believing* does, at least as far as determinate sentences are concerned.

Definition 8.5.6: Let ϕ be an objective formula, e an epistemic state, and δ a determinate sentence such that $e \models O\delta$. Suppose that $n_1, \ldots, n_k$, are all the names in ϕ or in δ, and that n' is some name that does not appear in ϕ or in δ.
$\text{RES}_K[\![\phi, e]\!]$ is defined by:

1. If ϕ has no free variables, then $\text{RES}_K[\![\phi, e]\!]$ is
   $\forall z(z = z)$, if $e \models K\phi$, and
   $\neg \forall z(z = z)$, otherwise.
2. If x is a free variable in ϕ, then $\text{RES}_K[\![\phi, e]\!]$ is
   $[((x = n_1) \wedge \text{RES}_K[\![\phi^x_{n_1}, e]\!]) \vee \ldots$
   $((x = n_k) \wedge \text{RES}_K[\![\phi^x_{n_k}, e]\!]) \vee$
   $((x \neq n_1) \wedge \ldots \wedge (x \neq n_k) \wedge \text{RES}_K[\![\phi^x_{n'}, e]\!^{n'}_x)].$

$\text{RES}_O[\![\phi, e]\!]$ is defined by:

1. If ϕ has no free variables, then $\text{RES}_O[\![\phi, e]\!]$ is
   $\forall z(z = z)$, if $e \models O\phi$, and
   $\neg \forall z(z = z)$, otherwise.
2. If x is a free variable in ϕ, then $\text{RES}_O[\![\phi, e]\!]$ is
   $[((x = n_1) \wedge \text{RES}_O[\![\phi^x_{n_1}, e]\!]) \vee \ldots$
   $((x = n_k) \wedge \text{RES}_O[\![\phi^x_{n_k}, e]\!]) \vee$
   $((x \neq n_1) \wedge \ldots \wedge (x \neq n_k) \wedge \text{RES}_O[\![\phi^x_{n'}, e]\!^{n'}_x)].$

Note that the definition of RES_K and RES_O are exactly like the old RES except that the implication $KB \models \phi$ is replaced by $e \models K\phi$ and $e \models O\phi$, respectively.

Given a determinate sentence δ, an epistemic state e with $e \models O\delta$, and an arbitrary wff α of $\mathcal{OL}$, $\|\alpha\|_e$ is the objective wff defined by

1. $\|\alpha\|_e = \alpha$, when α is objective
2. $\|\neg\alpha\|_e = \neg\|\alpha\|_e$
3. $\|(\alpha \wedge \beta)\|_e = (\|\alpha\|_e \wedge \|\beta\|_e)$
4. $\|\forall x\alpha\|_e = \forall x\|\alpha\|_e$
5. $\|K\alpha\|_e = \text{RES}_K[\![\|\alpha\|_e, e]\!]$
6. $\|O\alpha\|_e = \text{RES}_O[\![\|\alpha\|_e, e]\!]$

Again, the definition of $\|\cdot\|$ is exactly like the old one except that now we also reduce formulas of the form $O\alpha$.

Theorem 8.5.7: *For every determinate sentence δ there is an objective sentence ϕ such that $\models O\delta \equiv O\phi$.*

Proof: The proof is exactly analogous to the proof of the Representation Theorem. In particular, all the results of Section 7.2 and 7.3 carry over using the new definitions of RES and $\|\cdot\|$ in a straightforward way. Finally choose $\phi = \|\alpha\|_e$ where $e \models O\delta$. ∎

We will see in Theorem 9.6.5 of the next chapter that this property does not hold in general for non-determinate sentences.

In this chapter we have seen how introducing the concept of only-knowing allows us to fully characterize **ASK** and **TELL** within the logic itself. The logic $\mathcal{OL}$ has many other uses, which we will explore in subsequent chapters.

***

This then ends the first part of the book, which can be thought of as providing the basic concepts of a logic of knowledge bases. In the remaining chapters, we will touch on various more specialized topics which all build on the foundations we have laid out so far. We begin with a reconstruction and generalization using $\mathcal{OL}$ of autoepistemic reasoning, one of the main formalisms studied in the area of nonmonotonic reasoning, then consider an excursion into the proof theory of $\mathcal{OL}$, followed by an analysis of what it means to "only-know something about a subject matter," and then an in depth look at computational issues which involves, among other things, changing the underlying model of belief. Finally, we consider how to incorporate actions into our framework by amalgamating $\mathcal{OL}$ with the situation calculus.

8.6 Bibliographic notes

As we have seen, the key feature which distinguishes only-knowing from knowing (at least) is that both accessible and inaccessible worlds (e and its complement) are involved. This idea was independently developed by Humberstone [54] and later extended by Ben-David and Gafni [5]. Recently, Pratt-Hartmann [116] proposed what he calls *total knowledge*, which shares many of its properties with only-knowing. The semantics is based on sets of world states identical to ours except that beliefs are required to be true, that is, the actual world state is always considered possible. A sentence α is said to be total knowledge if α is known and every objective sentence which does not follow from knowing α is not known. As far as objective sentences are concerned, only-knowing and total knowledge basically coincide. In the general case, however, there are differences. Since these refer to properties of only-knowing treated in Chapter 9, we defer any further discussion of total knowledge to the end of that chapter.

Going back to the earlier work by Humberstone, Ben-David and Gafni, while they restrict themselves to the propositional case, they are in some sense more general than we are because they do not make the assumption of an underlying set of *all* worlds, that is, having a world for every interpretation of the atomic formulas. In fact, they allow general Kripke structures and consider modal logics other than K45. Allowing models with arbitrary sets of worlds, however, is problematic for only-knowing on intuitive grounds. For consider the case where we have just one world w where both p and q are true and w is the only accessible world. Then we have, for example, that both Op and Kq hold. This seems rather strange since only knowing p and, at the same time, knowing q seems incompatible with the intuitive reading of only-knowing.

To give only-knowing the right properties, then, it seems essential that the underlying models be large enough and contain worlds for every conceivable state of affairs. Note, however, that it is not at all obvious what constitutes a particular state of affairs. In our framework where we consider a single, fully introspective agent, it just so happens that a state of affairs can be identified with a world state. This is no longer the case when the agent is not fully introspective or when there are multiple agents. For example, consider the case of two agents A and B. From A's point of view, a state of affairs consists not just of facts about the world but also of B's beliefs about the world. This is because, as far as A is concerned, B's beliefs are just as objective for A as, say, the fact that Tina teaches Sue. Not surprisingly, modeling only-knowing for multiple agents is a complicated matter. Several approaches are discussed in [46, 68, 48]. In [16] Chen considers a specialized logic of only-knowing for two agents which allows the author to capture Gelfond's notion of *epistemic specifications* [40] within the framework of only-knowing.

In [49] Halpern and Moses define a concept of *minimal knowledge* in the propositional case which bears a striking resemblance to our notion of only-knowing. Roughly, given a sentence α, they define the corresponding epistemic state that only-knows α as the union of all sets of world states where α is known. α is called honest just in case α itself is known in this epistemic state. It is easily seen that every objective ϕ is honest. Indeed, for objective propositional formulas our notion of only-knowing coincides with that of Halpern and Moses. Also, just as there are sentences that cannot be only-known there are sentences that are dishonest, for example $Kp \vee Kq$. Despite those similarities there are differences as well. An obvious difference is that Halpern and Moses consider knowledge instead of belief, that is, $(K\alpha \supset \alpha)$ comes out valid or, equivalently, the real world is always assumed to be among the accessible world states. A much more surprising difference has to do with complexity. Recently, it was discovered [25] that reasoning about minimal knowledge is actually harder than reasoning about only-knowing (again, restricted to the propositional case).

Halpern and Moses' logic of minimal knowledge is only one example of a wide range of formalisms called nonmonotonic logics. Only-knowing has intimate connections to a number of these besides the one just mentioned. Since we will study one such connection in much more depth in the next chapter, we defer a discussion of the related literature to Section 9.8. Similarly, see Section 10.5 for literature on proof-theoretic and complexity issues regarding $\mathcal{OL}$.

8.7 Exercises

1. Generalize Theorem 8.3.1 to the case of non-finitely representable states as follows: Show that for any set of objective sentences Φ there is a unique epistemic state e such that for any objective ψ, $e \models K\psi$ iff Φ logically implies ψ.
2. Consider the statement "Only-knowing is closed under logical consequence." State this precisely as a theorem, and prove that it is true.
3. Give an example of a subjective σ such that $O\sigma$ is only satisfied by e_0, and another that is only satisfied by the inconsistent epistemic state.
4. Show that only-knowing is closed under introspection in the following sense: for any subjective σ,
$$\models O\alpha \wedge \sigma \supset O(\alpha \wedge \sigma).$$
Give an example of where this fails when O is replaced by K.
5. Show that for no α is it the case that $O\alpha$ is valid. Hint: consider finitely and infinitely representable states.
6. Show that for any falsifiable objective ϕ, $\models \neg O[\phi \vee K\phi]$.
7. Show that for any determinate KB and basic α, there is an objective ϕ with $\models (O\text{KB} \supset K(\alpha \equiv \phi))$.
8. Let KB $= \{\forall x((x \neq 3) \supset P(x)), \forall x(P(x) \supset KP(x))\}$. Show the validity of $O\text{KB} \equiv O[\forall x(P(x) \equiv (x \neq 3))]$.

Part II

9 Only-Knowing and Autoepistemic Logic

Up to now we have mainly focussed on knowledge bases (or arguments of *O*, for that matter) that uniquely determine an epistemic state. We saw that this restricted use of *O* is sufficient to completely characterize the interaction routines **ASK** and **TELL** developed earlier. While our intuitions about *O* are certainly strongest in the case of determinate sentences, going beyond them not only helps us deepen our understanding of *O*, but it also allows us to demonstrate a close connection between only-knowing and autoepistemic logic (AEL), originally introduced in the early eighties by R. Moore to capture certain forms of nonmonotonic or default reasoning.

Defaults are assumptions which are not based on facts about the world but rather on conventions, statistical information, and the like. For example, most people agree that birds generally fly. So, if presented with a particular bird called Tweety, it seems perfectly reasonable to assume that Tweety flies. Of course, later information may contradict this assumption, for example, if we find out that Tweety is a stuffed bird or an ostrich. In this case, we are more than willing to retract our previous belief about Tweety's flying ability. Notice that the use of defaults has the effect that the set of beliefs may grow *nonmonotonically* with the information obtained about the world. In other words, by adding new facts to our knowledge base we may be forced to retract beliefs held previously about the world. This is why reasoning by default is generally referred to as nonmonotonic reasoning, a term which also stands for the whole research area which has investigated the fundamental principles underlying this type of reasoning for over two decades.

AEL represents one branch of this endeavour. The idea, in a nutshell, is to interpret defaults such as *birds generally fly* epistemically. Roughly, one is willing to assume that a particular bird flies provided one's own knowledge about the world does not conflict with this assumption.

In the following, we will first demonstrate, by way of example, how autoepistemic reasoning is modeled in $\mathcal{OL}$. Then we will show how $\mathcal{OL}$ not only captures Moore's original ideas in a precise sense but also extends it substantially, mainly because we are using a more expressive language.

9.1 Examples of autoepistemic reasoning in $\mathcal{OL}$

To begin let us consider the following simple example, originally due to Moore.

> Suppose Bob is the oldest child in his family and someone asks him whether
> he has an older brother. Naturally he would answer no and, asked to explain

his reasoning, Bob may answer as follows: "If I had an older brother, I would certainly know about it. And since I do not know that I have an older brother, I conclude I do not have one."

Note what is happening here. Bob draws a conclusion not based on factual knowledge about the world but based on his ignorance (not knowing about an older brother), which is why this form of reasoning is called autoepistemic. The first sentence of Bob's explanation really expresses a default assumption. It is a quite reasonable one to make, but it can be defeated by new information.[1]

The logic $\mathcal{OL}$ allows us to formalize the example in a natural way. Let b stand for "Bob has an older brother" and let KB be Bob's knowledge base consisting of objective sentences such that $\not\models$ (KB $\supset b$). We can express the default as $\delta = (\neg Kb \supset \neg b)$. If we then assume KB $\wedge \delta$ is all Bob knows then we get the desired result, that is, $\neg b$ is believed.

Example 9.1.1: $(O(KB \wedge \delta) \supset K\neg b)$ is a valid sentence.

Proof: Let $e = \{w \mid w \models KB \wedge \neg b\}$. Clearly, $e \models K\neg b$. Let e^* be any set of worlds such that $e^* \models O(KB \wedge \delta)$. It suffices to show that $e^* = e$. To show that $e \subseteq e^*$, let $w \in e$. Then $w \models KB \wedge \neg b$ and, hence, $e^*, w \models KB \wedge \delta$, from which $w \in e^*$ follows. Conversely, let $w \in e^*$. Then $w \models KB$. Note that $e^* \models \neg Kb$. (For assume otherwise and let $w^* \notin e^*$ such that $w^* \models KB \wedge \neg b$. Then $e^*, w^* \models KB \wedge \delta$ and, hence, $w^* \in e^*$, a contradiction.) Then, since $e^*, w \models \delta$, we have $w \models \neg b$ and, hence, $w \in e$. ∎

An important characteristic of defaults is that the objective beliefs of an agent may change nonmonotonically if new information is added to the knowledge base. In the example, Bob initially knows KB and, by default, $\neg b$. When Bob's mom finally tells Bob the truth about the extent of his immediate family, Bob may add b to his knowledge base overriding his previous default belief. Formally, $(O(KB \wedge b \wedge \delta) \supset Kb)$ is valid, which can be derived using ordinary reasoning about K.

Let us now turn to more complex cases of defaults with quantifiers. Actually, we already saw examples of those in our previous discussion of determinate sentences (Section 8.5). Recall that we formalized the closed world assumption for a particular predicate P using the sentence $\gamma = (\forall x P(x) \supset KP(x))$. Note that we can rewrite γ as $(\forall x \neg KP(x) \supset \neg P(x))$, which is more suggestive of a default saying that P is a assumed to be false unless known otherwise. In this sense, the older-brother example is nothing more than an instance of the closed world assumption. Closed world reasoning, as it is commonly applied in databases, is perhaps the simplest form of default reasoning. But,

---

1 In fact, one of the authors has two older brothers whose existence was not revealed to him for a long time.

Only-Knowing and Autoepistemic Logic   145

of course, defaults can also be used to derive positive facts about the world. The classic example is about birds and their ability to fly. In particular, one would like to conclude that any bird such as the infamous *Tweety* can fly unless known otherwise. One way to express the appropriate default is by using sentences like

$$\forall x[Bird(x) \land \neg K\neg Fly(x) \supset Fly(x)]$$

within the scope of an O operator. If we let δ stand for this sentence, we obtain the following:

Example 9.1.2: Assume that KB = {$Bird(tweety)$}. Then the following sentences are valid:

1. $O(\text{KB} \land \neg Fly(tweety) \land \delta) \supset K\neg Fly(tweety)$
2. $O(\text{KB} \land Fly(tweety) \land \delta) \supset KFly(tweety)$
3. $O(\text{KB} \land \delta) \supset KFly(tweety)$.

Proof: (1) and (2) follow easily using the fact that $(O\alpha \supset K\alpha)$ is valid.

To show (3), let $e \models O(\text{KB} \land \delta)$. We first show that $e \models \neg K\neg Fly(tweety)$. Let w^* be any world state such that $w^* \models Bird(tweety) \land \forall x Fly(x)$. Then $e, w^* \models \text{KB} \land \delta$ and, hence, $w^* \in e$. Since $w^* \models Fly(tweety)$, we obtain $e \models \neg K\neg Fly(tweety)$.

Now we show that $e \models KFly(tweety)$, that is, for every $w \in e$, $w \models Fly(tweety)$. Let $w \in e$. Then $e, w \models (Bird(tweety) \land \neg K\neg Fly(tweety) \supset Fly(tweety))$. Clearly, $w \models Bird(tweety)$. By the above, we also have $e \models \neg K\neg Fly(tweety)$. Therefore, $w \models Fly(tweety)$. ∎

This example shows that Tweety's flying is indeed the default: if his flying ability is specified explicitly, then this works out properly (cases 1 and 2); otherwise, flying is taken as the default (case 3).

Note, however, that the proof uses the fact that there are worlds where all things fly, which is certainly true when KB = {$Bird(tweety)$}. However, this condition is too strong. We should be able to get the default for Tweety even if there are some flightless birds. So what happens if the KB implies of some bird that it is flightless? The answer is that the default still works properly, but for a slightly different reason. In the following, both *tweety* and *chilly* are meant to be distinct standard names.

Example 9.1.3: Assume that

KB = {$Bird(tweety), Bird(chilly), \neg Fly(chilly)$}.

Then

$$O(\text{KB} \land \delta) \supset K\forall x[Bird(x) \land (x \neq chilly) \supset Fly(x)]$$

is valid and thus, $(O(\text{KB} \wedge \delta) \supset \boldsymbol{K}Fly(tweety))$ is valid.

Proof: We leave the proof as an exercise. ∎

Note that the default belief that Tweety flies is based on the default belief that any bird other than Chilly flies. In fact, with no information at all, it will be assumed that all birds fly: the sentence $(O\delta \supset \boldsymbol{K}\forall x[Bird(x) \supset Fly(x)])$ is valid. In many applications this is too strong; we might not want to infer anything about distal unknown birds. One way to do this is to write the default as

$$\forall x[\boldsymbol{K}Bird(x) \wedge \neg \boldsymbol{K}\neg Fly(x) \supset Fly(x)].$$

This makes the default apply only to the *known* birds. However, it does have disadvantages compared to the previous form of default. For example, if KB = $\{\exists x(Bird(x) \wedge Yellow(x))\}$, then using this form of default, we would *not* conclude by default that there was a yellow bird that flies. Although its *existence* is known, the bird in question is not (yet) a known bird.

A nice property of the examples considered so far is that the KB together with the default is *determinate* as defined previously, that is, there is a unique corresponding epistemic state. In particular, the effect of the default is to add information to the KB, in this case, information about birds' flying ability. This "filling in the blanks" is precisely what one would expect from a default. Unfortunately the desired effect does not always obtain, and to see why, let us go back to our original birds-fly default δ, which applies to all birds, not just the known ones (although the same applies to the more restricted form). Note that so far it has always been the case that (1) the KB implies that Tweety is a bird and (2) it does not imply that Tweety is flightless. Unfortunately, these conditions are not sufficient for the default to go through. For suppose that

$$\text{KB} = \{Bird(tweety), Bird(chilly), (\neg Fly(chilly) \vee \neg Fly(tweety))\}.$$

Then the KB does not imply that Tweety is flightless, but it would be inappropriate to assume by default that she can fly, since by symmetry we could infer the same of Chilly, contradicting the fact that one of them is flightless. A similar complication occurs if

$$\text{KB} = \{Bird(tweety), \exists x(Bird(x) \wedge \neg Fly(x))\}.$$

Again, if we are prepared to infer that Tweety flies, by symmetry, we should be able to do likewise for any bird, and thus come to the conclusion that all birds fly, again contradicting the belief. The trouble with the two KBs above is that the default δ is actually believed to be false, that is, the sentence $(O\text{KB} \supset \boldsymbol{K}\neg\delta)$ is valid. In both cases the KB implies that there is a flightless bird but it does not specify which; so, if this is *all* that is known, then it *is* believed that there is a flightless bird whose identity is not known, which is $\neg\delta$. So what happens in these cases if we insist that $O[\text{KB} \wedge \delta]$ is true? That is, what happens when we believe KB and δ and nothing else, even though believing KB alone implies believing ¬δ? The answer, in short, is that (KB ∧ δ) is no longer determinate, that is, it fails to specify completely what is and is not believed. More specifically, we have:

Example 9.1.4: Let KB = {*Bird*(*tweety*), *Bird*(*chilly*), (¬*Fly*(*chilly*) ∨ ¬*Fly*(*tweety*))} and let *Exc*(*x*) be an abbreviation for *Bird*(*x*) ∧ ¬*Fly*(*x*). Then the sentence ***O***[KB ∧ δ] is logically equivalent to

$$O[\text{KB} \wedge \forall x(\textit{Exc}(x) \equiv x = \textit{tweety})] \vee O[\text{KB} \wedge \forall x(\textit{Exc}(x) \equiv x = \textit{chilly})].$$

Proof: To prove the if direction, let $e \models O[\text{KB} \wedge \forall x(\textit{Exc}(x) \equiv x = \textit{tweety})]$. (The other case is handled the same way.) We need to show that $e \models O[\text{KB} \wedge \delta]$. If $w \in e$, then clearly $w \models \text{KB}$. Also, $e, w \models \delta$ follows because all birds other than Tweety fly in w and Tweety is known not to fly. Conversely, let $w \notin e$ and assume that $w \models \text{KB}$. Thus $w \not\models \forall x(\textit{Exc}(x) \equiv x = \textit{tweety})$, that is, either there is an exceptional bird n other than Tweety or Tweety is not an exceptional bird. In the first case, $w \models \textit{Bird}(n) \wedge \neg \boldsymbol{K}\neg\textit{Fly}(n) \wedge \neg\textit{Fly}(n)$. In the second case, we have the same with n replaced by Chilly because KB requires one of Tweety and Chilly not to fly. In either case, $e, w \not\models \delta$, and we are done.

For the only-if direction, let $e \models O[\text{KB} \wedge \delta]$. First note that $e \models \boldsymbol{K}\neg\textit{Fly}(\textit{tweety}) \vee \boldsymbol{K}\neg\textit{Fly}(\textit{chilly})$ which follows by the default and the fact that either Tweety or Chilly does not fly. Thus let us assume that $e \models \boldsymbol{K}\neg\textit{Fly}(\textit{tweety})$. It suffices to show that $e = e^*$ with $e^* = \{w \mid w \models \text{KB} \wedge \forall x(\textit{Exc}(x) \equiv x = \textit{tweety})\}$. (The other case is handled the same way with Tweety replaced by Chilly.) Let $w \in e^*$. Since all birds other than Tweety fly at w and Tweety is known not to fly by assumption, we obtain $e, w \models \text{KB} \wedge \delta$ and, hence, $w \in e$. Conversely, suppose $w \in e$. By assumption, Tweety is known to be an exceptional bird at e and is therefore exceptional at w. Any bird n other than Tweety flies at w because of δ and the fact that $e \models \neg\boldsymbol{K}\neg\textit{Fly}(n)$, which follows from $e^* \subseteq e$. Hence $w \models \forall x(\textit{Exc}(x) \equiv x = \textit{tweety})$ and, therefore, $w \in e^*$. ∎

This says that only-knowing KB and the default is the same as only-knowing KB and that Tweety is the only flightless bird *or* only-knowing KB and that Chilly is the only flightless bird. But the KB and the default together are not sufficient to specify exactly what is believed; they *describe* what is believed, but do not determine it. They do, however, determine what is *common* to both epistemic states. For example,

$$O[\text{KB} \wedge \delta] \supset \boldsymbol{K}\forall x[\textit{Exc}(x) \supset (x = \textit{tweety} \vee x = \textit{chilly})]$$

is valid.

It is not hard to see that the default may even lead to an infinite number of compatible epistemic states. For instance, let KB = {∃*xExc*(*x*)}. The result here is that the sentence

$O[\text{KB} \wedge \delta] \equiv \exists y\, O[\text{KB} \wedge \forall x (Exc(x) \equiv (x = y))]$ is valid, by an argument similar to the one above. In other words, only knowing that birds fly by autoepistemic default and that there is an exceptional bird does not determine exactly what is believed; however, it only happens if, for some bird, all that is known is that this bird is the only exceptional one. In this case, there is a different epistemic state for each standard name.

Finally, instead of multiple epistemic states there is also the case, as we saw already in the previous chapter, that a sentence does not correspond to any epistemic state at all, that is, it can never be all that is known like $K\phi$, since $\models \neg O[K\phi]$ (see Corollary 8.3.6).

9.2 Stable sets and stable expansions

We now turn to the close relationship between only-knowing and Moore's original formulation of autoepistemic logic. There are two notions central to AEL, *stable sets* and *stable expansions*. We will give precise definitions below, but let us first look at these notions informally. Both have in common that they are syntactic characterizations of an agent's beliefs. A stable set simply states three basic conditions the beliefs of an ideal rational agent should satisfy: closure under logical consequence, positive and negative introspection. Stable expansions then define those sets of beliefs that are stable and in some sense derive from a set of assumptions A. In other words, a stable expansion describes the beliefs an ideal rational agent might hold provided her knowledge base consists of the sentences in A. The following properties will be established relating AEL and only knowing:

- Belief sets and stable sets coincide.
- The stable expansions of a sentence α are precisely those belief sets which result from only-knowing α.
- While AEL was originally only defined for a propositional language, $\mathcal{OL}$ provides a natural quantificational generalization.

The notion of stability depends on a definition of first-order consequence, so we should be clear about this concept first. The idea is simple: α is a first-order consequence of Γ when Γ implies α by virtue of the rules of ordinary first-order logic alone, that is, without using the rules for K or for O even over sentences containing these operators. One way to formalize this is to think of sentences like $K\alpha$ and $O\alpha$ as new atomic sentences so that there is no forced relationship between the truth value of, for example, $K\alpha$ and $\neg K \neg \neg \alpha$. Although the conjunction of these two is not satisfiable, this depends on the semantics of K, and so we want to say that it is first-order satisfiable.

More precisely, let θ be any function from sentences of the form $K\alpha$ or $O\alpha$ to $\{0, 1\}$, and w any world. We will say that a pair θ and w *first-order satisfies* a sentence α, which we write $\theta, w \models_{\text{FOL}} \alpha$ according to these rules:

Only-Knowing and Autoepistemic Logic

1. $\theta, w \models_{\text{FOL}} P(t_1, \ldots, t_k)$ iff $w[P(n_1, \ldots n_k)] = 1$, where $n_i = w(t_i)$;
2. $\theta, w \models_{\text{FOL}} t_1 = t_2$ iff $w(t_1)$ is the same name as $w(t_2)$;
3. $\theta, w \models_{\text{FOL}} \neg \alpha$ iff $\theta, w \not\models_{\text{FOL}} \alpha$;
4. $\theta, w \models_{\text{FOL}} \alpha \wedge \beta$ iff $\theta, w \models_{\text{FOL}} \alpha$ and $\theta, w \models_{\text{FOL}} \beta$;
5. $\theta, w \models_{\text{FOL}} \exists x \alpha$ iff for some n, $\theta, w \models_{\text{FOL}} \alpha_n^x$;
6. $\theta, w \models_{\text{FOL}} \boldsymbol{K}\alpha$ iff $\theta(\boldsymbol{K}\alpha) = 1$;
7. $\theta, w \models_{\text{FOL}} \boldsymbol{O}\alpha$ iff $\theta(\boldsymbol{O}\alpha) = 1$.

We will say that Γ is *first-order satisfiable* iff some θ and w first-order satisfies it. Finally, we will say that Γ *first-order implies* α, which we write $\Gamma \models_{\text{FOL}} \alpha$, iff $\Gamma \cup \{\neg \alpha\}$ is not first-order satisfiable. Clearly satisfiability implies first-order satisfiability, but not the converse.

We are now in a position to formally introduce stable sets and expansions and relate them to only-knowing. Since AEL only deals with basic sentences, we focus on those first, but we will already consider the full first-order case with quantifying-in. In the final section of this chapter, we will see that all definitions and results carry over naturally if we allow non-basic sentences as well.

9.3 Relation to stable sets

With first-order consequence the definition of a stable set is now very simple.

Definition 9.3.1: A set of basic sentences Γ is *stable* iff

1. If $\Gamma \models_{\text{FOL}} \alpha$, then $\alpha \in \Gamma$.[2]
2. If $\alpha \in \Gamma$, then $\boldsymbol{K}\alpha \in \Gamma$.
3. If $\alpha \notin \Gamma$, then $\neg \boldsymbol{K}\alpha \in \Gamma$.

Stability merely states in a rigorous way that beliefs are closed under perfect logical reasoning and introspection. Since we have been making these assumptions all along, it is clear that every basic belief set is also a stable set. Below we will show that the converse is also true, that is, stable sets correspond exactly to basic belief sets. But first we need a result stating that for certain sets of sentences satisfiability and first-order satisfiability coincide.

Definition 9.3.2: A set Σ is an <u>adjunct</u> of a set Γ iff $\Sigma = \{\boldsymbol{K}\alpha \mid \alpha \text{ is basic and } \alpha \in \Gamma\} \cup \{\neg \boldsymbol{K}\alpha \mid \alpha \text{ is basic and } \alpha \notin \Gamma\}$.

---

[2] In other words we are requiring Γ to be closed under first-order implication. Moore used propositional logical consequence since he only dealt with a propositional language.

Lemma 9.3.3: *If Σ is an adjunct of a basic belief set Γ, then for any subjective sentence σ, either $\Sigma \models \sigma$ or $\Sigma \models \neg\sigma$.*

Proof: Suppose Γ is a basic belief set for some maximal e and suppose that $e \models \sigma$. Then any maximal e' such that $e' \models \Sigma$ must have the same basic belief set as e. By Lemma 8.2.1, $e' \models \sigma$ follows and, consequently, $\Sigma \models \sigma$. The case with $\neg\sigma$ is analogous. ∎

Theorem 9.3.4: *Suppose Δ is a set of basic sentences that contains an adjunct to a stable set. Then Δ is satisfiable iff it is first-order satisfiable.*

Proof: The only-if direction is immediate. So suppose that Δ contains an adjunct to a stable set Γ and is first-order satisfiable, and that $\theta, w \models_{FOL} \Delta$. Define e as $\{w' \mid \theta, w' \models_{FOL} \Gamma\}$. We will show by induction that for any w' and any basic α, $e, w' \models \alpha$ iff $\theta, w' \models_{FOL} \alpha$.

This clearly holds for atomic sentences, equalities, and by induction, for negations, conjunctions, and quantifications. Now suppose that $\theta(K\alpha) = 1$. Therefore, $\neg K\alpha \notin \Delta$, and so $\alpha \in \Gamma$. Thus, for every $w' \in e$, $\theta, w' \models_{FOL} \alpha$ and so by induction, $e, w' \models \alpha$ and so, $e \models K\alpha$. Conversely, suppose that $\theta(K\alpha) = 0$. Therefore, $K\alpha \notin \Delta$, and so $\alpha \notin \Gamma$. But Γ is closed under first-order implication, so $\Gamma \cup \{\neg\alpha\}$ is first-order satisfiable. Therefore, there must be some θ^* and some w' such that $\theta^*, w' \models_{FOL} \Gamma \cup \{\neg\alpha\}$. But θ and θ^* can only differ on non-basic sentences since for every basic α, either $K\alpha \in \Gamma$ or $\neg K\alpha \in \Gamma$. Thus, $\theta, w' \models_{FOL} \Gamma \cup \{\neg\alpha\}$. This means that $w' \in e$, and so there is a $w' \in e$ such that $\theta, w' \models_{FOL} \neg\alpha$, and by induction $e, w' \models \neg\alpha$. Therefore, $e \models \neg K\alpha$.

Thus, for every w', $e, w' \models \alpha$ iff $\theta, w' \models_{FOL} \alpha$. This establishes that $e, w \models \Delta$, and so Δ is satisfiable. ∎

One simple consequence of this theorem is that it is not necessary to use first-order implication when dealing with (supersets of) adjuncts to stable sets:

Corollary 9.3.5: *Suppose Δ is a set of basic sentences that contains an adjunct to a stable set. Then for any basic α, $\Delta \models \alpha$ iff $\Delta \models_{FOL} \alpha$.*

Proof: Immediate from the theorem. ∎

Now we can show that stable sets and basic belief sets are one and the same.

Theorem 9.3.6: *Suppose Γ is a set of basic sentences. Then Γ is stable iff Γ is a basic belief set.*

Proof: The if direction is straightforward: the first condition is a result of the logical properties of a reasoner, and the last two are a result of its introspective capabilities.

Conversely, suppose Γ is stable. There are two cases. If Γ is satisfiable, then some $e, w \models \Gamma$. For any basic α, if $\alpha \in \Gamma$, then $\boldsymbol{K}\alpha \in \Gamma$, and so $e \models \boldsymbol{K}\alpha$; if $\alpha \notin \Gamma$, then $\neg \boldsymbol{K}\alpha \in \Gamma$, and so $e \models \neg \boldsymbol{K}\alpha$. Thus, $\alpha \in \Gamma$ iff $e \models \boldsymbol{K}\alpha$, and so Γ is a basic belief set for e. Suppose on the other hand that Γ is unsatisfiable. By properties (2) and (3) of stability, Γ must contain the adjunct to Γ. Then by Theorem 9.3.4, Γ is not first-order satisfiable. So for every basic α, $\Gamma \models_{\text{FOL}} \alpha$, and by definition of stability, $\alpha \in \Gamma$. Thus, Γ contains every basic sentence. It is therefore the basic belief set of the empty set of worlds. ∎

It has long been known that stable sets, when restricted to propositional sentences, are uniquely determined by their objective subsets. With quantifiers and, in particular, quantifying-in, this is no longer the case.[3]

Theorem 9.3.7: *Stable sets are in general not uniquely determined by their objective subsets.*

Proof: The result follows easily from Theorem 4.6.2, which says that there are two epistemic states e_1 and e_2 whose corresponding basic belief sets agree on all objective sentences but disagree on $\boldsymbol{K}\exists x[P(x) \wedge \neg \boldsymbol{K}P(x)]$. Since, by the previous theorem, stable sets and basic belief sets are one and the same, the theorem follows. ∎

9.4 Relation to stable expansions

Let us now turn to stable expansions. Roughly, a sentence γ belongs to a stable expansion of a set of basic sentences A if it follows from A using logical reasoning and introspection. Of course, we need to be clear about what we mean by introspection here. The trick is to assume we already know what the stable expansion is and use its adjunct as the characterization of the beliefs that can be inferred by introspection. γ is then simply a logical consequence of A and the adjunct. Formally, we obtain the following fixed-point definition.

Definition 9.4.1: A set of sentences Γ is a *stable expansion* of a set of basic sentences A iff Γ satisfies the fixed-point equation:

$$\Gamma = \{\gamma \mid \gamma \text{ is basic and } A \cup \{\boldsymbol{K}\beta \mid \beta \in \Gamma\} \cup \{\neg \boldsymbol{K}\beta \mid \beta \notin \Gamma\} \models_{\text{FOL}} \gamma\}.$$

---

3 If we disallow quantifying-in, we obtain the same results as in the propositional case.

The main result of this chapter says that the stable expansion of a sentence α and the basic belief sets that result from only-knowing α are one and the same.

Theorem 9.4.2: *For any basic α and any maximal set of worlds e, $e \models O\alpha$ iff the basic belief set of e is a stable expansion of $\{\alpha\}$.*

Proof: Let e be any maximal set of worlds with Γ as its basic belief set and Σ as the adjunct to Γ. Thus, $e \models \Sigma$. What we want to show is that $e \models O\alpha$ iff Γ is the set of basic sentences that are first-order implied by $\{\alpha\} \cup \Sigma$. Moreover, by Corollary 9.3.5, we can use full logical implication instead of first-order implication since Γ is a stable set. Thus, we need to show that

$$e \models O\alpha \quad \text{iff} \quad \text{for every basic } \beta, e \models K\beta \text{ iff } \{\alpha\} \cup \Sigma \models \beta.$$

First assume that $e \models O\alpha$. For the if part, assume that $\{\alpha\} \cup \Sigma \models \beta$. Now let w be any element of e. Since $e \models O\alpha$, $e, w \models \{\alpha\} \cup \Sigma$, and therefore, $e, w \models \beta$. Thus, for any $w \in e$, we have that $e, w \models \beta$, and so $e \models K\beta$.

For the only-if part, assume that $e \models K\beta$. To show that $\{\alpha\} \cup \Sigma \models \beta$, let e' be any maximal set of worlds and w be any world. If $e', w \models \{\alpha\} \cup \Sigma$, then $e' = e$ since Σ is an adjunct of the basic belief set for e by Lemma 9.3.3. Thus, $e, w \models \alpha$ and so $w \in e$, because $e \models O\alpha$. But if $w \in e$, then $e, w \models \beta$, since $e \models K\beta$. Thus for any e' and w, if $e', w \models \{\alpha\} \cup \Sigma$, then $e', w \models \beta$, and so $\{\alpha\} \cup \Sigma \models \beta$.

Now assume that $e \models K\beta$ iff $\{\alpha\} \cup \Sigma \models \beta$. First we need to show that $e \models K\alpha$, but this is immediate since clearly $\{\alpha\} \cup \Sigma \models \alpha$. Next we need to establish that if $e, w \models \alpha$ then $w \in e$. If $e, w \models \alpha$ then $e, w \models \{\alpha\} \cup \Sigma$, since $e \models \Sigma$. Now consider any β such that $e \models K\beta$. We have that $\{\alpha\} \cup \Sigma \models \beta$, and so $e, w \models \beta$. Therefore, by Theorem 6.1.1, we have that $e \approx (e + w)$ and so, because e is maximal, $w \in e$. Thus, for any w, if $e, w \models \alpha$ then $w \in e$, and so $e \models O\alpha$. ∎

So only-knowing a sentence means that what is believed is a stable expansion of that sentence (or, more intuitively, what is believed is derivable from that sentence using first-order logic and introspection alone). This theorem provides for the first time a semantic account for the notion of stable expansion. In addition, we have generalized the notion of a stable expansion to deal with a quantificational language with equality. To summarize, we have the following correspondences:

semantic	syntactic
believing	membership in a stable set
basic belief sets	stable sets
only believing	stable expansions

Only-Knowing and Autoepistemic Logic   153

One easy corollary to this theorem relates the number of stable expansions of a sentence to the number of sets of worlds where that sentence is all that is known.

Corollary 9.4.3: *A sentence α has exactly as many stable expansions as there are maximal sets of worlds where $O\alpha$ is true.*

Proof: By Theorem 8.2.1, the mapping between maximal sets of worlds and basic belief sets is bijective. By Theorem 9.3.6, beliefs sets are the stable sets. The correspondence then follows from the above theorem. ∎

What this says, among other things, is that our previous discussions of determinate and non-determinate sentences applies equally well to stable expansions.

9.5 Computing stable expansions

In the previous section, we saw that there was a one-to-one correspondence between the stable expansions of a formula and the epistemic states where the formula is all that is known. In this section, we examine a procedure for calculating these stable expansions or epistemic states in the propositional case. Specifically, we will return a set of objective formulas that represent each of the epistemic states where the given propositional formula is all that is known.

In the following, we will show that for any propositional $\beta \in \mathcal{OL}$, the formula $O\beta$ is equivalent to a finite disjunction of formulas of the form $O\psi$ where ψ is objective. In the process, we will need to substitute subwffs of the form $K\gamma$ or $O\gamma$ in β by either TRUE or FALSE. We begin by enumerating all subwffs $K\gamma_1, \ldots, K\gamma_k$, and $O\gamma_{k+1}, \ldots O\gamma_n$ that appear in β. In the proof below, we will let $\mathcal{OL}_\beta$ mean the subset of $\mathcal{OL}$ whose $K\gamma$ or $O\gamma$ subwffs appear in this list.

Definition 9.5.1: Let $v \in \{0, 1\}^n$. Then for any $\alpha \in \mathcal{OL}_\beta$, $\|\alpha\|_v$ is the objective formula that results from replacing a subwff $K\gamma_i$ or $O\gamma_i$ in α by TRUE if $v_i = 1$ and FALSE if $v_i = 0$.

Lemma 9.5.2: *Let e be an epistemic state, and suppose that $v \in \{0, 1\}^n$ satisfies $v_i = 1$ iff $e \models K\gamma_i$ (or $e \models O\gamma_i$). Then for any w, and any $\alpha \in \mathcal{OL}_\beta$, $e, w \models \alpha$ iff $w \models \|\alpha\|_v$, and consequently, $e \models O\alpha$ iff $e \models O\|\alpha\|_v$.*

Proof: By induction on the length of α. ∎

Lemma 9.5.3: *Suppose e and v are as above, and that for some $\alpha \in \mathcal{OL}_\beta$, we have that $e \models O\alpha$. Then for all $1 \leq i \leq k$, $v_i = 1$ iff $\models (\|\alpha\|_v \supset \|\gamma_i\|_v)$, and for all $k+1 \leq i \leq n$, $v_i = 1$ iff $\models (\|\alpha\|_v \equiv \|\gamma_i\|_v)$.*

Proof: For the first part with $i \leq k$, in the only-if direction, assume that $v_i = 1$. Now suppose that for any w, $w \models \|\alpha\|_v$. By the lemma above, $e \models O\|\alpha\|_v$, and so $w \in e$. Since $v_i = 1$, we have that $e \models K\gamma_i$, and therefore, $e, w \models \gamma_i$, and again by the same lemma, $w \models \|\gamma_i\|_v$. So $\models (\|\alpha\|_v \supset \|\gamma_i\|_v)$.

In the if-direction, assume that $\models (\|\alpha\|_v \supset \|\gamma_i\|_v)$, and suppose that w is any element of e. Since we have that $e \models O\|\alpha\|_v$, we get that $w \models \|\alpha\|_v$ and so $w \models \|\gamma_i\|_v$, and then by the above lemma, $e, w \models \gamma_i$. Thus, $e \models K\gamma_i$, and so $v_i = 1$.

The second part of the proof with $i > k$ is analogous. ∎

Lemma 9.5.4: *Assume that $v \in \{0, 1\}^n$ and that for the given β we have that for all $1 \leq i \leq k$, $v_i = 1$ iff $\models (\|\beta\|_v \supset \|\gamma_i\|_v)$, and for all $k+1 \leq i \leq n$, $v_i = 1$ iff $\models (\|\beta\|_v \equiv \|\gamma_i\|_v)$. Let $e = \Re[\![\|\beta\|_v]\!]$. Then for any $\alpha \in \mathcal{OL}_\beta$ and any w, we have that $e, w \models \alpha$ iff $w \models \|\alpha\|_v$, and so $e \models O\beta$.*

Proof: By induction on the length of α. For atoms, negations and conjunctions, the argument is clear. If α is $K\gamma_i$, then $e, w \models \alpha$ iff for every $w' \in e$, we have that $e, w' \models \gamma_i$ iff (by induction) for every $w' \in e$, we have that $w' \models \|\gamma_i\|_v$. Since $e = \Re[\![\|\beta\|_v]\!]$, this happens iff $\models (\|\beta\|_v \supset \|\gamma_i\|_v)$, iff $v_i = 1$ iff $\|\alpha\|_v =$ TRUE iff $w \models \|\alpha\|_v$. The final case with $O\gamma_i$ is analogous. ∎

Theorem 9.5.5: *For any formula $\beta \in \mathcal{OL}$ and any epistemic state e, $e \models O\beta$ iff there is a $v \in \{0, 1\}^n$ such that $e = \Re[\![\|\beta\|_v]\!]$ and where for all $1 \leq i \leq k$, $v_i = 1$ iff $\models (\|\beta\|_v \supset \|\gamma_i\|_v)$, and for all $k+1 \leq i \leq n$, $v_i = 1$ iff $\models (\|\beta\|_v \equiv \|\gamma_i\|_v)$.*

Proof: In the if direction, we can define the v using e as in Lemma 9.5.2, and then apply Lemma 9.5.3. The only-if direction is an immediate consequence of Lemma 9.5.4. ∎

Corollary 9.5.6: *For any $\beta \in \mathcal{OL}$, there are objective wffs $\psi_1, \ldots, \psi_m$, $m \geq 0$ such that $\models O\beta \equiv (O\psi_1 \vee \cdots \vee O\psi_m)$.*

Proof: Let S be the set of all objective wffs of the form $\|\beta\|_v$ where $v \in \{0, 1\}^n$ and for all $1 \leq i \leq k$, $v_i = 1$ iff $\models (\|\beta\|_v \supset \|\gamma_i\|_v)$, and for all $k+1 \leq i \leq n$, $v_i = 1$ iff

---

Input: any propositional formula $\beta \in \mathcal{OL}$;
Output: a set of objective formulas $\psi_1, \ldots, \psi_m$ satisfying
$$\models \boldsymbol{O}\beta \equiv (\boldsymbol{O}\psi_1 \vee \cdots \vee \boldsymbol{O}\psi_m).$$

Procedure
    /* Assume that β has subwffs $\boldsymbol{K}\gamma_1, \ldots, \boldsymbol{K}\gamma_k, \boldsymbol{O}\gamma_{k+1}, \ldots, \boldsymbol{O}\gamma_n$. */
    $S \leftarrow \{\}$
    for $v \in \{0,1\}^n$ **do**
        if for all $1 \leq i \leq k$, $v_i = 1$ iff $\models (\|\beta\|_v \supset \|\gamma_i\|_v)$
            and for all $k+1 \leq i \leq n$, $v_i = 1$ iff $\models (\|\beta\|_v \equiv \|\gamma_i\|_v)$
        then $S \leftarrow S \cup \{\|\beta\|_v\}$
    end
    return S
end

Figure 9.1: Calculating stable expansions

$\models (\|\beta\|_v \equiv \|\gamma_i\|_v)$. Then by the theorem, if $e \models \boldsymbol{O}\beta$, then for some $\psi \in S$, $e \models \boldsymbol{O}\psi$. Furthermore, if $\psi \in S$, and $e = \mathfrak{R}[\![\psi]\!]$, then again by the theorem, $e \models \boldsymbol{O}\beta$. Thus we have that $\boldsymbol{O}\beta$ is logically equivalent to $\vee\{\boldsymbol{O}\psi \mid \psi \in S\}$. ∎

We can also see looking at the proof of this corollary that the m in question can be no larger that 2^n where n is the number of subwffs of the form $\boldsymbol{K}\gamma$ or $\boldsymbol{O}\gamma$ that appear in β.

This then suggests a procedure for generating the epistemic states that satisfy a given propositional formula $\boldsymbol{O}\beta$ by generating a finite set of objective formulas that represent all that is known in each of these states.[4] The procedure appears in Figure 9.1. Because of Corollary 9.4.3, this procedure also generates the stable expansions of any propositional formula. More precisely, the objective formulas returned by the procedure represent the epistemic states whose basic belief sets are the stable expansions.

Finally, the theorem leads us to the conclusion that propositional $\mathcal{OL}$ is reducible, in the sense of Section 4.6: it is possible to reduce any propositional formula involving perhaps nested $\boldsymbol{K}$ or $\boldsymbol{O}$ operators to an equivalent one where the $\boldsymbol{K}$ and the $\boldsymbol{O}$ only dominate objective formulas. The proof of this is left as an exercise.

9.6 Non-reducibility of $\mathcal{OL}$

We have seen in Chapter 4 that $\mathcal{KL}$ is irreducible, that is, there are sentences such as $\boldsymbol{K}[\exists x P(x) \wedge \neg \boldsymbol{K} P(x)]$ with nested occurrences of $\boldsymbol{K}$ which are not equivalent to any sentence without nested $\boldsymbol{K}$'s. Of course, the same holds in $\mathcal{OL}$ as well since it subsumes $\mathcal{KL}$. But what about sentences of the form $\boldsymbol{O}\alpha$? Do we obtain reducibility at least for this special

---

4 This will not work in the first-order case since as in the example on page 148, a formula $\boldsymbol{O}\beta$ can be satisfied by an infinite set of epistemic states.

1. $\forall xyz[R(x,y) \wedge R(y,z) \supset R(x,z)]$
   R is transitive.
2. $\forall x \neg R(x,x)$
   R is irreflexive.
3. $\forall x[KP(x) \supset \exists y.R(x,y) \wedge KP(y)]$
   For every known instance of P, there is another one that is R related to it.
4. $\forall x[K \neg P(x) \supset \exists y.R(x,y) \wedge K \neg P(y)]$
   For every known non-instance of P, there is another one that is R related to it.
5. $\exists x KP(x) \wedge \exists x K \neg P(x)$
   There is at least one known instance and known non-instance of P.
6. $\exists x \neg KP(x)$
   There is something that is not known to be an instance of P.
7. $\forall x KP(x) \supset P(x)$.
   Every known instance of P is a P.
8. $\forall x K \neg P(x) \supset \neg P(x)$.
   Every known non-instance of P is not a P.

Figure 9.2: A sentence unsatisfiable in finite states

case? We saw in the previous section that we do get reducibility when α has no quantifiers. With quantifiers, the answer, in short, is no, but finding an appropriate irreducible sentence is not as straightforward as one might think. For example, obvious candidates like $O[\exists x P(x) \wedge \neg KP(x)]$ and $O[\exists x P(x) \wedge \neg OP(x)]$ are both equivalent to $O \exists x P(x)$ and hence reducible (see Exercise 6).

To show that only-knowing does not reduce, we choose a sentence which is almost identical to the sentence π in Figure 6.1 on page 105, which was used to show that finitely representable epistemic states are not sufficient to capture $\mathcal{KL}$. Let ζ be the conjunction of the sentences of Figure 9.2, which differs from π only in that there are two additional conjuncts (7) and (8).

Our first task is to show that ζ can be all that is known. To this end, let Ω be the set $\{{}^\#1, {}^\#3, {}^\#5, \ldots\}$ and let us call a standard name *odd* if it is in Ω and *even* otherwise. Let e be the set of world states w which satisfy the following conditions:

a) w satisfies all of the following objective sentences:
   $$\{P({}^\#1), \neg P({}^\#2), P({}^\#3), \neg P({}^\#4), \ldots\};$$
b) w satisfies conjuncts (1) and (2) stating that R is transitive and irreflexive;
c) for every even n there are infinitely many even standard names m which are R-related to n, that is, for which $w \models R(n,m)$;
d) for every odd n there are infinitely many odd standard names which are R-related to n.

Lemma 9.6.1: $e \models O\zeta$.

Only-Knowing and Autoepistemic Logic

Proof: It is easy to see that for every $w \in e$, $e, w \models \zeta$. Given our particular choice of R (conditions (b)–(d)), the argument is very similar to the one used to show the satisfiability of π. Note also that the conjuncts (7) and (8) are clearly satisfied because both $K[\forall x KP(x) \supset P(x)]$ and $K[\forall x K\neg P(x) \supset \neg P(x)]$ are valid sentences. Now consider an arbitrary world state w not in e. Then it violates one of the conditions (a)–(d). We will show that, in each case, one of the conjuncts of ζ is falsified by w. If w violates condition (a), then w does not satisfy an even P or w satisfies an odd P, that is, either conjunct (7) or (8) turns out false. If condition (b) is violated, then clearly either (1) or (2) is false. Now consider the case where (c) is violated. Then there is an even n and at most finitely many even $m_1, \ldots, m_k$ such that $w \models R(n, m_i)$, and for all other even m, $w \not\models R(n, m)$. We claim that there must be some $m^* \in \{m_1, \ldots, m_k\}$ such that $w \not\models R(m^*, m)$ for all even m. For assume otherwise, that is, for every m_i there is an even m'_i such that $w \models R(m_i, m'_i)$. Then, by the transitivity of R, we also have $w \models R(n, m'_i)$. Hence $m'_i \in \{m_1, \ldots, m_k\}$. However, this is only possible if there is a cycle, that is, w satisfies all of $\{R(m_i, m_{j_1}), R(m_i, m_{j_2}), \ldots, R(m_i, m_{j_k})\}$ and $m_{j_k} = m_i$ for some i. But then $w \models R(m_i, m_i)$, contradicting the irreflexivity of R. Given that there is an even name m^* such that for all even names m, $w \not\models R(m^*, m)$, conjunct (3) of ζ is clearly not satisfied. Similarly, the case where condition (d) is violated implies that conjunct (4) is not satisfied. Therefore, for every $w \notin e$, $e, w \not\models \zeta$ and $e \models O\zeta$ follows. Finally, note that e is also a maximal set because any $w \notin e$ falsifies a known basic sentence, namely ζ. ∎

Lemma 9.6.2: *For any e such that $e \models O\zeta$, both the set of known instances of P and the set of known non-instances are infinite.*

Proof: If we assume that there are only finitely many known instances of P, say, $m_1, \ldots, m_k$, then the assumption that conjunct (3) of ζ is satisfied at every $w \in e$ leads to a contradiction with the irreflexivity of R, using an argument similar to the one in the previous proof. The case of the known non-instances of P is symmetric. ∎

Lemma 9.6.3: *Let e be any epistemic state such that $e \models O\zeta$. Then for all objective ϕ, $e \not\models O\phi$.*

Proof: Assume otherwise, that is, suppose there is an objective ϕ such that $e \models O\phi$. Then e is finitely represented by ϕ. By Lemma 6.5.3, either the set of known instances of P is finite or the set of known non-instances, contradicting Lemma 9.6.2. ∎

Lemma 9.6.4: *Let $e \models O\zeta$ and let q be a 0-ary predicate symbol not occurring in ζ. Let $e^* = e \cap \{w \mid w \models q\}$. Then for any w and any α which does not mention q and whose subformulas $O\phi$ are restricted to objective ϕ, $e, w \models \alpha$ iff $e^*, w \models \alpha$.*

Proof: The proof is by induction on the structure of α. The lemma clearly holds for objective sentences and, by induction, for $\neg$, $\vee$, and $\exists$. Let $e \models K\alpha$. Then for all $w \in e$, $e, w \models \alpha$ and, by induction, for all $w \in e$, $e^*, w \models \alpha$. Since $e^* \subseteq e$, $e^* \models K\alpha$ follows. Now suppose $e \not\models K\alpha$. Then there is a $w \in e$ such that $e, w \not\models \alpha$. By induction, $e^*, w \not\models \alpha$. If $w \in e^*$ we are done. Otherwise, by the construction of e^*, $w \models \neg q$. Since q does not appear in ζ, it is easy to see that there must be a $\overline{w} \in e$ which is exactly like w except that $\overline{w} \models q$. Then $\overline{w}$ is in e^* and $e^*, \overline{w} \not\models \alpha$ because q does not appear in α. Hence $e^* \not\models K\alpha$. Finally, let us consider $O\phi$. Since ϕ is objective, $e^* \models \neg O\phi$ because e^* knows q and q does not occur in ϕ. Also, $e \models \neg O\phi$ because of Lemma 9.6.3. Hence, $e \models O\phi$ iff $e^* \models O\phi$. ∎

Theorem 9.6.5: *$O\zeta$ is not logically equivalent to any sentence where modal operators occur nested.*

Proof: Assume, to the contrary, that there is an α without nested modal operators such that $\models O\zeta \equiv \alpha$. Let e be any maximal set of world states such that $e \models O\zeta$ and let $e^* = e \cap \{w \mid w \models q\}$, where q is a 0-ary predicate symbol occurring nowhere in ζ or α.

First we show that e^* itself is maximal. For that it suffices to show that for any $w \notin e^*$, $e^*, w \not\models \gamma$ for some basic γ such that $e^* \models K\gamma$. If $w \not\models q$ we are done because $e^* \models Kq$. Otherwise $w \models q$ and, since $w \notin e^*$, $w \notin e$. Since $e \models O\zeta$ by assumption, $e, w \not\models \zeta$. Also, since ζ does not mention q, $e^*, w \not\models \zeta$ follows from Lemma 9.6.4. By the same lemma, $e^* \models K\zeta$ and we are done.

Continuing with the main proof of the theorem, since $\models O\zeta \equiv \alpha$ and $e \models O\zeta$ by assumption, we have $e, w \models \alpha$ for any w. Then, since any occurence of O in α applies only to an objective formula, $e^*, w \models \alpha$ by Lemma 9.6.4. Now consider a w in e which is not in e^*. (Such w clearly exists.) Then $e, w \models \zeta$ and thus, by Lemma 9.6.4, $e^*, w \models \zeta$. Therefore, $e^* \not\models O\zeta$, contradicting our assumption that α and $O\zeta$ are equivalent. ∎

9.7 Generalized stability

So far, the two main results relating $\mathcal{OL}$ to Moore's autoepistemic logic, Theorem 9.3.6 and Theorem 9.4.2, have only dealt with basic sentences or sentences like $O\alpha$, where α is basic. However, the generalization to deal with arbitrary sentences is not difficult. First define the *generalized belief set* of e to be the set of all sentences α (basic or not) such that $e \models K\alpha$. Then we have the following:

Theorem 9.7.1: *A set of sentences Γ is a generalized belief set iff Γ is a generalized stable set, that is, it satisfies the following conditions:*

1. *If $\Gamma \models \alpha$, then $\alpha \in \Gamma$.*[5]
2. *If $\alpha \in \Gamma$, then $K\alpha \in \Gamma$.*
3. *If $\alpha \notin \Gamma$, then $\neg K\alpha \in \Gamma$.*

Proof: The proof is identical to that of Theorem 9.3.6, except without the diversion via Theorem 9.3.4 to handle first-order implication. ∎

So to convert a belief set to a generalized belief set, we need only close it under implication (rather than just first-order implication).

Dealing with $O\alpha$ in general is also straightforward:

Theorem 9.7.2: *For any α and any maximal e, $e \models O\alpha$ iff the generalized belief set of e is a generalized stable expansion of α, that is, the generalized belief set Γ satisfies*

$$\Gamma \text{ is the set of implications of } \{\alpha\} \cup \{K\beta \mid \beta \in \Gamma\} \cup \{\neg K\beta \mid \beta \notin \Gamma\}.$$

Proof: The proof is the same as that of Theorem 9.4.2, except again without the diversion through Theorem 9.3.4 to handle first-order implication, and a generalized belief set is used here. However, by Lemma 9.3.3, belief sets completely determine the generalized belief sets. ∎

This, then, is perhaps the most succinct characterization of only-knowing: α is all that is known iff every belief follows logically from α and the basic subjective facts.

9.8 Bibliographic notes

Autoepistemic logic was originally developed by Moore [110, 109], who was inspired by earlier work on nonmonotonic modal logic by McDermott and Doyle [105, 106]. The idea of stable sets goes back to Stalnaker and appeared in a note already in 1980, but was published only much later in [139]. There have been a number of proposals to syntactically characterize the stable expansions of a given set of assumptions, for example [135, 137,

---

5 This closure under full implication is the only change to the definition of stability.

100, 113, 142]. Ours differs perhaps in that it yields a rather simple algorithm (Figure 9.1). Corollary 9.5.6 was independently obtained by Waaler [142] (see also [132]).

Except for [15], who relates $\mathcal{OL}$ to Lifschitz' logic of minimal belief and negation as failure [90], we know of no work which directly relates $\mathcal{OL}$ to other nonmonotonic formalisms. Of course, given the exact correspondence with AEL, comparative studies involving AEL can be imported directly. For example, the connection between AEL and Reiter's default logic [120] has been thoroughly investigated. Early proposals translating default logic into AEL such as [58, 93, 112, 99, 141] all required some modification of AEL by eliminating certain *ungrounded* stable expansions (see the next section for more on this issue). Gottlob [43] finally established that a faithful translation from default logic into standard AEL is possible, but that it cannot be done in a modular way, that is, by translating every default separately. Also, it is well known that the reverse translation sometimes is not possible at all. For example, there is no default theory corresponding to only-knowing ($Kp \supset p$).

It should be noted that the above comparisons are either explicitly restricted to the propositional case or at least rule out quantifying-in. In fact, as far as we know only Konolige [59, 60] and Lifschitz [90] address nonmonotonic reasoning with quantifying-in apart from us. While there clearly are similarities, there are also significant differences, in particular, regarding the use of names. For example, while Lifschitz requires there to be a name for every individual, the name need not be unique. Moreover, there is no restriction on the cardinality of the domain. Konolige even allows for individuals which have no name at all. While certainly interesting, a detailed analysis of what these differences amount to remains largely open.

Finally, let us go back to the concept of total knowledge [116], which we discussed already briefly at the end of the previous chapter and which coincides with only-knowing in the case of objective sentences. Interestingly, in contrast to the strong correspondence between AEL and only-knowing, no such correspondence seems to exist between AEL and total knowledge. To see the difference, consider again the "older-brother" default rule $\delta = (\neg Kb \supset \neg b)$. As we have seen, only-knowing δ implies knowing $\neg b$. Requiring δ to be total knowledge, however, is inconsistent. While total knowledge is perhaps not useful for default reasoning, it does have the nice property that total knowledge of an arbitrary sentence always reduces to total knowledge of an objective sentence.

9.9 Where do we go from here?

The connection to autoepistemic reasoning is the first chapter of Part II of this book, which covers special topics extending our basic logic of knowledge bases in various ways. Naturally, there are many open research issues which are only touched upon lightly or not at all. For this reason, we include a paragraph or more in this and the following chapters to draw attention to some of the open issues.

As was already hinted at in the previous section, AEL sometimes yields expansions which may be considered ungrounded. The standard example is the sentence $\alpha = (Kp \supset p)$. Intuitively, α can be read as "assume p provided you already know p." Clearly, if α is all that is known, there seems to be little if any reason to believe p. And indeed, there is an expansion of α where nothing is believed apart from the valid sentences (and those obtained by introspection). However, there is also a second expansion which contains p. Since there really is no good reason to believe p based only on α,[6] such expansions have been called ungrounded. To remedy the situation, various notions of groundedness of a stable expansion have been proposed and AEL was modified so that only grounded expansions obtain. Of course, $\mathcal{OL}$ suffers from the same problems as AEL. It is an open problem whether there are restricted forms of only-knowing which would also eliminate ungrounded expansions.

Drawing on the work by Konolige and Lifschitz on quantifying-in, which we discussed in the previous section, it seems interesting to investigate extensions of $\mathcal{OL}$ where the isomorphism between the standard names and the universe of discourse is given up. With luck this would yield versions of only-knowing with exact correspondences to the above proposals.

Finally, while some of the similarities and differences between only-knowing and total knowledge have been established, their exact relationship is still unclear.

9.10 Exercises

1. Show that $(Kp \supset p)$ has two stable expansions.
2. Let KB = {$Bird(tweety), Bird(chilly), \neg Fly(chilly)$}. We assume that both *tweety* and *chilly* are standard names. Show that $(O(\text{KB} \wedge \delta) \supset KFly(tweety))$ is valid, where δ is the birds-fly default.
   Hint: Use the proof of Example 9.1.2, part 3, but with $\forall x(x \neq chilly) \supset Fly(x)$ instead of $\forall x Fly(x)$.
3. Let KB = {$\exists x Exc(x)$}, where $Exc(x)$ stands for $Bird(x) \wedge \neg Fly(x)$ as before. Show that $O[\text{KB} \wedge \delta] \equiv \exists y O[\text{KB} \wedge \forall x(Exc(x) \equiv (x = y))]$ is valid, that is, there are infinitely many epistemic states compatible with only-knowing KB.
4. Let $\delta_K = \forall x[KBird(x) \wedge \neg K\neg Fly(x) \supset Fly(x)]$, that is, the default about flying birds

---

6 One possible reason is to read α as saying that it is sufficient to believe p for p to be true, as when believing p makes p true.

only applies to known birds, as discussed on page 146. Let
KB = {*Bird*(*tweety*), *Bird*(*best_friend*(*tweety*)), ¬*Fly*(*best_friend*(*tweety*))}.
Show whether (O(KB $\wedge$ δ_K) $\supset$ $K$$Fly$(*tweety*)) is valid.

5. Show that for any propositional formula α there is an equivalent one α' where the K and the O only dominate objective formulas..

6. Show that both $O[\exists x P(x) \wedge K P(x)]$ and $O[\exists x P(x) \wedge O P(x)]$ are logically equivalent to $O \exists x P(x)$.

10 On the Proof Theory of $\mathcal{OL}$

We already saw proof theoretic characterizations of $\mathcal{L}$ and $\mathcal{KL}$ and found them useful because they gave us an independent account of the valid sentences of the respective logics. In this chapter, we consider an axiom system for $\mathcal{OL}$. We demonstrate its usefulness by syntactically deriving some of the default conclusions of the previous chapter and prove their completeness for the propositional case. Unfortunately, they are not complete for the whole language, and we will discuss why this is so. In addition, the axioms are not even recursively enumerable, that is, we do not have a proof theory in the traditional sense. We will see that this feature is inescapable in $\mathcal{OL}$.

10.1 Knowing at least and at most

To better analyze only-knowing and for this chapter only, it is convenient to consider $\boldsymbol{O}$ not as a primitive notion but to define it in terms of $\boldsymbol{K}$ and a new operator $\boldsymbol{N}$ in the following way. One way to read $\boldsymbol{O}\alpha$ is to say that α is believed and nothing more, whereas $\boldsymbol{K}\alpha$ says that α is believed, and perhaps more. In other words, $\boldsymbol{K}\alpha$ means that α *at least* is believed to be true. A natural dual to this is to say that α *at most* is believed to be false, which we write $\boldsymbol{N}\alpha$. The idea behind introducing this operator is that $\boldsymbol{O}\alpha$ would then be *definable* as $(\boldsymbol{K}\alpha \wedge \boldsymbol{N}\neg\alpha)$, that is, at least α is believed and at most α is believed.[1] So, *exactly* α is believed. In other words, we are taking $\boldsymbol{K}$ to specify a lower bound on what is believed (since there may be other beliefs) and $\boldsymbol{N}$ to specify an upper bound on beliefs (since there may be fewer beliefs).[2] What is actually believed must lie between these two bounds.

These bounds can be seen most clearly when talking about objective sentences. Given an epistemic state as specified by a maximal set of world states e, to say that $\boldsymbol{K}\phi$ is true wrt e is to say that e is a subset of the states where ϕ is true. By symmetry then, $\boldsymbol{N}\neg\phi$ will be true when the set of states satisfying ϕ are a subset of e. The fact that e must contain all of these states means that nothing else can be believed that would eliminate any of them. This is the sense in which no more than ϕ is known. Finally, as before, $\boldsymbol{O}\phi$ is true iff both conditions hold and the two sets coincide.

This leads us to the precise definition of $\boldsymbol{N}\alpha$:

6'. $e, w \models \boldsymbol{N}\alpha$ iff for every w', if $e, w' \not\models \alpha$ then $w' \in e$.

from which the original constraint 6 on $\boldsymbol{O}\alpha$ follows trivially.

---

1 Although using negation with the $\boldsymbol{N}$ operator appears perhaps to be needlessly complex, it will indeed simplify matters later.
2 A slightly simplistic way of saying this is that $\boldsymbol{K}\alpha$ means that what is actually believed is of the form $(\alpha \wedge \beta)$, whereas $\boldsymbol{N}\alpha$ means what is actually believed is of the form $(\neg\alpha \vee \beta)$.

So what are the properties of believing at most that α is false? It is very easy to show that if α is valid, then $N\alpha$ will be valid too, if $N\alpha$ and $N(\alpha \supset \beta)$ are both true, then so is $N\beta$, and if some subjective σ is true, then so is $N\sigma$. In other words, remarkably enough, N behaves like an ordinary belief operator: it is closed under logical implication and exhibits perfect introspection. This is most clearly seen by rephrasing very slightly the definition of N and comparing it to that of K:

5. $e, w \models K\alpha$ iff   for every $w' \in e$, $e, w' \models \alpha$.
6'. $e, w \models N\alpha$ iff   for every $w' \notin e$, $e, w' \models \alpha$.[3]

Letting $\bar{e}$ stand for the set of states not in e, we have

6'. $e, w \models N\alpha$ iff   for every $w' \in \bar{e}$, $e, w' \models \alpha$.

So N is like a belief operator with one important difference: we use the complement of e. In possible-world terms, we range over the *inaccessible* possible world states. In other words, K and N give us two belief-like operators: one, with respect to e, and one with respect to $\bar{e}$.

In these terms, the relation between the two belief operators is that the accessible world states they range over must be *disjoint* (empty intersection) and *exhaustive* (universal union). As it turns out, only the exhaustiveness property is used in the axiomatization. In fact, we will show as part of the propositional completeness proof that there is an equivalent semantics where the set of world states considered for K and N may overlap.

In the following, a *subjective* sentence is understood as in $\mathcal{KL}$ except that any occurrence of K (but not necessarily all) can be replaced by N. For example, the sentence $(\forall x \forall z(x = y) \supset KNP(x, y))$ is considered subjective since its truth depends only on the epistemic state and its complement. The axioms for $\mathcal{OL}$ are then given as follows:[4]

1. $L\alpha$, where α is an instance of an axiom of $\mathcal{L}$ (with the proviso on specialization).
2. $L(\alpha \supset \beta) \supset L\alpha \supset L\beta$.
3. $\forall x L\alpha \supset L\forall x\alpha$.
4. $\sigma \supset L\sigma$, where σ is subjective.
5. The N vs. K axiom:
   $(N\phi \supset \neg K\phi)$, where ϕ is any objective sentence such that $\not\models \phi$.
6. The definition of O: $O\alpha \equiv (K\alpha \wedge N\neg\alpha)$.

The first thing to notice is that N, taken by itself, has precisely the same properties as K, that is, K and N can both be thought of as ordinary belief operators except, of course, that there is a strong connection between the two. For one, Axiom 4 expresses that both K

---

3 Note that phrased this way, α rather than $\neg\alpha$ is required to be true. This is why negation was used for N.
4 In the following axioms, **L** is used as a modal operator standing for either K or N. Multiple occurrences of **L** in the same axiom should be uniformly replaced by K or by N.

On the Proof Theory of $\mathcal{OL}$ 165

and N are perfectly and mutually introspective. For example, ($K\alpha \supset NK\alpha$) is an instance of 4. If we think of K and N as two agents, then this says that each agent has perfect knowledge of what the other knows. The other and more interesting connection between the two is, of course, Axiom 5, which is valid because K and N together range over all possible world states.

It is not hard to show that all these axioms are sound:

Theorem 10.1.1: *If a sentence of $\mathcal{OL}$ is derivable, then it is valid.*

Proof: The proof proceeds by a standard induction on the length of a derivation, just like in the case $\mathcal{KL}$. Here we only show the base case for the N vs. K axiom. Thus let ϕ be an objective sentence such that $\not\models \phi$ and e a maximal set of world states such that $e \models N\phi$. Then for all world states not in e, $w \models \phi$. Since $\not\models \phi$, there must be a world state w' such that $w' \models \neg\phi$. Moreover, w' must be in e, from which $e \models \neg K\phi$ follows. ∎

10.2 Some example derivations

Before turning to the issue of completeness, let us go through several derivations in order to get a better feel for the axioms. The examples are the defaults about birds, which we already proved semantically in Chapter 9. Syntactically deriving defaults is particularly instructive, since they exhibit very nicely the use of Axiom 5 and, more generally, the power of the proof theory.

In the derivations to follow, we will use a natural-deduction-style argument with three justifications: (1) the definition of O in terms of N and K, (2) the axiom relating N to K for objective sentences, and (3) $\mathcal{KL}$, which we normally do not analyze further. When writing $\mathcal{KL}$ as a justification we really mean Axioms 2–4, which include the axioms of $\mathcal{KL}$ with K replaced by N.

In the following we consider the examples about flying birds with the default

$$\delta = \forall x[Bird(x) \land \neg K\neg Fly(x) \supset Fly(x)].$$

Example 10.2.1: If KB = $\{Bird(tweety)\}$ then $(O(\text{KB} \land \delta) \supset KFly(tweety))$ is a theorem. (See Example 9.1.2 for the semantic proof.)

Proof:

1. $O(\text{KB} \land \delta)$ Assumption.
2. $K(\text{KB} \land \delta)$ 1;defn. of O.

3. $(K\neg K\neg Fly(tweety) \supset KFly(tweety))$	2;$\mathcal{KL}$.
4. $(\neg K\neg Fly(tweety) \supset KFly(tweety))$	3;$\mathcal{KL}$.
5. $N\neg(\text{KB} \wedge \delta)$	1;defn. of O.
6. $N(\text{KB} \supset \exists x \neg Fly(x))$	5;$\mathcal{KL}$.
7. $\neg K(\text{KB} \supset \exists x \neg Fly(x))$	6;N vs. K.
8. $\neg K \neg Fly(tweety)$	7;$\mathcal{KL}$.
9. $KFly(tweety)$	4,8;$\mathcal{KL}$.

Most of the uses of $\mathcal{KL}$ here are direct: lines 6 and 8 involve propositional reasoning within a modal operator; line 9 is the result of ordinary propositional logic; line 3 requires distributing the K over a conjunction, then over an implication; and line 4 uses the $\mathcal{KL}$ axiom $(\neg K \neg Fly(tweety) \supset K \neg K \neg Fly(tweety))$ to obtain the final formula. Note that line 7 is the only place where we need to make use of the special connection between K and N (Axiom 5). The derivation is valid because $(\text{KB} \supset \exists x \neg Fly(x))$ clearly is falsifiable. ∎

Example 10.2.2: Let

$$\text{KB} = \{Bird(tweety), Bird(chilly), (\neg Fly(chilly) \vee \neg Fly(tweety))\}$$

and $Exc(x)$ be an abbreviation for $Bird(x) \wedge \neg Fly(x)$. The sentence $O[\text{KB} \wedge \delta]$ is logically equivalent to

$$(O[\text{KB} \wedge \forall x(Exc(x) \equiv x = tweety)] \vee O[\text{KB} \wedge \forall x(Exc(x) \equiv x = chilly)]).$$

Proof: In the following syntactic derivation we will occasionally write a line σ and then follow it by $K\sigma$ or $N\sigma$. This is *not* using the rule: from α infer $K\alpha$ which, in general, is only sound when α is valid. Rather it is an application of *modus ponens*, together with the axiom $(\sigma \supset K\sigma)$, which holds whenever σ is subjective. Also, for clarity, we will expand some of the steps that depend only on properties of $\mathcal{KL}$. The first part of the proof (only-if direction) has two stages: we first establish that $K\neg Fly(tweety) \vee K\neg Fly(chilly)$ is provable given the assumption $O[\text{KB} \wedge \delta]$; then we show that each of the disjuncts of the desired conclusion is derivable from either $K\neg Fly(tweety)$ or $K\neg Fly(chilly)$.

1. $O[\text{KB} \wedge \delta]$	Assumption.
2. $N[\text{KB} \supset \exists x(Exc(x) \wedge \neg K\neg Fly(x))]$	1;defn. of O.
3. $K[\text{KB} \wedge \forall x(Exc(x) \supset K\neg Fly(x))]$	1;defn. of O.
4. $K(Exc(tweety) \vee Exc(chilly))$	3;$\mathcal{KL}$.
5. $K(K\neg Fly(tweety) \vee K\neg Fly(chilly))$	3,4;$\mathcal{KL}$.
6. $K\neg Fly(tweety) \vee K\neg Fly(chilly)$	5;$\mathcal{KL}$.

On the Proof Theory of $\mathcal{OL}$

Now what we will do is show that

$$\mathbf{K}\neg Fly(tweety) \supset \mathbf{O}[KB \wedge \forall x(Exc(x) \equiv (x = tweety))]$$

is a theorem, and so, by an analogous derivation, we have that

$$\mathbf{K}\neg Fly(chilly) \supset \mathbf{O}[KB \wedge \forall x(Exc(x) \equiv (x = chilly))]$$

is a theorem, in which case, the required disjunction follows immediately from Line 6.

7. $\mathbf{K}\neg Fly(tweety)$ Assumption.
8. $(\mathbf{K}\forall x(x = tweety \supset Exc(x)))$ 3,7;$\mathcal{KL}$.
9. $\mathbf{NK}\neg Fly(tweety)$ 7;$\mathcal{KL}$.
10. $\mathbf{N}[KB \supset \exists x(\neg Fly(x) \wedge (x \neq tweety))]$ 2,9;$\mathcal{KL}$.
11. $\neg \mathbf{K}[KB \supset \exists x(\neg Fly(x) \wedge (x \neq tweety))]$ 10;$\mathbf{N}$ vs. $\mathbf{K}$.
12. $(\forall x(x \neq tweety) \supset \neg \mathbf{K}\neg Fly(x))$ 11;$\mathcal{KL}$.
13. $\mathbf{K}[\forall x(x \neq tweety) \supset \neg \mathbf{K}\neg Fly(x)]$ 12;$\mathcal{KL}$.
14. $\mathbf{K}[\forall x(x \neq tweety) \supset \neg Exc(x)]$ 3,13;$\mathcal{KL}$.
15. $\mathbf{K}[KB \wedge \forall x(Exc(x) \equiv (x = tweety))]$ 8,14;$\mathcal{KL}$.
16. $\mathbf{N}[KB \supset \exists x(Exc(x) \wedge (x \neq tweety))]$ 2,7;$\mathcal{KL}$.
17. $\mathbf{N}[KB \supset \neg \forall x(Exc(x) \equiv (x = tweety))]$ 16;$\mathcal{KL}$.
18. $\mathbf{O}[KB \wedge \forall x(Exc(x) \equiv (x = tweety))]$ 15,17;defn. of $\mathbf{O}$.

This completes the first part of the proof. Note that except for line 11 all the reasoning involved requires only the axioms of $\mathcal{KL}$ (for both $\mathbf{K}$ and $\mathbf{N}$). Line 11 requires Axiom 5, which is applicable because $(KB \supset \exists x(\neg Fly(x) \wedge (x \neq tweety)))$ is not valid simply by letting Tweety be the only non-flying bird.

We now proceed to the only-if part. What we will do is show that

$$\mathbf{O}[KB \wedge \forall x(Exc(x) \equiv (x = tweety))] \supset \mathbf{O}[KB \wedge \delta]$$

is a theorem. Because this proof applies equally well to Chilly, the disjunction of the two possibilities gives us the desired conclusion.

1. $\mathbf{O}[KB \wedge \forall x(Exc(x) \equiv (x = tweety))]$ Assumption.
2. $\mathbf{K}[KB \wedge \forall x(Exc(x) \equiv (x = tweety))]$ 1;defn. of $\mathbf{O}$.
3. $\mathbf{K}\neg Fly(tweety)$ 2;$\mathcal{KL}$.
4. $(\mathbf{K}\forall x(Exc(x) \supset \mathbf{K}\neg Fly(x)))$ 2,3;$\mathcal{KL}$.
5. $\mathbf{K}[KB \wedge \delta]$ 2,4;$\mathcal{KL}$.
6. $\mathbf{N}[KB \supset \neg \forall x(Exc(x) \equiv (x = tweety))]$ 1;defn. of $\mathbf{O}$.
7. $\mathbf{N}[KB \supset \exists x Exc(x)]$ $\mathcal{KL}$.
8. $\mathbf{N}[KB \supset \exists x(x \neq tweety \wedge Exc(x))]$ 6,7;$\mathcal{KL}$.

9. $N[\text{KB} \supset \exists x(x \neq \textit{tweety} \wedge \neg \textit{Fly}(x))]$ 8;$\mathcal{KL}$.
10. $\neg K[\text{KB} \supset \exists x(x \neq \textit{tweety} \wedge \neg \textit{Fly}(x))]$ 9;N vs. K.
11. $\neg K[\exists x(x \neq \textit{tweety} \wedge \neg \textit{Fly}(x))]$ 10;$\mathcal{KL}$.
12. $(\forall x(x \neq \textit{tweety} \supset \neg K \neg \textit{Fly}(x)))$ 11;$\mathcal{KL}$.
13. $(N \forall x(x \neq \textit{tweety} \supset \neg K \neg \textit{Fly}(x)))$ 12;$\mathcal{KL}$.
14. $N[\text{KB} \supset \exists x(\textit{Exc}(x) \wedge \neg K \neg \textit{Fly}(x))]$ 8,13;$\mathcal{KL}$.
15. $O[\text{KB} \wedge \delta]$ 5,14;defn. of O.

Note that there is again only one place in the proof where we need Axiom 5 (line 10), and it is exactly the same as in the first part. This completes the proof. ∎

10.3 Propositional completeness

We now turn to the issue of completeness of the axiom system in the propositional case. The proof uses the standard technique of constructing satisfying models for maximally consistent sets. As in the case of $\mathcal{KL}$, not every maximally consistent set is satisfiable. In $\mathcal{KL}$ this was due mainly to technical reasons, having to do with quantification and standard names. Interestingly, the reasons here are not only entirely different, which is not surprising, but they also lead to a deeper understanding of the underlying semantics. As we already mentioned, it is sufficient for Axiom 5 to be sound if K and N cover all world states. Whether or not the two sets are disjoint or not seems to be neither sufficient nor necessary. In fact, it turns out that there are maximally consistent sets which are only satisfiable if we allow the set of world states considered by K to overlap with the set of world states considered by N. In a moment, we will consider just such an "overlapping" semantics. Not only are all of the axioms of $\mathcal{OL}$ (restricted to the propositional case) sound, but the valid sentences of the overlapping semantics are precisely the same as in our original semantics. What this tells us is that, while the disjointness property maybe more intuitively appealing, it cannot be captured axiomatically. Indeed, insisting on disjointness needlessly complicates the completeness proof since we would have to show that we can restrict ourselves to a subset of the maximally consistent sets. While this can be done, we will instead prove completeness for the equivalent overlapping semantics, which is straightforward given the known techniques for modal logics. Finally, the overlapping semantics has another revealing feature in that it does not require maximal sets of world states. So, as a corollary, we get that our original semantics yields the same valid sentences if we allow arbitrary epistemic states, not just maximal ones.

We begin with the definition of the overlapping semantics for propositional $\mathcal{OL}$. Let e and e' be two sets of world states. If $e \cup e' = e_0$, then we call (e, e') an *exhaustive pair*. Define a new satisfaction relation $\models^x$ that is exactly like $\mathcal{OL}$'s except for K- and N-formulas.

Let (e_K, e_N) be an exhaustive pair. Then
1. $e_K, e_N, w \models^x K\alpha$ if $e_K, e_N, w' \models^x \alpha$ for all $w' \in e_K$
2. $e_K, e_N, w \models^x N\alpha$ if $e_K, e_N, w' \models^x \alpha$ for all $w' \in e_N$.

Note that K and N are now treated in a completely symmetric way.

A sentence α is called *x-valid* ($\models^x \alpha$) iff $e_K, e_N, w \models^x \alpha$ for all world states w and all exhaustive pairs (e_K, e_N). α is called *x-satisfiable* iff $\neg\alpha$ is not x-valid.

Theorem 10.3.1: *The axioms are sound with respect to $\models^x$.*

The proof is straightforward and omitted. Note that for Axiom 5 to be sound it suffices that e_K and e_N together cover all world states. In particular, it does not matter whether or not the two sets overlap.

It is clear that the notions of satisfiability and validity coincide for both semantics when restricted to objective formulas. To prove that this is true in general, we need the following lemmas.

Lemma 10.3.2: *Let ϕ and ψ be objective formulas such that $\phi \wedge \neg\psi$ is satisfiable. Then $(K\neg\phi \supset \neg N\psi)$ and $(N\neg\phi \supset \neg K\psi)$ are both valid and x-valid.*

Proof: The proof relies only on the fact that the world states considered for K together with those for N cover all world states, which holds in either semantics. Here we only consider the nonstandard case.

Let $e_K, e_N, w \models^x K\neg\phi$. Then e_N contains every world state that satisfies ϕ. Hence, by assumption, there is a $w' \in e_N$ such that $w' \models^x \phi \wedge \neg\psi$, that is, $e_K, e_N, w \models^x \neg N\psi$.

$(N\neg\phi \supset \neg K\psi)$ is handled in a completely symmetric way. ∎

Next we show that every propositional sentence is provably equivalent to one without nested modal operators. Let us write $\mathbf{L}$ to stand for either K or N. As before, we write $\vdash \alpha$ for α is provable from the axioms.

Lemma 10.3.3: $\vdash \mathbf{L}(\phi \vee \sigma) \equiv (\mathbf{L}\phi \vee \sigma)$.

Proof: Since $\vdash (\sigma \supset \mathbf{L}(\phi \vee \sigma))$ because σ is subjective, and $\vdash (\mathbf{L}\phi \supset \mathbf{L}(\phi \vee \sigma))$, then clearly $\vdash ((\mathbf{L}\phi \vee \sigma) \supset \mathbf{L}(\phi \vee \sigma))$. Conversely, since $\vdash (\mathbf{L}(\alpha \supset \beta) \supset (\mathbf{L}\alpha \supset \mathbf{L}\beta))$ in general, we have that $\vdash (\mathbf{L}(\phi \vee \sigma) \supset (\mathbf{L}\phi \vee \neg\mathbf{L}\neg\sigma))$. But since σ is subjective, $\vdash (\neg\mathbf{L}\neg\sigma \supset \sigma)$. Thus, $\vdash (\mathbf{L}(\phi \vee \sigma) \supset (\mathbf{L}\phi \vee \sigma))$. ∎

Lemma 10.3.4: *Every sentence is provably equivalent to one where K or N operators apply only to objective sentences.*

Proof: Consider a subformula $L\alpha$. Using the rules of standard propositional logic, put α into conjunctive normal form so that $\vdash (\alpha \equiv \bigwedge(\phi_i \vee \sigma_i))$, where we have separated the subjective and objective parts. By induction, assume that each σ_i is in the correct form. Then,[5] $\vdash (L\alpha \equiv \bigwedge L(\phi_i \vee \sigma_i))$ and so, by the above lemma, $\vdash (L\alpha \equiv \bigwedge(L\phi_i \vee \sigma_i))$. One level of nesting of L has been eliminated. By applying this to all subformulas, the correct sentence will be obtained. ∎

Theorem 10.3.5: *For all propositional sentences α, α is x-satisfiable iff α is satisfiable.*

Proof: The only-if direction is straightforward. If $e, w \models \alpha$ then simply let $e_K = e$ and let e_N be the complement of e.

To prove the if direction, let $e_K, e_N, w \models^x \alpha$. We need to find a maximal set of world states e such that $e, w \models \alpha$. By Theorem 10.3.1 and Lemma 10.3.4, we can assume, without loss of generality, that $\alpha = \bigvee \alpha_m$, where each α_m has the form

$$\alpha_m = \phi \wedge \bigwedge Kr_i \wedge \bigwedge \neg Ks_j \wedge \bigwedge Nt_k \wedge \bigwedge \neg Nu_l$$

for objective ϕ, r_i, s_j, t_k, and u_l. Then $e_K, e_N, w \models^x \alpha_m$ for some m. It suffices to find a maximal set e such that $e, w \models \alpha_m$.

Let $\alpha^* = (\bigwedge t_k \supset (p \wedge \bigwedge r_i))$ for some atom p not occurring in α and let $e = \{w \mid w \models \alpha^*\}$. e is obviously maximal and $e \models O\alpha^*$ or, equivalently, $e \models K\alpha^* \wedge N\neg\alpha^*$. To show that $e, w \models \alpha_m$ we need to prove that each of the conjuncts is satisfied.

$e, w \models \phi$: Follows immediately because ϕ is objective and $e_K, e_N, w \models^x \alpha_m$.

$e, w \models Kr_i$: Let $w' \in e$. Then $w' \models \neg \bigwedge t_k \vee (p \wedge \bigwedge r_i)$. If $w' \models \bigwedge t_k$, then $w' \models r_i$. Otherwise, $w' \models \neg \bigwedge t_k$. Since $N \bigwedge t_k \wedge Kr_i$ is x-satisfiable, by Lemma 10.3.2, $\models (\neg \bigwedge t_k \supset r_i)$. Hence $w' \models r_i$.

$e, w \models \neg Ks_j$: Since $(e_K, e_N, w) \models^x K \bigwedge r_i \wedge \neg Ks_j$, $\not\models (\bigwedge r_i \supset s_j)$. Since p does not occur in any of the r_i and s_j, $\not\models ((p \wedge \bigwedge r_i) \supset s_j)$. Thus, by Lemma 10.3.2, $\models (N\neg(p \wedge \bigwedge r_i) \supset \neg Ks_j)$, from which $e \models \neg Ks_j$ follows.

$e, w \models Nt_k$: Follows immediately from $e \models N\neg\alpha^*$.

$e, w \models \neg Nu_l$: Since $e_K, e_N, w \models^x N \bigwedge t_k \wedge \neg Nu_l$, $\not\models (\bigwedge t_k \supset u_l)$ and, therefore, $\not\models ((\bigwedge t_k \wedge \neg p) \supset u_l)$, which implies $(\not\models \neg\alpha^* \supset u_l)$. Thus, by Lemma 10.3.2, $\models (K\alpha^* \supset \neg Nu_l)$, from which $e \models \neg Nu_l$ follows. ∎

---

5 We repeatedly use the fact here and later that $\vdash \bigwedge L\alpha_i \equiv L \bigwedge \alpha_i$.

Since a sentence is valid (x-valid) iff its negation is not satisfiable (x-satisfiable) we immediately get:

Corollary 10.3.6: $\models \alpha$ iff $\models^x \alpha$.

An interesting consequence of the theorem is that whether or not we use maximal sets of world states has no effect on the valid sentences, at least as far as the propositional subset of $\mathcal{OL}$ is concerned.

Corollary 10.3.7: $\models \alpha$ iff $e, w \models \alpha$ for all world states w and epistemic states e (including nonmaximal e).

Proof: The if direction holds immediately since maximal sets of world states are special epistemic states.

For the converse, assume $\models \alpha$ and let w be a world state and e an arbitrary epistemic state. Let $\bar{e}$ denote the complement of e. By Theorem 10.3.6, $\models^x \alpha$ and hence $e, \bar{e}, w \models^x \alpha$, from which $e, w \models \alpha$ follows immediately. ∎

Theorem 10.3.8: For all propositional sentences α, if $\models^x \alpha$ then $\vdash \alpha$.

Proof: The approach is very similar to the one taken in the completeness proof of $\mathcal{KL}$ (Theorem 4.5.1). Recall that to establish completeness it suffices to show that every consistent sentence is satisfiable, which in turn is established by showing that any maximally consistent sets that contains α is satisfiable. Of course, (maximal) consistency now refers to the axioms of $\mathcal{OL}$.

Let $\mathcal{C}_0$ be the set of all maximally consistent sets of propositional sentences of $\mathcal{OL}$. For $\Gamma \in \mathcal{C}_0$, define $\Gamma/K = \{\alpha \mid K\alpha \in \Gamma\}$ and $\Gamma/N = \{\alpha \mid N\alpha \in \Gamma\}$. We then define

- $\mathcal{C}_K^\Gamma = \{\Gamma' \in \mathcal{C}_0 \mid \Gamma/K \subseteq \Gamma'\}$,
- $\mathcal{C}_N^\Gamma = \{\Gamma' \in \mathcal{C}_0 \mid \Gamma/N \subseteq \Gamma'\}$.

If we view maximally consistent sets as world states, then $\mathcal{C}_K^\Gamma$ and $\mathcal{C}_N^\Gamma$ represent the world states accessible from Γ for K and N, respectively. The following lemma reflects the fact that K and N are both fully and mutually introspective (Axiom 4).

Lemma 10.3.9: If $\Gamma' \in \mathcal{C}_K^\Gamma \cup \mathcal{C}_N^\Gamma$, then $\mathcal{C}_K^{\Gamma'} = \mathcal{C}_K^\Gamma$ and $\mathcal{C}_N^{\Gamma'} = \mathcal{C}_N^\Gamma$.

Proof: We prove the lemma for $\Gamma' \in \mathcal{C}_K^\Gamma$. The case $\Gamma' \in \mathcal{C}_N^\Gamma$ is completely symmetric. To show that $\mathcal{C}_K^\Gamma = \mathcal{C}_K^{\Gamma'}$, it clearly suffices to show that $\Gamma/K = \Gamma'/K$. Let $\alpha \in \Gamma/K$. Then

$K\alpha \in \Gamma$ and also $KK\alpha \in \Gamma$ by Axiom 4. Thus $K\alpha \in \Gamma'$ (since $\Gamma' \in \mathcal{C}_K^\Gamma$ implies that $\Gamma/K \subseteq \Gamma'$) and, hence, $\alpha \in \Gamma'/K$.

For the converse, let $\alpha \in \Gamma'/K$. Thus, $K\alpha \in \Gamma'$. Assume that $\alpha \notin \Gamma/K$. Then $\neg K\alpha \in \Gamma$ (since Γ is a maximally consistent set) and, therefore, $K\neg K\alpha \in \Gamma$, from which $\neg K\alpha \in \Gamma'$ follows, a contradiction.

The proof that $\mathcal{C}_N^{\Gamma'} = \mathcal{C}_N^\Gamma$ proceeds the same way, that is, we show that $\Gamma/N = \Gamma'/N$. Let $\alpha \in \Gamma/N$. Then $N\alpha \in \Gamma$ and also $KN\alpha \in \Gamma$ by Axiom 4. Hence $N\alpha \in \Gamma'$, so $\alpha \in \Gamma'/N$.

For the converse, let $\alpha \in \Gamma'/N$. Thus, $N\alpha \in \Gamma'$. Assume that $\alpha \notin \Gamma/N$. Then $\neg N\alpha \in \Gamma$ and also $K\neg N\alpha \in \Gamma$, from which $\neg N\alpha \in \Gamma'$ follows, a contradiction. ∎

As in the completeness proof for $\mathcal{KL}$, each maximally consistent set Γ is mapped into the world state w_Γ with $w[p] = 1$ iff $p \in \Gamma$ for all atomic propositions p.

For any $\Gamma \in \mathcal{C}_0$, let $e_K^\Gamma = \{w_{\Gamma'} \mid \Gamma' \in \mathcal{C}_K^\Gamma\}$ and $e_N^\Gamma = \{w_{\Gamma'} \mid \Gamma' \in \mathcal{C}_N^\Gamma\}$.

Lemma 10.3.10:

(a) (e_K^Γ, e_N^Γ) is an exhaustive pair.

(b) For all α, we have $\alpha \in \Gamma$ iff $e_K^\Gamma, e_N^\Gamma, w_\Gamma \models^x \alpha$.

Proof: For part (a), to show that (e_K^Γ, e_N^Γ) is an exhaustive pair, we must show that $e_K^\Gamma \cup e_N^\Gamma$ consists of all world states. By way of contradiction, suppose there is a world state w not in $e_K^\Gamma \cup e_N^\Gamma$. Let $F_w = \{p \mid p \text{ is an atom and } w \models p\} \cup \{\neg p \mid p \text{ is an atom and } w \models \neg p\}$. $F_w \cup \Gamma/K$ cannot be consistent, for otherwise there would be some $\Gamma' \in \mathcal{C}_K^\Gamma$ that contains F_w, which would mean that $w \in e_K^\Gamma$. Similarly $F_w \cup \Gamma/N$ cannot be consistent. Thus, there must be formulas $\phi_1, \phi_2, \phi_3, \phi_4$ such that ϕ_1 and ϕ_2 are both conjunctions of a finite number of formulas in F_w, ϕ_3 is the conjunction of a finite number of formulas in Γ/L, and ϕ_4 is the conjunction of a finite number of formulas in Γ/N, and both $\phi_1 \wedge \phi_3$ and $\phi_2 \wedge \phi_4$ are inconsistent. Thus, we have $\vdash (\phi_3 \supset \neg \phi_1)$ and $\vdash (\phi_4 \supset \neg \phi_2)$. Using standard modal reasoning, we have $\vdash (K\phi_3 \supset K\neg \phi_1)$ and $\vdash (N\phi_4 \supset N\neg \phi_2)$. Since $K\psi \in \Gamma$ for each conjunct ψ of ϕ_3, standard modal reasoning shows that $K\phi_3 \in \Gamma$. Similarly, we have $N\phi_4 \in \Gamma$. Since Γ is a maximally consistent set, both $K\neg\phi_1$ and $N\neg\phi_2$ are in Γ. Since $\vdash (K\neg\phi_1 \supset K(\neg\phi_1 \vee \neg\phi_2))$ and $\vdash (N\neg\phi_2 \supset N(\neg\phi_1 \vee \neg\phi_2))$, it follows that both $K(\neg\phi_1 \vee \neg\phi_2)$ and $N(\neg\phi_1 \vee \neg\phi_2)$ are in Γ. But this contradicts Axiom 5, since $\phi_1 \wedge \phi_2$ is a propositionally consistent objective formula.

For part (b), the proof proceeds by induction on the structure of α. The statement holds trivially for atomic propositions, conjunctions, and negations. In the case of $K\alpha$, we proceed by the following chain of equivalences:

$K\alpha \in \Gamma$
iff   for all $\Gamma' \in \mathcal{C}_K^\Gamma$, we have $\alpha \in \Gamma'$
iff   for all $\Gamma' \in \mathcal{C}_K^\Gamma$, we have $e_K^{\Gamma'}, e_N^{\Gamma'}, w_{\Gamma'} \models^x \alpha$ (by induction)
iff   for all $w_{\Gamma'} \in e_K^\Gamma$, we have $e_K^\Gamma, e_N^\Gamma, w_{\Gamma'} \models^x \alpha$ (by Lemma 10.3.9)
iff $(e_K^\Gamma, e_N^\Gamma, w_\Gamma) \models^x K\alpha$.

The case $N\alpha$ is completely symmetric. ∎

The completeness result now follows easily. Let α be a consistent formula and Γ a maximally consistent set of sentences containing α. $e_K^\Gamma, e_N^\Gamma, w_\Gamma \models^x \alpha$ then follows immediately from Lemma 10.3.10. ∎

Finally, since, by Corollary 10.3.6, validity and x-validity are one and the same, we obtain

Corollary 10.3.11: *For all propositional sentences α, if $\models \alpha$ then $\vdash \alpha$.*

10.4 Incompleteness

As already mentioned in the introduction, the axioms are incomplete for the full language, that is, there are sentences which are valid in $\mathcal{OL}$ but which cannot be derived. We will not prove this result here, but rather discuss the ideas behind it in an informal way.

The reasons for the incompleteness can essentially all be traced back to Axiom 5:

$(N\phi \supset \neg K\phi)$, where ϕ is any objective sentence such that $\not\models \phi$.

To begin with, note that the axiom is already problematic for reasons other than completeness. Proof theories normally require axioms to be recursive, that is, it should be decidable whether any given sentence is an instance of an axiom. Clearly, Axiom 5 violates this requirement since the set of non-valid sentences in $\mathcal{L}$ is not even recursively enumerable. As we will see below, this deficiency is really inescapable in $\mathcal{OL}$, that is, no complete axiom system can be recursive.

The second and perhaps more serious problem with Axiom 5 is that it is simply too weak. In a nutshell, only considering objective non-valid sentences is just not good enough in a logic where nested beliefs do not reduce to non-nested ones, a property of $\mathcal{KL}$ and hence of $\mathcal{OL}$, which we proved in Chapter 4.[6] In fact, a sentence quite similar to the one which we used to show irreducibility can be employed to prove incompleteness. Let

$\zeta = \exists x[P(x) \wedge \neg KP(x)] \vee \exists x[\neg P(x) \wedge KP(x)]$.

---

6 In the propositional completeness proof this problem did not arise since propositional nested beliefs indeed reduce.

Thus ζ says that either there is a P which is not believed to be a P, or there is a non-P which is believed to be a P. We then obtain the following result.

Lemma 10.4.1: $(N\zeta \supset \neg K\zeta)$ *is valid.*

Proof: Suppose $e \models N\zeta$. Let $A = \{n \mid n$ is a standard name and $e \models KP(n)\}$ and let w be a world state such that $w \models P(n)$ iff $n \in A$. It is easy to see that $e, w \models \neg \zeta$. Since $e \models N\zeta$, it must be the case that $w \in e$. Thus, $e \models \neg K\zeta$. ∎

While valid, it can be shown that the sentence is not derivable from the axioms by devising a slightly different semantics where all of the axioms of $\mathcal{OL}$ are sound, but where $(N\zeta \supset \neg K\zeta)$ is not valid. Incompleteness then follows immediately because no non-valid sentence can be derived from sound axioms.

Of course, the question remains what a complete axiomatization would look like. We already remarked that the set of instances of Axiom 5 is not recursively enumerable (r.e.). Even without knowing what a complete axiomatization might look like, it is easy to see that it cannot be recursive.

Lemma 10.4.2: *Every complete axiomatization of* $\mathcal{OL}$ *is non-recursive.*

Proof: Suppose there were a recursive complete axiomatization of $\mathcal{OL}$. Then the set of non-valid objective formulas would be r.e., since we could generate them by generating all the objective formulas ϕ such that $(N\phi \supset \neg K\phi)$ is provable. Since the set of non-valid objective formulas is co-r.e., this is a contradiction. ∎

Given the non-recursive nature of the axioms, there is, in a sense, a trivial solution to the problem simply by declaring every valid sentence an axiom. Of course, such a "proof theory" is completely useless since it does not give us any new insights into the logic. Instead we would expect axioms to be natural in that they at least have a compact representation. We do not know whether there is such a natural proof-theoretic account of the logic, at least within first-order modal logic. As the following results suggest, if there is one, it may be hard to find.

Recall that the incompleteness proof proceeds by showing that, for a particular basic formula ζ, the formula $(N\zeta \supset \neg K\zeta)$ is valid yet not provable from the axioms. The latter formula almost looks like an instance of Axiom 5. It is not, of course, since 5 would apply only if the formula ζ were objective. The obvious idea, namely to strengthen Axiom 5 by allowing it to range over all non-valid basic sentences, can easily be dismissed. For example, consider the subjective sentence $KP(n)$ for some predicate P and standard name n. $KP(n)$ is obviously not valid, yet $(NKP(n) \supset \neg KKP(n))$ is not valid. In fact, $NKP(n) \equiv KKP(n)$ is easily derivable from the axioms (using 2) and is therefore valid.

But what about basic sentences that are not subjective like the sentence ζ used above? In other words, do we obtain a complete axiomatization if we replace Axiom 5 by the following Axiom 5'?

(5'.) $(N\alpha \supset \neg K\alpha)$, where α is a basic non-subjective sentence such that $\not\models \alpha$.

Since ζ is basic, non-subjective, and not valid, the offending sentence $(N\zeta \supset \neg K\zeta)$ would now come out trivially as a theorem. Unfortunately, 5' does not solve the problem either, since restricting the axiom to non-subjective basic sentences is still unsound. To see this, consider the formula

$$\xi = \forall x (P(x) \supset KP(x)),$$

which is obviously not valid. However,

Lemma 10.4.3: $(N\xi \supset \neg K\xi)$ *is not valid.*

Proof: Let e_P consist of all world states w such that $w \models \forall x P(x)$. Clearly e_P is maximal. We now show that $e_P \models K\xi \wedge N\xi$. It is easy to see that $e_P, w \models (K(\forall x P(x)) \supset \xi)$ for all world states w. Since $e_P \models K(\forall x P(x))$, it follows that $e_P, w \models \xi$ for all world states w. This means that $e_P \models K\xi \wedge N\xi$. Hence $(N\xi \supset \neg K\xi)$ is not valid. ∎

Although, as we just showed, $(N\xi \supset \neg K\xi)$ is not valid, there is a sense in which it just misses being valid. As we now show, the only time it fails to be valid is when every standard name is known not to satisfy P (as was the case for the set e_P of world states considered in Lemma 10.4.3).

Lemma 10.4.4: $(\neg K(\forall x P(x)) \supset (N\xi \supset \neg K\xi))$ *is valid.*

Proof: Let e be any maximal set of world states such that $e \models \neg K(\forall x P(x)) \wedge N\xi$. Since $e \models \neg K(\forall x P(x))$, there is a standard name n^* such that $e \models \neg KP(n^*)$. Since $e \models N\xi$, it follows that for all $w' \notin e$, we have $e, w' \models (\forall x (P(x) \supset KP(x)))$. In particular, this means that for all $w' \notin e$, we must have $w' \models \neg P(n^*)$. Thus, there must be some $w \in e$ such that $w \models P(n^*)$. Clearly $e, w \models \neg\xi$, so $e \models \neg K\xi$, as desired. ∎

These lemmas suggest that it may not be easy to find an extension of Axiom 5 that would cover the counterexample, let alone lead to a complete axiomatization.

10.5 Bibliographic notes

In [86] an alternative completeness proof for the propositional case is given, which uses the original semantics of $\mathcal{OL}$. Recently, Rosati [125, 126] investigated the computational complexity of only-knowing in the propositional case. He presents an algorithm which decides satisfiability in propositional $\mathcal{OL}$ in nondeterministic polynomial time using an NP-oracle for propositional satisfiability. In other words, the problem is in Σ_2^p (the second level of the polynomial hierarchy). Rosati also shows that the problem is in fact Σ_2^p-complete, which follows easily because of the close connection between $\mathcal{OL}$ and AEL and a result by Gottlob [42], who showed that determining whether a formula has a stable expansion is Σ_2^p-hard.

The alternative semantics for $\mathcal{OL}$ with overlapping sets of world states was first introduced in [48]. The proof that the axioms are incomplete in the first-order case appeared in [47]. Most of the material of Section 10.4 is taken from there.

10.6 Where do we go from here?

The most immediate open question remaining is whether there is a finite set of axioms which is complete for all of $\mathcal{OL}$. Besides that, it seems also interesting to ask whether there are classes of sentences for which the given axioms are actually complete. Clearly, this is true for sentences of $\mathcal{KL}$ or those mentioning only N. But what about sentences involving both K and N (or O). For example, what about sentences of the form (OKB $\supset$ $K\alpha$), where KB is objective or, more generally, where KB is a determinate sentence?

10.7 Exercises

1. Give a syntactic derivation of the older-brother example (Example 9.1.1 on page 144), that is, show that ($O[\text{KB} \wedge (\neg Kb \supset \neg b)] \supset K\neg b$) follows from the axioms.

2. Do the same for Example 9.1.3 on page 145.

3. Consider KB = $\{P(^\#1), P(^\#2), \forall x(P(x) \supset KP(x))\}$. In Chapter 8 we proved that only-knowing KB amounts to the closed-world assumption for P, that is, ($O\text{KB} \supset K\forall x(P(x) \equiv [(x = ^\#1) \vee (x = ^\#2)]))$ is valid. Show that the sentence follows from the axioms. (See Lemma 8.5.2 on page 135 for a semantic proof.)

4. In the propositional case, the soundness of Axiom 5 depends on the fact that there are *infinitely many* atomic propositions.

    (a) Give an example where Axiom 5 fails in the case where there are only finitely many atomic propositions in the language.

(b) Show that in this case O can be expressed using K alone.

Hint: Use the fact that each world state can be represented completely by a sentence as long as there are only finitely many propositions.

11 Only-Knowing-About

As we have seen, only-knowing is a useful notion to completely characterize the functionality of **ASK** and **TELL** within the logic itself. While nothing prevents us from using O as part of a query or a sentence to be added to the knowledge base, doing so does not seem terribly interesting, quite in contrast to the use of K. For example, who would want to know if α is all the KB knows? Most likely the answer would be false anyway since, for all practical purposes, queries are much too small to cover the contents of the whole knowledge base. Similarly, telling the KB about all it knows or does not know seems to be of little value, if any.

On the other hand, if we were able to qualify only-knowing by focusing on a specific subject matter, things would be quite different. Rather than asking whether a sentence is all that is known, it seems much more interesting to ask whether a sentence is all that is known *about a certain subject matter*. Ideally we would like to pose queries of the sort "tell me all you know about Toronto" or "what do you know about John and Mary's relationship?" At this point, however, we need to be a lot more modest, since our results on this subject are restricted to the propositional case. Nevertheless, even in its limited form, interesting issues emerge that warrant its treatment here.

In the next section, we develop the logic of *only-knowing-about* by suitably extending only-knowing, and discuss some of its properties. We will show how **ASK** and **TELL** can be adapted to deal with the extended expressiveness. Finally, we will indicate how only-knowing-about can be used to give precise meaning to certain forms of relevance.

11.1 The logic of only-knowing-about

Before we can make sense of only-knowing-about, we need to be clear about what we mean by a *subject matter*. Since we are only dealing with a propositional language, a reasonable choice for a subject matter seems to be a set of atomic sentences. For example, if KB $= (p \vee q) \wedge r$, we might be interested in what the KB knows about p, which in this case amounts to $(p \vee q)$. While we could allow arbitrary sets of atoms as subject matters, we confine ourselves to *finite* sets. The reason is mainly convenience, since it allows us to get away with countably many symbols to refer to the different subject matters.

The language $\mathcal{OL}^a$ is obtained from $\mathcal{OL}$ by taking its propositional subset and adding an infinite number of modal operators $O\langle \pi \rangle$ to it, one for each finite set of atomic sentences π. $O\langle \pi \rangle \alpha$ should then be read as "all the agent knows about π is α." As usual, modal operators are allowed to be arbitrarily nested so that we can form sentences like $KO\langle \pi_1 \rangle p \wedge \neg O\langle \pi_1 \rangle O\langle \pi_2 \rangle q$. When referring to a specific subject matter, we often omit the curly brackets. For example, for atoms p and q, we write $\boldsymbol{O}\langle p, q\rangle$ instead of $\boldsymbol{O}\langle\{p, q\}\rangle$.

Clauses, which are disjunctions of *literals*,[1] play a central role in our formalization of only-knowing-about. To ease the presentation below, we introduce a few conventions about clauses. Given an atomic proposition p and a clause c, we say that c *mentions* p just in case either p or $\neg p$ occurs in c. It is often convenient to identify a clause with the set of literals occurring in the clause. A clause c is contained in a clause c' ($c \subseteq c'$) if every literal in c occurs in c'. We write $c \subsetneq c'$ instead of $c \subseteq c'$ and $c' \nsubseteq c$.

11.1.1 A formal semantics

To get a better sense of what we mean by only-knowing-about, let us consider the following example:

$$KB = (cows \supset mammals) \wedge (mammals \supset animals).$$

Although the sentence is purely propositional, we should think of the atoms as representing sets with the obvious connotation. When asked what is known about mammals given KB, the only reasonable answer seems to be: all of KB. Similarly, one would not hesitate to say that the KB knows nothing about fish. But what about cows? Clearly, ($cows \supset mammals$) is known about them. It is less clear whether ($mammals \supset animals$) should count as well. If we did not know that cows were mammals, the answer would clearly be no. With that knowledge, however, ($mammals \supset animals$) conveys some information about cows, namely that cows are animals. Here we take the stance that only the directly cow-related information that follows from KB is considered knowledge about the bovine species, which eliminates ($mammals \supset animals$) but includes ($cows \supset animals$). Hence all KB knows about cows is ($cows \supset mammals$) $\wedge$ ($cows \supset animals$).[2]

For the general case of only-knowing-about, consider an arbitrary epistemic state e and a subject matter π. To get at what is only-known about π, we first collect in a set $\Gamma_{e,\pi}$ all those smallest clauses which are known at e and which mention the subject matter (formal definitions are given below). We claim that this represents the information that is directly π-related and contained in e. Then we consider a new epistemic state $e|_\pi$ which satisfies only-knowing $\Gamma_{e,\pi}$. We say that e satisfies only-knowing α about π just in case $e|_\pi$ satisfies only-knowing α. Intuitively, $e|_\pi$ can be understood as the epistemic state that results after the agent forgets everything that is not relevant to π.

Definition 11.1.1: Let e be a set of world states and π a subject matter.

---

[1] A literal is either an atom or its negation.

[2] One could view ($mammals \supset animals$) as being about cows in light of the fact that this piece of information is essential in inferring that cows are animals from KB. However, this view would introduce a distinct syntactic flavour into the definition of only-knowing-about, which is counter to the knowledge level analysis we have been following so far.

Only-Knowing-About

1. A clause c is called *e-minimal* iff $e \models Kc$ and for all clauses $c' \subsetneq c$, $e \not\models Kc'$.
2. A clause c is called *e-p-minimal* iff c is e-minimal and, in addition, c mentions p.
3. $\Gamma_{e,\pi} = \{c \mid c$ is an e-p-minimal clause for some $p \in \pi.\}$
4. $e|_\pi = \{w \mid w \models c$ for all $c \in \Gamma_{e,\pi}\}$.

Note that $e|_\pi$ is a superset of e and, hence, what is known about the world at $e|_\pi$ is always included in what is known at e. By restricting ourselves to e-p-minimal clauses, we rule out clauses that mention the subject matter but do not really tell us anything about it. For example, let the subject matter be p and assume all we know is q, that is, $e = \{w \mid w \models q\}$. Then we certainly also know $(p \vee q)$, which is not e-minimal because q is known as well. While $(p \vee q)$ mentions the subject matter p, it does so, in a sense, only accidentally, since it does not convey what is really known about p, namely nothing. The only e-minimal clause mentioning p is $(p \vee \neg p)$, which gives us the right information.

Given these definitions of what it means to forget irrelevant things, we obtain the following semantic definition of only-knowing-about.

Definition 11.1.2: $e \models O\langle\pi\rangle\alpha$ iff $e|_\pi \models O\alpha$

Logical implication, validity, and satisfiability are defined as usual.

11.1.2 Some properties of only-knowing-about

In the following, we present examples that demonstrate that the new concept $O\langle\pi\rangle$ has reasonable properties. We first discuss them informally and defer the proofs to the end of this subsection. Throughout, p and q are assumed to be distinct atoms.

1. If $\models \alpha \equiv \beta$, then $\models O\langle\pi\rangle\alpha \equiv O\langle\pi\rangle\beta$.
   If all I believe about π is α, then the syntactic form of α is immaterial.
2. $\models (O\langle\pi\rangle\phi \supset K\phi)$ for objective ϕ.
   While the fact that knowing ϕ follows from only-knowing ϕ about a subject matter should not come as a surprise, it is perhaps more surprising why this property fails in general when ϕ is not objective. For example,
   $(O\langle p\rangle\neg Kq \supset K\neg Kq)$ is not valid.
   The reason is as follows: First note that $O\langle p\rangle\neg Kq$ is just another way of saying that absolutely nothing is known about p. In fact, $e \models O\langle p\rangle\neg Kq$ iff $e \models O\langle p\rangle\text{TRUE}$, that is, $e|_p = e_0$, the set of all world states. It is in the context of e_0 that q is not known. On the other hand, it may very well be the case that q is known at e, only this information is forgotten once we focus on what is known about p. Hence, what may

seem counterintuitive at first, is not all that unreasonable once we notice that any K within the scope of an $O\langle\pi\rangle$ refers to the epistemic state which results from forgetting any information not related to π.

3. $\models (O\langle p\rangle(p \vee q) \supset (\neg Kp \wedge \neg Kq))$.
This is because if I knew either p or q, then $(p \vee q)$ would be contingent information and thus, in either case, would not capture what I really know about p. Similarly,

$\models (O\langle p\rangle(p \vee q) \supset (\neg K\neg p \wedge \neg K\neg q))$.
This is because if I knew $\neg q$, I would also know p because I know $(p \vee q)$. Thus $(p \vee q)$ would once again not capture what I really know about p. (A similar argument shows that knowing $\neg p$ is also inconsistent with only-knowing $(p \vee q)$ about p.)

4. $\models (O\langle p, q\rangle p \supset O\langle q\rangle\text{TRUE})$.
If all I know about p and q is p, then I know nothing about q.

5. $\models \neg O\langle p\rangle q$.
In other words, something totally independent of p cannot be all I know about p. For example, it does not make sense to say that all I know about Tweety is that Black Forest cake is tasty.

6. Let α be such that $\not\models \neg\alpha$ and $\not\models \neg O\langle\pi\rangle\alpha$. Then $\not\models (O\langle\pi\rangle\alpha \supset O\alpha)$.
This says that, except for trivial cases, what is only-known about a specific subject matter does not determine what is only-known. For example, if all I know about Tweety is that it is a bird, then I most likely know other things as well, for example, that Black Forest cake is tasty. Note that the implication goes through if $\neg\alpha$ is valid. In this case the only epistemic state e which satisfies only-knowing α is the empty set of world states. Then $e|_\pi$ is also empty because for every $p \in \pi$ both p and $\neg p$ are in $\Gamma_{e,\pi}$. While the converse sometimes holds (for example, $\models (O(p \vee q) \supset O\langle p\rangle(p \vee q))$), it fails in general, as it should. For example, $\not\models (O(p \wedge q) \supset O\langle p\rangle(p \wedge q))$. For if $e \models O(p \wedge q)$ then $e|_p = \{w \mid w \models p\}$ and, therefore, $e \not\models O\langle p\rangle(p \wedge q)$.

7. We have already seen that only-knowing captures autoepistemic reasoning. The following example indicates that the weaker assumption of only-knowing-about in fact suffices for this purpose.
Let p denote the proposition *Tweety flies*. Given the assumption that Tweety flies unless known otherwise, autoepistemic reasoning lets us conclude that Tweety flies. Formally, $\models (O(\neg K\neg p \supset p) \supset Kp)$. Later, if we discover that Tweety indeed does not fly, we retract our previous conclusion, that is, $\models (O(\neg p \wedge (\neg K\neg p \supset p)) \supset \neg Kp)$. However, it is intuitively clear that we need not require that the assumption is all we know to get the desired conclusion, but that the weaker requirement that this is all we know *about Tweety* suffices. And indeed, our formalization gives us just that. (The subject matter *about Tweety* in this case is simply $\{p\}$.)

$\models O\langle p\rangle(\neg K\neg p \supset p) \supset Kp$.

$\models O\langle p\rangle(\neg p \wedge (\neg K\neg p \supset p)) \supset K\neg p$.

8. The following example concerns the case of sentences with multiple stable expansions. Again, the weaker requirement of only-knowing-about suffices to make the same distinctions as regular autoepistemic logic.

$\models O\langle p\rangle(Kp \supset p) \equiv [O\langle p\rangle p \vee O\langle p\rangle\text{TRUE}]$.

In other words, believing only $(Kp \supset p)$ about p is the same as believing either only p or nothing about p.

Proofs:

1. If $\models \alpha \equiv \beta$, then $\models O\langle\pi\rangle\alpha \equiv O\langle\pi\rangle\beta$.

   Let $\models \alpha \equiv \beta$ and let e be a set of world states such that $e \models O\langle\pi\rangle\alpha$. We need to show that $e \models O\langle\pi\rangle\beta$. (The reverse direction is completely symmetric and is omitted.)
   Since, by assumption, $e|_\pi \models O\alpha$, we obtain that for all world states w, $w \in e|_\pi$ iff $e|_\pi, w \models \alpha$. Since $\models \alpha \equiv \beta$, this is the same as for all w, $w \in e|_\pi$ iff $e|_\pi, w \models \beta$ and, therefore, $e|_\pi \models O\beta$, from which $e \models O\langle\pi\rangle\beta$ follows.

2. $\models (O\langle\pi\rangle\phi \supset K\phi)$ for objective ϕ.

   Let $e \models O\langle\pi\rangle\phi$, that is, $e|_\pi \models O\phi$ where $e|_\pi = \{w \mid w \models \Gamma_{e,\pi}\}$. Since ϕ is objective, $\Gamma_{e,\pi} \models \phi$ follows. Hence, since every element of $\Gamma_{e,\pi}$ is known at e, so is ϕ.

3. $\models O\langle p\rangle(p \vee q) \supset (\neg Kp \wedge \neg Kq)$.

   Let $e \models O\langle p\rangle(p \vee q)$. Then $e|_p \models O(p \vee q)$, that is, $e|_p = \{w \mid w \models (p \vee q)\}$. Suppose $e \models Kp$. Since p is e-minimal, we have $p \in \Gamma_{e,p}$ and hence $e|_p \not\models Kp$, a contradiction. Now suppose $e \models Kq$. Then none of the clauses in $\Gamma_{e,p}$ mentions q because q is e-minimal. Therefore $\Gamma_{e,p} \not\models (p \vee q)$, contradicting the assumption that $e|_p \models K(p \vee q)$.

   $\models O\langle p\rangle(p \vee q) \supset (\neg K\neg p \wedge \neg K\neg q)$.

   Let $e \models O\langle p\rangle(p \vee q)$. Suppose $e \models K\neg p$. Then, since $e \models K(p \vee q)$, $e \models Kq$ follows, contradicting $\models (O\langle p\rangle(p \vee q) \supset \neg Kq)$ from above. The proof of the other case is completely analogous.

4. $\models O\langle p, q\rangle p \supset O\langle q\rangle\text{TRUE}$.

   Let $\pi = \{p, q\}$ and $e \models O\langle\pi\rangle p$. Suppose, to the contrary, that $e \not\models O\langle q\rangle\text{TRUE}$. Then there is a $c \in \Gamma_{e,q}$ which mentions q and c is not a tautology. Since $q \in \pi$, c is also in $\Gamma_{e,\pi}$. Note that c does not mention p because p itself is e-minimal. Yet, by assumption, $e|_\pi = \{w \mid w \models p\}$, that is $e|_\pi \not\models Kc$, a contradiction.

5. $\models \neg O\langle p\rangle q$.

   Assume, to the contrary, that there is an e such that $e \models O\langle p\rangle q$. Then $e|_p \models Oq$, that is, $e|_p = \{w \mid w \models q\}$. Since $e \models Kq$, q is e-minimal and no e-minimal clause that mentions p contains q as a literal. Thus $\Gamma_{e,p} \not\models q$, that is, $e|_p$ contains a world state which does not satisfy q, a contradiction.

6. Let α be such that $\not\models \neg\alpha$ and $\not\models \neg O\langle\pi\rangle\alpha$. Then $\not\models (O\langle\pi\rangle\alpha \supset O\alpha)$.
   Let $e \models O\langle\pi\rangle\alpha$, that is, $e|_\pi \models O\alpha$. Now let p be an atom which does not occur in α or π and let $e' = e \cap \{w \mid w \models p\}$. Then $e'|_\pi = e|_\pi$. Therefore, $e'|_\pi \models O\langle\pi\rangle\alpha$. On the other hand, it is easy to see that $e' \not\models O\alpha$. For assume otherwise. Since $\not\models \neg\alpha$ by assumption, $e' \neq \{\}$. Thus let $w \in e'$, that is, $e', w \models \alpha$. Let $w_{\bar{p}}$ be just like w except that $w_{\bar{p}} \models \neg p$. Clearly, since α does not mention p, we have $e', w_{\bar{p}} \models \alpha$ and, therefore $w_{\bar{p}} \in e'$, contradicting the assumption that $e' \models Kp$.

7. $\models O\langle p\rangle(\neg K\neg p \supset p) \supset Kp$.
   Let $e \models O\langle p\rangle(\neg K\neg p \supset p)$. Then, by definition, $e|_p \models O(\neg K\neg p \supset p)$. Similar to the older-brother example (see Example 9.1.1 on page 144) we have that the only epistemic state where $(\neg K\neg p \supset p)$ is only-known is $e|_p = \{w \mid w \models p\}$, that is, $e|_p \models Kp$. Since $e \subseteq e|_p$, $e \models Kp$ holds as well.

   $\models O\langle p\rangle(\neg p \wedge (\neg K\neg p \supset p)) \supset K\neg p$.
   Similar to the previous case, if we assume that $e \models O\langle p\rangle(\neg p \wedge (\neg K\neg p \supset p))$, then $e|_p \models O(\neg p \wedge (\neg K\neg p \supset p))$ and thus $e|_p = \{w \mid w \models \neg p\}$, which implies that $e \models K\neg p$.

8. $\models O\langle p\rangle(Kp \supset p) \equiv [O\langle p\rangle p \vee O\langle p\rangle \text{TRUE}]$.
   Let $e \models O\langle p\rangle(Kp \supset p)$. Then $e|_p \models O(Kp \supset p)$. It follows from Exercise 1 of Chapter 9 that $e|_p \models Op \vee O\text{TRUE}$. Hence $e \models O\langle p\rangle p \vee O\langle p\rangle \text{TRUE}$.
   To prove $\models ([O\langle p\rangle p \vee O\langle p\rangle \text{TRUE}] \supset O\langle p\rangle(Kp \supset p))$, let us first assume that $e \models O\langle p\rangle p$. Then $e|_p \models Op$ and, by the same exercise as before, $e|_p \models O(Kp \supset p)$, from which $e \models O\langle p\rangle(Kp \supset p)$ follows. By a similar argument, $e \models O\langle p\rangle \text{TRUE}$ implies $e \models O\langle p\rangle(Kp \supset p)$. ∎

The following theorem characterizes cases where reasoning from only-knowing-about is the same as reasoning from only-knowing. This is interesting for at least two reasons. For one, reducing only-knowing-about to only-knowing has the advantage that the latter is simpler conceptually and better understood. The theorem also seems to support the way only-knowing is used colloquially. We often say "all I know is..." and draw inferences based on that. Of course, what we really mean is "all I know *about such and such* is..." The theorem, in a sense, justifies our carelessness in leaving out the about-part in that the inferences remain the same as long as we restrict ourselves to queries that mention only the subject matter at hand.

Theorem 11.1.3: *For any formula α let π_α be the subject matter which consists of all* atoms occurring in α. Let ϕ and ψ be objective formulas. Then $\models (O\langle \pi_\phi \rangle \phi \supset K\psi)$ iff $\models (O\phi \supset K\psi)$.

Proof: Let $\models (O\langle \pi_\phi \rangle \phi \supset K\psi)$ and $e \models O\phi$. It is not hard to show that $e|_{\pi_\phi} = e$ in this case. (We leave the proof as an exercise.) Then $e \models K\psi$ follows immediately. Conversely, assume that $\models (O\phi \supset K\psi)$ and let $e \models O\langle \pi_\phi \rangle \phi$. Then $e|_{\pi_\phi} \models O\phi$, from which $e|_{\pi_\phi} \models K\psi$ follows by assumption. Since ϕ is objective and $e \subseteq e|_{\pi_\phi}$, we obtain $e \models K\psi$. ∎

11.1.3 Prime implicates

Given an objective knowledge base, all that is known about a subject matter can be characterized succinctly using *prime implicates*. For any objective ϕ, the prime implicates of ϕ are simply the smallest clauses (in terms of set inclusion) that are logically implied by ϕ. Formally:

Definition 11.1.4: Let ϕ be an objective sentence. A clause c is called a prime implicate of ϕ iff

1. $\models (\phi \supset c)$ and
2. for all $c' \subsetneq c$, $\not\models (\phi \supset c')$.

For any atom p, let $\mathcal{P}(\phi, p) = \{c \mid c \text{ is a prime implicate of } \phi \text{ mentioning } p\}$, $\mathcal{P}(\phi, \pi) = \bigcup_{p \in \pi} \mathcal{P}(\phi, p)$, and $\mathcal{P}(\phi) = \bigcup_{p \in \phi} \mathcal{P}(\phi, p)$, where $p \in \phi$ stands for p occurs in ϕ. We sometimes use these sets as formulas and mean the conjunction of the clauses contained in them.

For example, if $\phi = (p \vee q) \wedge (p \vee r \vee s) \wedge \neg s$, then $\mathcal{P}(\phi, p) = \{(p \vee q), (p \vee r), (p \vee \neg p)\}$. In cases where p is not mentioned in ϕ at all, $\mathcal{P}(\phi, p) = \{(p \vee \neg p)\}$.

It is easy to see that for any given ϕ and p, $\mathcal{P}(\phi, p)$ and $\mathcal{P}(\phi)$ are finite assuming we identify clauses with sets of literals and hence eliminate redundancies. What is known about π relative to a sentence (background theory) ϕ has a simple characterization in terms of prime implicates of ϕ, as the following lemma and theorem show.

Lemma 11.1.5: Let $e \models O\phi$. Then for any atom p, $c \in \mathcal{P}(\phi, p)$ iff c is an e-p-minimal clause.

Proof: Since $e \models O\phi$, $e = \{w \mid w \models \phi\}$. Hence for any clause c, $e \models Kc$ iff $\models (\phi \supset c)$. Therefore the e-minimal clauses are precisely the prime implicates of ϕ. In particular, c is e-p-minimal iff $c \in \mathcal{P}(\phi, p)$. ∎

Theorem 11.1.6: $\models (O\phi \supset O\langle\pi\rangle\psi)$ iff $\models (\psi \equiv \mathcal{P}(\phi, \pi))$.

Proof: Let $e \models O\phi$ and let $\psi^* = \mathcal{P}(\phi, \pi)$. We show that $e \models O\langle\pi\rangle\psi^*$. Together with the fact that a sentence that is only-known about π is unique up to equivalence, the theorem follows.

To show that $e \models O\langle\pi\rangle\psi^*$, consider $e|_\pi = \{w \mid w \models c \text{ for all } c \in \Gamma_{e,\pi}\}$ with $\Gamma_{e,\pi} = \{c \mid c \text{ is an } e\text{-}p\text{-minimal clause for some } p \in \pi\}$. By Lemma 11.1.5, $\Gamma_{e,\pi} = \{c \mid c \in \mathcal{P}(\phi, p) \text{ for some } p \in \pi\}$. Therefore, $e|_\pi \models O\psi^*$ and thus $e \models O\langle\pi\rangle\psi^*$. ∎

In other words, given all that is known is ϕ, then what is only-known about π is characterized by the prime implicates of ϕ that mention atoms in π.

11.2 ASK and TELL

We will now consider versions of the **ASK**- and **TELL**-operations, where we allow sentences containing the operators $O\langle\pi\rangle$ as arguments.[3] For the rest of this section we assume that KB is a finite set of objective sentences and α is a sentence in $\mathcal{OL}^a$ that does not mention O.

Semantically, the story is exactly the same as before except for the interface language. In other words, we define **ASK** and **TELL** as in Definitions 5.2.1 and 5.5.1.

Definition 11.2.1: For any epistemic state e,

$$\text{ASK}[\alpha, e] = \begin{cases} yes & \text{if } e \models K\alpha \\ no & \text{otherwise} \end{cases}$$

$$\text{TELL}[\alpha, e] = e \cap \{w \mid e, w \models \alpha\}$$

In order to obtain a representation theorem similar to Theorem 7.4.1 we need to extend the definitions of RES and $\|\cdot\|$ to account for $O\langle\pi\rangle\alpha$.

Definition 11.2.2: Let ϕ be an objective sentence. Then

$$\text{RES}_{O\langle\pi\rangle}[\phi, \text{KB}] = \begin{cases} \text{TRUE} & \text{if } \models \mathcal{P}(\text{KB}, \pi) \equiv \phi \\ \text{FALSE} & \text{otherwise} \end{cases}$$

Definition 11.2.3: $\|O\langle\pi\rangle\alpha\|_{\text{KB}} = \text{RES}_{O\langle\pi\rangle}[\|\alpha\|_{\mathcal{P}(\text{KB}, \pi)}, \text{KB}]$

---

[3] We could easily accommodate all of $\mathcal{OL}^a$ as the interface language. We leave out O because it is simpler and, as we said in the beginning of this chapter, there really seems very little need from a practical point of view to ask or tell a KB about all it knows.

Only-Knowing-About   187

Note the change from $\|\cdot\|_{KB}$ to $\|\cdot\|_{\mathcal{P}(KB,\pi)}$ on the right hand side of the definition. This reflects the fact that any modal operator in the scope of $O\langle\pi\rangle$ must be evaluated relative to what is known about π given KB.

Lemma 11.2.4: $\Re[\![KB]\!] \models O\langle\pi\rangle\|\alpha\|_{\mathcal{P}(KB,\pi)}$ *iff* $\models \|\alpha\|_{\mathcal{P}(KB,\pi)} \equiv \mathcal{P}(KB,\pi)$

Proof: Since $\|\alpha\|_{\mathcal{P}(KB,\pi)}$ is objective, the proof follows immediately from the fact that $\Re[\![KB]\!] \models O KB$ together with Theorem 11.1.6. ∎

Lemma 11.2.5: *For any world state* w, $\Re[\![KB]\!], w \models \alpha$ *iff* $w \models \|\alpha\|_{KB}$.

Proof: The proof is by induction on the structure of α. We only need to prove the case for $O\langle\pi\rangle\alpha$, since the rest has already been taken care of by Lemma 7.3.2.

$\Re[\![KB]\!], w \models O\langle\pi\rangle\alpha$
iff $\Re[\![\mathcal{P}(KB,\pi)]\!], w \models O\alpha$
iff $\forall w'[w' \in \Re[\![\mathcal{P}(KB,\pi)]\!]$ iff $\Re[\![\mathcal{P}(KB,\pi)]\!], w' \models \alpha]$
iff $\forall w'[w' \in \Re[\![\mathcal{P}(KB,\pi)]\!]$ iff $\Re[\![\mathcal{P}(KB,\pi)]\!], w' \models \|\alpha\|_{\mathcal{P}(KB,\pi)}]$ by induction
iff $\Re[\![\mathcal{P}(KB,\pi)]\!], w \models O\|\alpha\|_{\mathcal{P}(KB,\pi)}$
iff $\Re[\![KB]\!], w \models O\langle\pi\rangle\|\alpha\|_{\mathcal{P}(KB,\pi)}$
iff $\models \|\alpha\|_{\mathcal{P}(KB,\pi)} \equiv \mathcal{P}(KB,\pi)$ by Lemma 11.2.4
iff $\text{RES}_{O\langle\pi\rangle}[\|\alpha\|_{\mathcal{P}(KB,\pi)}, KB] = \text{TRUE}$
iff $w \models \|O\langle\pi\rangle\alpha\|_{KB}$. ∎

As in the case of $\mathcal{KL}$ (Theorem 7.4.1), the previous lemma yields a representation theorem for **ASK** and **TELL**, where this time the interaction language allows us to talk about only-knowing-about.

Theorem 11.2.6:
    Let KB be any finite set of objective sentences and α be any sentence of $\mathcal{OL}^a$. Then:
    1. **TELL**$[\alpha, \Re[\![KB]\!]] = \Re[\![(KB \wedge \|\alpha\|_{KB})]\!]$.
    2. **ASK**$[\alpha, \Re[\![KB]\!]] = yes$ *iff* KB $\models \|\alpha\|_{KB}$.

11.3 Relevance

The study of relevance has gained considerable attention recently in areas as diverse as machine learning and knowledge representation. Clearly only-knowing-about, by its very nature, has something to do with relevance. For if all I know about Tweety is that Tweety is a bird, then I seem to know something relevant about Tweety. In this section, we will elaborate on this connection and demonstrate how only-knowing-about can be employed to give precise answers to questions like: when is a sentence (or theory) relevant to a set of propositions, or, when is one set of propositions relevant to another given some background theory?

To motivate the following definitions, we begin with an example:

$$\phi = (rain \supset wet) \land (sprinkler\_on \supset wet).$$

Here *rain*, *wet*, and *sprinkler_on* stand for the propositions *"it rains," "the ground is wet,"* and *"the sprinkler is on,"* respectively. The sentence ϕ clearly seems to tell us something relevant about the proposition *rain* (as well as *sprinkler_on* and *wet*). However, ϕ seems irrelevant to anything else like *Jack_is_happy*. When we look a little closer, not everything ϕ conveys is relevant to *rain* because the sentence contains the extraneous information (*sprinkler_on* ⊃ *wet*). On the other hand, everything ϕ tells us is about the ground being wet and, hence, we want to call ϕ *strictly* relevant to *wet*.

It also makes sense to talk about relevance between propositions relative to a background theory. For example, given ϕ, *rain* should count as relevant to *wet* since the latter is true whenever the former is. On the other hand, there really is no connection between *rain* and *sprinkler_on*, since they are either forced to both be false by some other condition (the ground not being wet) or their truth values can vary independently of each other. Therefore, *rain* and *sprinkler_on* are not relevant to each other given ϕ.

If we view relevance with regard to what a sentence (or theory) tells us about the world, then logically equivalent formulations should not change the relevance relation. For example, given $\phi' = (rain \supset wet) \land rain$, *rain* and *wet* are not considered relevant to each other, since ϕ' is logically equivalent to (*wet* ∧ *rain*).

With only-knowing and only-knowing-about, it turns out to be easy and natural to give precise meaning to the notions of relevance which we just introduced. We say that a sentence ϕ is relevant to p just in case it is impossible to know nothing about p assuming ϕ is all that is known. Going back to our initial example, ϕ is then found to be relevant to *rain* because knowing only ϕ implies that we know something nontrivial about *rain*, namely (*rain* ⊃ *wet*). Note the importance of assuming that *only* ϕ is believed and nothing else. For if we merely require ϕ to be believed, we do not rule out believing *wet* as well, in which case $\phi \land wet$ reduces to *wet* and all relevance to *rain* disappears.

In the following, we make these and other notions of relevance precise, including a syntactic characterization in terms of prime implicates in each case.

Varieties of relevance

We begin by defining what it means for a sentence to be relevant to some subject matter. The intuition behind ϕ being relevant to π is that ϕ must contain nontrivial information about π. Our logic allows us to express this directly.

Definition 11.3.1: An objective sentence ϕ is <u>relevant</u> to a subject matter π iff $\models (O\phi \supset \neg O\langle\pi\rangle \text{TRUE})$.

In other words, if we let $e = \mathfrak{R}[\![\phi]\!]$, then for ϕ to be relevant to π, there must be e-minimal clauses which mention some atom in π.

Example 11.3.2: It is easy to verify that, while $\neg p$ and $(p \supset q) \wedge (q \supset r)$ are relevant to p, $(q \supset r)$ and $(p \supset (q \supset p))$ are not.

Theorem 11.3.3: ϕ is relevant to π iff there is some $\gamma \in \mathcal{P}(\phi)$ such that $\not\models \gamma$ and γ mentions some $p \in \pi$.

Proof: The theorem follows immediately from Theorem 11.1.6. ∎

While the previous definition, in a sense, only requires part of the sentence to be about π, we can be even more restrictive and require that everything ϕ tells us is about π in a relevant way and hence arrive at the notion of strict relevance.

Definition 11.3.4: An objective sentence ϕ is <u>strictly relevant</u> to a subject matter π iff $\not\models \phi$ and $O\langle\pi\rangle\phi$ is satisfiable.

Example 11.3.5: $(p \supset q) \wedge (q \supset r)$ is not strictly relevant to p because $(q \supset r)$ is not about p. However, $(p \equiv q) \wedge (q \supset r)$ is strictly relevant to p. This time $(q \supset r)$ is recognized as being about p since p and q are assumed to be equivalent.

Theorem 11.3.6: Let ϕ be an objective sentence such that $\not\models \phi$. Then the following statements are equivalent.

1. ϕ is strictly relevant to π.
2. $\models O\phi \supset O\langle\pi\rangle\phi$.
3. $\models \phi \equiv \bigwedge_{p \in \pi} \bigwedge_{\gamma \in \mathcal{P}(\phi, p)} \gamma$.

Proof: The equivalence of (2) and (3) follows immediately from Theorem 11.1.6. We now show that (1) iff (2). To prove the if direction it suffices to show that $O\langle\pi\rangle\phi$ is satisfiable because ϕ is not valid by assumption. Also, $O\phi$ is satisfiable for any objective ϕ. Since $\models (O\phi \supset O\langle\pi\rangle\phi)$ by assumption, $O\langle\pi\rangle\phi$ is satisfiable as well. Conversely, let $O\langle\pi\rangle\phi$ be satisfiable. Hence there is an e such that $e|_\pi \models O\phi$, that is, $e|_\pi = \{w \mid w \models \phi\}$. To show that $\models (O\phi \supset O\langle\pi\rangle\phi)$, let $e^* \models O\phi$. Since ϕ is objective, $e^* = e|_\pi$. It is not hard to show that $e|_\pi = (e|_\pi)|_\pi$ for any e (Exercise 2). Thus $e^* = e^*|_\pi$ and, therefore, $e^*|_\pi \models O\phi$, which immediately implies $e^* \models O\langle\pi\rangle\phi$. ∎

Next we want to express that a subject matter is relevant to another, relative to some background theory.

Definition 11.3.7: Let π_1 and π_2 be sets of atoms and ϕ and ψ objective sentences such that $\models (O\phi \supset O\langle\pi_2\rangle\psi)$. π_1 is relevant to π_2 with respect to ϕ iff $\pi_1 \cap \pi_2 \neq \{\}$ or ψ is relevant to π_1.

We denote "π_1 is relevant to π_2 with respect to ϕ" by $R_\phi(\pi_1, \pi_2)$.

In other words, π_1 is relevant to π_2 if whatever is known about π_2 contains some nontrivial information about π_1.

Example 11.3.8: Let $\phi = (s \supset w) \wedge (r \supset w)$. Since $\models (O\phi \supset O\langle w\rangle\phi)$ and $\models (O\phi \supset O\langle s\rangle(s \supset w))$, we obtain immediately that s is relevant to w. Similarly, r is relevant to w. However, s is not relevant to r, since $\models (O\phi \supset O\langle r\rangle(r \supset w))$ and $\models (O(r \supset w) \supset O\langle s\rangle\text{TRUE})$.

Theorem 11.3.9: Let π_1 and π_2 be disjoint sets of atoms. $R_\phi(\pi_1, \pi_2)$ iff there is a $\gamma \in \mathcal{P}(\phi)$ such that γ mentions atoms from both π_1 and π_2.

Proof: Let ψ be such that $\models (O\phi \supset O\langle\pi_2\rangle\psi)$.

If direction: Let $\gamma \in \mathcal{P}(\phi)$ such that γ mentions both π_1 and π_2. Then γ is also a prime implicate of ψ. Since γ mentions atoms from both π_1 and π_2, γ is not a tautology. Thus, by Theorem 11.3.3, ψ is relevant to π_1.

Only-if direction: Let $R_\phi(\pi_1, \pi_2)$. Then ψ is relevant to π_1 and, by Theorem 11.3.3, there is a $c \in \mathcal{P}(\psi)$ such that c mentions some atom in π_1. Also $\models \psi \equiv \bigwedge_{p \in \pi_2} \bigwedge_{\gamma \in \mathcal{P}(\phi, p)} \gamma$. Assume none of these γ mentions π_1. Then none of the prime implicates of ψ mentions atoms from π_1, a contradiction.

The relation $R_\phi(\pi_1, \pi_2)$ is obviously reflexive by definition. While symmetry is not obvious from the definition, it nevertheless follows immediately from Theorem 11.3.9, that is, if π_1 is relevant to π_2 with respect to ϕ, then π_2 is relevant to π_1. Note, however, that transitivity does not hold. Example 11.3.8 provides a counterexample. Just because *sprinkler_on* is relevant to *wet* and *wet* is relevant to *rain* does not mean that *sprinkler_on* is relevant to *rain*.

For the next and last definition of relevance, it is convenient to extend our language once again and allow operators that refer to what is known about π without requiring that this is all that is known about π. Naturally, we use operators of the form $\boldsymbol{K}\langle\pi\rangle$ for this purpose and the semantics is just what we would expect:

$$e, w \models \boldsymbol{K}\langle\pi\rangle\phi \quad \text{iff} \quad e|_\pi, w \models \boldsymbol{K}\phi.$$

The reader can easily verify that $(\boldsymbol{O}\langle q\rangle[(p \supset q) \wedge (q \supset r)] \supset \boldsymbol{K}\langle q\rangle(p \supset q))$ is valid according to this definition. Note that, while $\boldsymbol{O}\langle q\rangle(p \supset r)$ is not satisfiable because the sentence in question does not even mention q, this is no longer the case when considering knowing-about instead. In fact, $(\boldsymbol{O}\langle q\rangle[(p \supset q) \wedge (q \supset r)] \supset \boldsymbol{K}\langle q\rangle(p \supset r))$ is valid.

With knowing-about we can now strengthen our previous definition of relevance with respect to sets of atoms π_1 and π_2 by requiring that whatever is known about π_1 is also known about π_2.

Definition 11.3.10: π_1 is <u>subsumed</u> by π_2 with respect to ϕ ($\pi_1 \prec_\phi \pi_2$) iff $\models (\boldsymbol{O}\phi \supset (\boldsymbol{K}\langle\pi_1\rangle\psi \supset \boldsymbol{K}\langle\pi_2\rangle\psi))$ for all objective ψ. π_1 and π_2 are equivalent with respect to ϕ ($\pi_1 \approx_\phi \pi_2$) iff $\pi_1 \prec_\phi \pi_2$ and $\pi_2 \prec_\phi \pi_1$.

As an example, consider $\phi = (cows \supset mammals) \wedge (cows \supset animals)$. Then cows are subsumed by mammals, that is, whatever I know about cows is also known about mammals.

It is easy to see that $\models (\phi \supset (p \equiv q))$ implies $p \approx_\phi q$. Note, however, that the converse does not hold. For example, $p \approx_\phi q$ holds even for $\phi = (p \supset q)$. With Theorem 11.1.6 we obtain the following

Theorem 11.3.11: $\pi_1 \prec_\phi \pi_2$ iff $\models (\mathcal{P}(\phi, \pi_2) \supset \mathcal{P}(\phi, \pi_1))$.

Proof: Note that, by Theorem 11.1.6, $e|_{\pi_1} = \{w \mid w \models \mathcal{P}(\phi, \pi_1)\}$ and $e|_{\pi_2} = \{w \mid w \models \mathcal{P}(\phi, \pi_2)\}$.

To prove the only-if direction, let $\pi_1 \prec_\phi \pi_2$, that is, for all objective ψ, $\models (\boldsymbol{O}\phi \supset (\boldsymbol{K}\langle\pi_1\rangle\psi \supset \boldsymbol{K}\langle\pi_2\rangle\psi))$. Let $e \models \boldsymbol{O}\phi$ and assume, to the contrary, that $\not\models (\mathcal{P}(\phi, \pi_2) \supset \mathcal{P}(\phi, \pi_1))$. Then there is a w^* such that $w \models \mathcal{P}(\phi, \pi_2)$ but $w^* \not\models \mathcal{P}(\phi, \pi_1)$. In other words, there is a $w^* \in e|_{\pi_2}$ such that $w^* \not\models \mathcal{P}(\phi, \pi_1)$. Since $e \models \boldsymbol{O}\langle\pi_1\rangle\mathcal{P}(\phi, \pi_1)$, we obtain immediately that $e \models \boldsymbol{K}\langle\pi_1\rangle\mathcal{P}(\phi, \pi_1)$ and hence, by assumption, that $e \models \boldsymbol{K}\langle\pi_2\rangle\mathcal{P}(\phi, \pi_1)$, from which $w^* \models \mathcal{P}(\phi, \pi_1)$ follows, a contradiction.

To prove the if direction, let $\models (\mathcal{P}(\phi, \pi_2) \supset \mathcal{P}(\phi, \pi_1))$. We need to show that for all objective ψ, $\models (\mathbf{O}\phi \supset (\mathbf{K}\langle\pi_1\rangle\psi \supset \mathbf{K}\langle\pi_2\rangle\psi))$. Let $e \models \mathbf{O}\phi \wedge \mathbf{K}\langle\pi_1\rangle\psi$ for an arbitrary objective ψ. Then $e|_{\pi_1} \models \mathbf{K}\psi$ and, therefore, $\models (\mathcal{P}(\phi, \pi_1) \supset \psi)$. Given our assumption, we obtain that $\models (\mathcal{P}(\phi, \pi_2) \supset \psi)$, which implies $e|_{\pi_2} \models \mathbf{K}\psi$ and thus $e \models \mathbf{K}\langle\pi_2\rangle\psi$. ∎

11.4 Bibliographic notes

Only-knowing-about was first introduced in [66], which also discusses a close connection between only-knowing-about and abductive explanations as they are typically computed by an ATMS [124]. In a nutshell, to find out what is only-known about p, one only needs to collect all the explanations of p and $\neg p$. Only-knowing-about was later extended to multiple agents in [69]. This has the advantage that agents themselves may become a subject so that one can talk about what Jack knows about Jill, among other things. The material relating only-knowing-about to relevance is adapted from [71]. A more general account of our particular brand of relevance is given in [73], which includes considering weaker models of belief of the sort we are about to study starting in the next chapter. There are various connections between our notion of relevance and others found in the literature. Perhaps the most prominent one is *conditional independence*, which was introduced by Darwiche and Pearl [20] as a symbolic analogue of probabilistic conditional independence. To see what is behind this concept, let us consider the following variant of the sprinkler example, which is originally due to Darwiche and Pearl: $\phi = \{(rain \vee sprinkler\_on) \equiv wet\}$. Given this background information, *rain* and *sprinkler_on*, taken by themselves, are considered independent, since knowing the truth value of one does not allow us to infer the other. However, if we know the truth value of *wet*, then the two become dependent with respect to ϕ. It turns out that conditional independence can be completely captured within our framework using the definition of $R_\phi(\pi_1, \pi_2)$. The details can be found in [71, 73], which also includes a detailed account of the connections to Lin and Reiter's version of relevance which they define in terms of "forgetting" certain information in a given knowledge base [91].

11.5 Where do we go from here?

Besides the definitions of relevance we considered in this chapter others seem plausible as well. For example, we could replace *relevant* by *strictly relevant* in Definition 11.3.7. Theorem 11.3.9 could then be strengthened accordingly (see Exercise 4). Whether there are other, less obvious candidates with interesting properties remains to be seen.

Perhaps the most pressing issue, however, is the need to find a first-order generalization of only-knowing-about. As we already hinted at in the beginning, we really would like to make sense of questions like "what do you know about Toronto." To begin with, it is not at all clear what should be considered a subject matter, since we have more primitives to choose from such as standard names, terms, or predicates.[4] Perhaps even more troubling is the fact that the idea of using minimal clauses does not seem to generalize to the first-order case. For example, if all that is known is $\exists x\, P(x)$, then we simply cannot represent what is known as a set of clauses. The problem is, of course, the existential quantifier. One fallback position could be to rule out such cases and, say, consider only objective knowledge bases without existential quantifiers. Then we do obtain a clausal representation, since every sentence with only universal quantifiers can be represented by a set of propositional sentences where the variables are replaced by all combinations of standard names. While we could then in principle apply our original definition of only-knowing-about, new problems arise. For example, we would be forced to deal with an infinite set of clauses and it does not seem obvious how to represent what is only-known about a subject matter, if it is finitely representable at all.

11.6 Exercises

1. Let ϕ be objective with $e \models O\phi$. Show that $e|_{\pi_\phi} = e$. (See the proof of Theorem 11.1.3, where this fact is used.)
2. Show that for any π and e, $e|_\pi = (e|_\pi)|_\pi$.
3. Consider versions of **ASK** and **TELL**, where the operator $K\langle\pi\rangle$ is allowed as part of the query. In particular, modify Definitions 11.2.2 and 11.2.3 appropriately and prove Lemma 11.2.5 for the extended language.
4. Consider the strengthening of Definition 11.3.7 where *relevant* is replaced by *strictly relevant*. Similar to Theorem 11.3.9, formulate and prove the corresponding characterization in terms of prime implicates.

---

4 Actually, even in the propositional case, there is the possibility to choose arbitrary sentences as the subject matter.

12 Avoiding Logical Omniscience

So far in our considerations, the "symbol level" has played a relatively minor role. In as much as we cared at all about representing knowledge, we focused on results like the Representation Theorem of Chapter 7, which deals with the existence of finite representations, but more or less neglects the computational cost of manipulating them, with one exception. We were glad to see that **ASK** and **TELL** can be characterized completely using first-order reasoning alone, even though the interaction language is modal. While we did not say so explicitly, one motivation for avoiding modal reasoning is that it is much less understood and, in general, much harder computationally than non-modal reasoning. However, this is only partly good news, since it is well known that first-order reasoning alone is already undecidable. In particular, any faithful implementation of **ASK** and **TELL** is bound to run forever on certain inputs. While this may be acceptable under certain circumstances such as proving mathematical theorems, it clearly is not when it comes to commonsense reasoning tasks, which has been the main motivation behind our work.

If we look at the problem in terms of the properties of belief, then one reason for bad computational behaviour seems to be that beliefs are closed under logical implication, that is, whatever follows from the knowledge base must be believed. In the extreme, even if the knowledge base is empty, the agent still believes every valid sentence of the logic. As already mentioned in Chapter 4, this assumption is known as *logical omniscience*, which seems clearly unacceptable for real resource-bounded agents. One way out of this dilemma would be to severely limit the expressiveness of the representation language as is done in relational databases, for example. Here we will consider the case where we leave the language unchanged and look for ways to reduce the set of beliefs so that they become more manageable. Not restricting the language in any way is perhaps closer to human reasoning. After all, we have no trouble dealing with highly complex representation languages such as English or German, to choose two at random. What we have trouble with is inferring everything that is implicit in those representations. Of the many reasons why humans find this part hard, one is certainly limited resources. After all, we cannot spend all day thinking.

A naive approach to dealing with this problem in logic would be to simply use a first-order theorem prover with explicit resource limitations to compute the answer to a query. For example, if the theorem prover returns the correct answer within time *t* we are happy; otherwise the computation stops and *unknown* is returned as an answer. While easy to implement, this simple approach has a severe disadvantage. For any given resource bound it is no longer clear what the corresponding epistemic state of the agent is. The problem is that the only characterization of what the agent believes would be the current implementation of the inference mechanism. As a result, it would be impossible, in general, to predict the behaviour of the system because the only way to find out whether a sentence is believed would be to actually ask the system and wait for the reply. If one cares about predictability, as we do, one needs to look at limiting the reasoning power in more principled ways.[1]

To do that, we advocate a *semantic* approach, which looks at ways of modifying the model theory in appropriate ways. In the context of a possible-world model of belief, as in our case, the general idea is actually quite simple. Reasoning, in an abstract sense, means determining whether a sentence belongs to a given epistemic state. If, as we have argued above, this problem is too hard in general, then we must somehow remove beliefs from the epistemic state until we are left with a manageable set of them. Now recall that an epistemic state is represented by a set of world states e. This means, in particular, that beliefs can be removed by adding world states to e.[2] Just adding world states, however, does not solve our problem yet because logical omniscience has not gone away, that is, all valid sentences of the logic are believed at all times and beliefs are closed under logical implication. What is needed to overcome the logical omniscience hurdle semantically is to allow more possibilities than we have in the models considered so far. In the following we propose using four-valued extensions of world states for this purpose. Allowing such nonstandard world states as part of epistemic states will reduce the set of beliefs enough to break logical omniscience with a considerable computational pay-off.

Working within a modal logic has the advantage that we need not switch to an entirely different logic when specifying a limited reasoner. All we need to do is add a new notion of belief to what we already have. We call this new notion *explicit belief* and denote it by a new modality B. What we previously simply called belief (K) now becomes *implicit belief*. When an agent's knowledge is represented by a finite KB, the explicit beliefs should be thought of as those which follow from the KB in a (yet to be defined) simple way, and the implicit beliefs are those which may take the full power of first-order logic to compute. Naturally, we expect the two forms to be connected in that every explicit belief is also implicitly believed. In other words, ($B\alpha \supset K\alpha$) should be valid.

In $\mathcal{OL}$, the complexity of reasoning about the implicit beliefs of a KB is measured by the complexity of determining the validity of sentences of the form ($O\text{KB} \supset K\alpha$). Analogously, the validity problem of sentences of the form ($O\text{KB} \supset B\alpha$) now models limited reasoning using explicit belief.

In this chapter, we will consider the simple case where both KB and α are objective. In that case, asking whether ($O\text{KB} \supset B\alpha$) is valid is equivalent to asking whether ($B\text{KB} \supset$

---

[1] We could make the above approach a little less dependent on the implementation by counting inference steps instead of measuring computing time, but the principle problem would not go away since the beliefs would still depend on the particular algorithm.

[2] Of course, some beliefs are also added, namely those about what we no longer believe.

$B\alpha$) is valid. Since sentences of the latter form play such a central role in the following, we frequently use the term *belief implications* to refer to them.

We begin our journey into the world of non-classical logic gently by ignoring quantifiers for now. This allows us to introduce key features of our approach in a simple setting. As we will see later, the first-order version has some complications in stock which would only distract us at this point.

12.1 The propositional case

Let $\mathcal{BL}$ be the propositional subset of $\mathcal{KL}$ with an additional modal operator B. We ignore beliefs about beliefs and assume that neither K nor B appear nested. In other words, for any (sub-)sentence $K\phi$ or $B\phi$, ϕ is assumed to be objective.

The key idea of the semantics of $\mathcal{BL}$ is that of a *situation*. Intuitively one may think of a situation as a slice of the world as it presents itself to an observer. For example, imagine an office with a desk, chairs, a work station on the desk, shelves, papers etc. While an office contains a whole lot of things, compared to the whole world, many (if not most) things are irrelevant to offices such as the price of tea in China or the height of the CN Tower. To reflect these distinctions, situations do not assign a truth value to every proposition, only to the ones that matter. Furthermore, some of the facts about the office may be ambivalent to the observer. For example, there may be evidence that a particular paper is somewhere in the office since it was seen there just a few days ago. On the other hand, there may also be evidence that it is not there since it cannot be found anywhere. To reflect this intuition we allow that a situation supports both the truth and the falsity of a proposition.

In summary, situations extend the usual notion of a world state in two ways. For one, not every proposition has a truth value and, for another, some propositions may have both true- and false-support. In other words, situations have a four-valued semantics.

Definition 12.1.1: A situation s is a mapping which assigns each propositional variable a subset of $\{0, 1\}$.

$\{1\}$ means that there is only true-support, $\{0\}$ that there is only false-support, $\{\}$ that there is neither true- nor false-support, and $\{0, 1\}$ that there is both true- and false-support. Naturally, we identify $\{0\}$ and $\{1\}$ with the usual notions of truth and falsity. Hence a situation is called a world state just in case every proposition is assigned either $\{1\}$ or $\{0\}$.

The main feature to keep in mind about situations is that there is no longer a connection between the true- and false-support of an atomic proposition, that is, if a proposition has no true-support, then it does not necessarily have false-support and vice versa. The following semantics generalizes this feature to arbitrary objective sentences and provides the key ingredient for modeling what we called shallow reasoning in the introduction of this chapter.

In order to keep the true- and false-support of sentences independent, we need to define them separately and denote them by $\models_T$ and $\models_F$, respectively. Let p be an atomic proposition, α and β arbitrary sentences in $\mathcal{BL}$, ϕ an objective sentence, s a situation, and e a set of situations.

1. $e, s \models_T p$ iff $1 \in s[p]$;
   $e, s \models_F p$ iff $0 \in s[p]$;
2. $e, s \models_T \neg\alpha$ iff $e, s \models_F \alpha$;
   $e, s \models_F \neg\alpha$ iff $e, s \models_T \alpha$;
3. $e, s \models_T \alpha \vee \beta$ iff $e, s \models_T \alpha$ or $e, s \models_T \beta$;
   $e, s \models_F \alpha \vee \beta$ iff $e, s \models_F \alpha$ and $e, s \models_F \beta$;
4. $e, s \models_T \boldsymbol{B}\phi$ iff for all $s' \in e$, $e, s' \models_T \phi$;
   $e, s \models_F \boldsymbol{B}\phi$ iff $e, s \not\models_T \boldsymbol{B}\phi$;
5. $e, s \models_T \boldsymbol{K}\phi$ iff for all world states $w \in e$, $e, w \models_T \phi$;
   $e, s \models_F \boldsymbol{K}\phi$ iff $e, s \not\models_T \boldsymbol{K}\phi$.

Since we are interested in situations primarily in the context of explicit belief and to obtain regular notions of satisfiability and validity for this logic, we define them with respect to world states and epistemic states only. Given a world state w, an epistemic state e, and a sentence α of $\mathcal{BL}$, we say that α is *true* at w and e if $e, w \models_T \alpha$ and *false* if $e, w \not\models_T \alpha$. (It is easy to show that $e, w \not\models_T \alpha$ iff $e, w \models_F \alpha$.) α is said to be valid ($\models\alpha$) if α is true in all world states w and epistemic states e. α is satisfiable if $\neg\alpha$ is not valid.

Before we turn to the properties of $\mathcal{BL}$ in detail, let us consider some sets of sentences that are satisfiable and show that explicit belief indeed does not suffer from logical omniscience.

1. $\{\boldsymbol{B}p, \boldsymbol{B}(p \supset q), \neg \boldsymbol{B}q\}$ This shows that beliefs are not closed under implication.
2. $\{\neg\boldsymbol{B}(p \vee \neg p)\}$ A valid sentence need not be believed.
3. $\{\boldsymbol{B}p, \neg\boldsymbol{B}(p \wedge (q \vee \neg q))\}$ A logical equivalent to a belief need not be believed.
4. $\{\boldsymbol{B}p, \boldsymbol{B}\neg p, \neg \boldsymbol{B}q\}$ Beliefs can be inconsistent without every sentence being believed.

To see why (1) is satisfiable, consider $e = \{s\}$, where s supports both the truth and falsity of p and neither the truth nor the falsity of q, that is, $s[p] = \{0, 1\}$ and $s[q] = \{\}$. Then $s \models_T p \wedge (p \supset q)$ yet $s \not\models_T q$. Given the definition of $\boldsymbol{B}$, e clearly satisfies all the sentences in (1). We leave the proof of the other cases as an exercise. Note that while all of the above sets are satisfiable, none of them would be if we replaced all occurrences of $\boldsymbol{B}$ by $\boldsymbol{K}$.

12.1.1 A proof theory

The above sets show what freedom the logic allows in terms of belief; to demonstrate that the logic does impose reasonable constraints on belief, we must look at the valid sentences of $\mathcal{BL}$. We will present these in terms of a proof theory for $\mathcal{BL}$ that is both sound and complete with respect to the above semantics. As in the case of $\mathcal{KL}$ the main reason why we consider a proof theory here is that it does provide an elegant and vivid way to examine the valid sentences of $\mathcal{BL}$ (especially those using $\boldsymbol{B}$).[3]

As for the objective part of the logic of implicit belief, there is nothing new. We simply inherit the properties given by $\mathcal{KL}$ restricted to the propositional case and non-nested $\boldsymbol{K}$'s.

1. all instances of propositional tautologies.
2. $\boldsymbol{K}\phi$, where ϕ is a tautology.
3. $\boldsymbol{K}\phi \wedge \boldsymbol{K}(\phi \supset \psi) \supset \boldsymbol{K}\psi$.

The connection between explicit and implicit belief is as expected, that is, every explicit belief is implicitly believed, but not necessarily vice versa.

4. $\boldsymbol{B}\phi \supset \boldsymbol{K}\phi$.

For explicit belief we have to dream up a set of axioms stating what has to be believed when something else is. In other words, we need a set of axioms of the form ($\boldsymbol{B}\phi \supset \boldsymbol{B}\psi$), for various ϕ and ψ. Fortunately, most of this work has already been done for us in relevance logic, in particular regarding its fragment *tautological entailment*.

Using our own notation, tautological entailment has the following simple semantics:

Definition 12.1.2: Let ϕ and ψ be objective sentences. Then ϕ tautologically entails ψ ($\phi \longrightarrow \psi$) iff for all situations s,

1. if $s \models_T \phi$ then $s \models_T \psi$,
2. if $s \models_F \psi$ then $s \models_F \phi$.

It turns out that the first of the two conditions suffices to characterize tautological entailment. (For symmetric reasons, the second condition alone would do as well.)

Lemma 12.1.3: $\phi \longrightarrow \psi$ *iff for all situations s, if $s \models_T \phi$ then $s \models_T \psi$.*

Proof: Given any situation s, let s^* be a situation that is identical everywhere to s unless $s[p]$ is empty in which case $s^*[p] = \{0, 1\}$, or $s[p] = \{0, 1\}$ in which case $s^*[p]$ is

---

[3] We could imagine constructing a decision procedure for $\mathcal{BL}$ directly from the above without even passing through a proof theory at all. Such a decision procedure, after all, is what counts when building a system that reasons with $\mathcal{BL}$.

empty. A simple induction argument shows that s^* maintains this property over non-atomic sentences. Then, assuming the first condition above is satisfied for all situations, if $s \not\models_F \phi$, then regardless of whether $s \models_T \phi$ or not, $s^* \models_T \phi$ and, hence, $s^* \models_T \psi$, guaranteeing that $s \not\models_F \psi$. ■

With that we obtain a straightforward connection between explicit belief and tautological entailment.

Theorem 12.1.4: $\models (B\phi \supset B\psi)$ iff $\phi \longrightarrow \psi$.

Proof: To prove the if direction, let $e, s \models_T B\phi$. Then for all $s' \in e$, $s' \models_T \phi$. Hence, by assumption, for all $s' \in e$, $s' \models_T \psi$, that is, $e, s \models_T B\psi$.

To prove the only-if direction, let $s^* \models_T \phi$. Let $e_{max} = \{$situation $s \mid s \models_T \phi\}$. Clearly, $s^* \in e$. Also, by assumption, $e_{max} \models_T B\psi$, from which $s^* \models_T \psi$ follows immediately. ■

What this theorem tells us is that $\mathcal{BL}$ subsumes a fragment of relevance logic: questions of tautological entailment can be reduced to questions of belief in $\mathcal{BL}$. Moreover, we get this relevance logic without having to give up classical logic or the normal interpretation of $\supset$ and the other connectives.

So all that is needed to specify the constraints satisfied by belief is to borrow the characterization of tautological entailment in relevance logic.[4] One such account is the following:

1. $B(\phi \wedge \psi) \equiv B(\psi \wedge \phi)$.
   $B(\phi \vee \psi) \equiv B(\psi \vee \phi)$.
2. $B(\phi \wedge (\psi \wedge \eta)) \equiv B((\phi \wedge \psi) \wedge \eta)$.
   $B(\phi \vee (\psi \vee \eta)) \equiv B((\phi \vee \psi) \vee \eta)$.
3. $B(\phi \wedge (\psi \vee \eta)) \equiv B((\phi \wedge \psi) \vee (\phi \wedge \eta))$.
   $B(\phi \vee (\psi \wedge \eta)) \equiv B((\phi \vee \psi) \wedge (\phi \vee \eta))$.
4. $B\neg(\phi \vee \psi) \equiv B(\neg\phi \wedge \neg\psi)$.
   $B\neg(\phi \wedge \psi) \equiv B(\neg\phi \vee \neg\psi)$.
5. $B\neg\neg\phi \equiv B\phi$.
6. $B\phi \wedge B\psi \equiv B(\phi \wedge \psi)$.
   $B\phi \vee B\psi \supset B(\phi \vee \psi)$.
   From $((B\phi \vee B\psi) \supset B\eta)$, infer $(B(\phi \vee \psi) \supset B\eta)$.

This particular axiomatization emphasizes that belief must respect properties of the logical operators such as commutativity, associativity, distributivity, and De Morgan's laws.

---

4 One necessary constraint on belief not expressible in relevance logic is that believing two sentences implies believing their conjunction.

The main observation about the proof theory of $\mathcal{BL}$ is that nothing forces *all* the logical consequences of what is believed to be believed (as in Axiom 1 and 3, above, for implicit belief), although *some* consequences have to be believed (e.g., a double negation of a sentence must be believed if the sentence itself is).

Another way to understand the axiomatization is as constraints on the individuation of beliefs. For example, $(\phi \vee \psi)$ is believed iff $(\psi \vee \phi)$ is because these are two lexical notations for the *same* belief. In this sense, it is not that there is an automatic inference from one belief to another, but rather two ways of describing a single belief.

This, in itself, does not *justify* the axioms, however. It is easy to imagine logics of belief that are different from this one, omitting certain of the above constraints or perhaps adding additional ones. Indeed, there is not much to designing a proof theory with any collection of constraints on belief. The interesting fact about *this* particular set of axioms, however, is that it corresponds so nicely to an independently motivated semantic theory. Specifically, we have the following result:

Theorem 12.1.5: *A sentence of $\mathcal{BL}$ is a theorem of the above logic iff it is valid.*

Proof: The soundness of $\mathcal{BL}$ follows by induction on the length of a proof, given that each axiom can easily be seen to be valid, and that *modus ponens* and the inference rule(s) from relevance logic preserve validity.

To prove completeness, we show that any consistent sentence α is satisfiable. For that we first extend the set $\{\alpha\}$ to what is called a *downward saturated* set. A set of sentences Γ is downward saturated if (a) for every $\neg\neg\gamma \in \Gamma$, $\gamma \in \Gamma$, (b) for every $(\gamma \wedge \delta) \in \Gamma$, both γ and δ are in Γ, and (c) for every $(\gamma \vee \delta) \in \Gamma$, either γ or δ is in Γ. Note that sentences dominated by a *B* or *K* operator are treated as atomic. Let us call these *modal atoms*. The idea is then to find an appropriate w and e such that w satisfies all the literals in Γ and e satisfies all the modal atoms. Since Γ is downward saturated, a simple induction establishes that such e and w satisfy all of Γ and, in particular, α.

Finding an appropriate w is easy: just pick any world state which satisfies all the literals in Γ, which must exist by the completeness of propositional logic. The more difficult part is to find an appropriate e. Here we lean on the fact that we have completely characterized tautological entailment so that if $\phi \longrightarrow \psi$ then $(\boldsymbol{B}\phi \supset \boldsymbol{B}\psi)$ is a theorem. Taking this for granted, let us consider the subset of all modal atoms in Γ, which can be written as

$$\Sigma = \{\boldsymbol{B}\phi_i\} \cup \{\neg\boldsymbol{B}\psi_i\} \cup \{\boldsymbol{K}\eta_i\} \cup \{\neg\boldsymbol{K}\zeta_i\}.$$

Let us call a situation *s incoherent* if it supports both the truth and falsity of some

propositional letter. We claim that the following epistemic state satisfies Σ:
$$e = \bigcup_{\psi_i}\{\text{incoherent } s \mid s \models_{\text{T}} \bigwedge \phi_j, s \not\models_{\text{T}} \psi_i\} \cup$$
$$\bigcup_{\zeta_i}\{\text{world state } w \mid w \models_{\text{T}} \bigwedge \phi_j, w \models_{\text{T}} \bigwedge \eta_k, w \not\models_{\text{T}} \zeta_i\}.$$

The first claim is that for each ψ_i, there is a situation s in e that fails to support it. For if there were not, then every incoherent situation supporting all the ϕ_j would also support ψ_i. But this would mean that the conjunction of the ϕ_j tautologically entailed ψ_i, since if there was a situation that supported all the ϕ_j but failed to support ψ_i, a situation that was incoherent (by virtue of some propositional letter that appeared nowhere else) would have to do the same. But this entailment relation would mean that

$$\boldsymbol{B}(\bigwedge \phi_j) \supset \boldsymbol{B}\psi_i$$

was a theorem which, by the axiom for conjunctive belief, would imply that

$$\bigwedge \boldsymbol{B}\phi_j \supset \boldsymbol{B}\psi_i$$

was a theorem also. But this would mean Σ was inconsistent, contradicting our assumption.

The second claim is that for each ζ_i, there is a world state in e that fails to support it. If not, then every world state supporting all the ϕ_j and the η_k also supports ζ_i. But these situations are just the regular propositional truth value assignments. So, by the completeness of propositional logic, we would get that

$$\bigwedge \phi_j \wedge \bigwedge \eta_k \supset \zeta_i$$

is a theorem, and so

$$\bigwedge \boldsymbol{K}\phi_j \wedge \bigwedge \boldsymbol{K}\eta_k \supset \boldsymbol{K}\zeta_i$$

is a theorem also, since $\boldsymbol{K}$ is closed under *modus ponens* and every axiom of propositional logic is known. Finally, since it is an axiom that every belief is known, we would get that

$$\bigwedge \boldsymbol{B}\phi_j \wedge \bigwedge \boldsymbol{K}\eta_k \supset \boldsymbol{K}\zeta_i$$

is a theorem, meaning that Σ was inconsistent, again contradicting our assumption.

The next thing to notice about e is that by definition each of its elements supports the truth of every ϕ_i and, moreover, that each of its world states by definition supports the truth of every η_i.

In summary, the epistemic state we have defined satisfies all of Σ: for each $\boldsymbol{B}\phi_i$ we have that every situation in e supports ϕ_i, for each $\neg \boldsymbol{B}\psi_i$, we have some element of e that does not support ψ_i, for each $\boldsymbol{K}\eta_i$, we have every world state in e supporting η_i, and for each $\neg \boldsymbol{K}\zeta_i$, we have a world state that fails to support ζ_i. Thus e is an epistemic state which satisfies all the modal atoms in Γ. Together with any world state w satisfying all the literals in Γ we have that e and w satisfies all of Γ because the set is downward saturated. Since $\alpha \in \Gamma$, α is satisfied as well. Thus, any consistent sentence is satisfiable, and so any valid sentence is provable. ∎

12.1.2 Computing explicit belief

Now that we have an idea of how explicit and implicit belief compare logically, let us now turn to the computational side. In particular, we are interested in the following question: given that the sentences in a KB are explicitly believed, how difficult is it to determine whether any sentence ϕ is believed, either explicitly or implicitly. More precisely, we are interested in comparing the computational complexity of deciding the validity of formulas of the form (BKB $\supset B\phi$) and (BKB $\supset K\phi$).[5] Equivalently, this amounts to comparing the complexity of tautological entailment and logical implication. (We leave it as an exercise to show that (BKB $\supset K\phi$) is valid iff KB logically implies ϕ.)

We begin with a somewhat surprising negative result, namely that there is no difference in the complexity of logical implication and tautological entailment if KB and ϕ are arbitrary objective sentences.

Theorem 12.1.6: *Let* KB *and ϕ be objective sentences. Then determining whether* KB *logically implies ϕ and whether* KB *tautologically entails ϕ are both co-NP complete.*

Proof: To see why the first problem is co-NP-complete, note that a special case of the problem (namely whether or not the KB logically implies $(p \wedge \neg p)$) holds exactly if the KB is unsatisfiable. But satisfiability of sentences is NP-complete and so, our original problem is the complement of a problem at least as hard as an NP-complete one. Moreover, the problem is not too hard since it is a special case of a test for validity. Thus, the problem is co-NP-complete.

Let us now see why the second problem is also co-NP-complete. The key observation is that logical implication can actually be expressed as a special case of tautological entailment. In particular, if Σ is the set of all atoms occurring in KB and ϕ we obtain that KB logically implies ϕ iff KB $\wedge \bigwedge_{p \in \Sigma} (p \vee \neg p)$ tautologically entails $\phi \vee \bigvee_{p \in \Sigma} (p \wedge \neg p)$. We defer the proof to Corollary 12.2.2, where this result falls out as a special case when considering first-order explicit beliefs. Hence tautological entailment is at least as hard as a co-NP complete problem. To see why it is in co-NP, let us consider its complement, that is, the problem of determining whether KB *does not* tautologically entail ϕ. There we need to find a situation s such that $s \models_T$ KB and $s \not\models_T \phi$. This can be done by guessing an appropriate assignment of the atoms in Σ and then checking (in linear time) the true-support of KB and ϕ. This problem is clearly in NP and, therefore, tautological entailment is in co-NP. ∎

---

[5] The question whether (KKB $\supset B\phi$) is valid has a trivial answer: no explicit belief follows from implicitly believing KB. The proof is left as an exercise.

However, this negative result for tautological entailment does not hold when the KB and ϕ are in conjunctive normal form:

Theorem 12.1.7: *Suppose KB and ϕ are propositional sentences in conjunctive normal form. Determining if KB tautologically entails ϕ has an $O(mn)$ algorithm, where $m = |KB|$ and $n = |\phi|$.*

Proof: To see how the algorithm for tautological entailment works, let $\phi_1, \ldots, \phi_n$, and $\psi_1, \ldots \psi_m$ be disjunctions of literals. First note that $\bigwedge \phi_i \longrightarrow \bigwedge \psi_j$ iff $\bigwedge \phi_i \longrightarrow \psi_j$ for each ψ_j. The claim we prove below is that for every ψ_j, $\bigwedge \phi_i \longrightarrow \psi_j$ exactly when there is an ϕ_i whose literals are a subset of those of ψ_j. This gives us an algorithm of the proper order, since we only have to traverse all the ϕ_i for every ψ_j.

First suppose that the literals of some ϕ_i are a subset of those of ψ_j. This means that for any situation s, if $s \models_T \phi_i$ then $s \models_T \psi_j$. But then if $s \models_T \bigwedge \phi_i$, then $s \models_T \psi_j$. Thus, the conjunction of the ϕ_i entails ψ_j.

Next suppose that for each ϕ_i there is a literal λ_i that is in ϕ_i but not ψ_j. Define a situation s as follows: $s \models_T p$ iff $p = \lambda_i$, for some i, and $s \models_F p$ iff $\neg p = \lambda_i$, for some i. Then, for any literal λ, $s \models_T \lambda$ iff $\lambda = \lambda_i$, for some i. Consequently, $s \not\models_T \psi_j$ but for each i, $s \models_T \phi_i$. Thus, the conjunction of ϕ_i does not entail ψ_j. ∎

Corollary 12.1.8: *Assume KB and ϕ are as above. Then, in the worst case, deciding if*
a) $\models (\mathbf{B}KB \supset \mathbf{K}\phi)$ is very difficult.
b) $\models (\mathbf{B}KB \supset \mathbf{B}\phi)$ is relatively easy.

What this amounts to is that if we consider answering questions of a given fixed size, the time it takes to calculate what the KB believes will grow *linearly* at worst with the size of the KB, but the time it takes to calculate the implications of what the KB believes will grow *exponentially*[6] at worst with the size of the KB.

12.2 The first-order case

Given this dramatic difference in complexity between explicit and implicit belief, the obvious question, of course, is whether there is a similar difference in the first-order case. The answer, in short, is yes, but the story turns out to be much more complicated than one might expect. The problem is, as we will see in a moment, that a straightforward first-order

---

6 More precisely, it will grow faster than any polynomial function, unless P equals NP.

generalization of explicit belief still leads to an undecidable reasoner. How could this be when an agent cannot even apply the rule of *modus ponens*? To understand intuitively what else could be responsible for intractability, recall that we identified *reasoning by cases* as the main culprit for bad computational behaviour. While we have ruled out one source by introducing situations, which decouple the meaning of literals from their negations, existential quantifiers introduce another source, which needs to be taken care of as well. For example, assume our KB consists only of the sentence $P(a) \vee P(b)$. Then any world state or situation the agent considers possible supports either $P(a)$ or $P(b)$. Hence in any *case* there is some individual which has the property P. In other words, $\exists x P(x)$ is believed given KB, a property which we refer to as *existential generalization*.[7] Note that this has nothing to do with the relation between literals and their negation. So situations alone will not help us in eliminating this form of reasoning by cases.

The solution is to restrict the interpretation of an existential quantifier within explicit beliefs in a way analogous to its treatment in intuitionistic logic. Essentially we require that $\exists x P(x)$ be believed only if we find a term t (also called a *witness*) such that $P(t)$ is believed. Note that, in the example, we would not be able to find such a witness since all we know is that either a or b is a P, but we do not know which one. Hence $\exists x P(x)$ would not be believed under the new interpretation of $\exists$.

In the following we will formalize this idea in detail. We begin by introducing first-order situations and prove that the obvious generalization of the semantics of explicit belief indeed leads to undecidability. We then modify the interpretation of $\exists$ within **B** and prove various properties including the decidability of belief implication. In addition to nested beliefs, we also ignore equality for now. Both restrictions will be lifted in the next chapter.

Situations in the first-order case behave exactly like world states as far as the interpretation of terms is concerned, that is, every term is assigned a unique co-referring standard name. Only predicates are given a four-valued semantics. Formally, a situation s is a mapping from the primitive terms and primitive formulas into the standard names and subsets of $\{0, 1\}$ such that

$s[t] = n$ for some standard name n, where t is a primitive term.

$s[p] \subseteq \{0, 1\}$, where p is a primitive formula.

Note that world states are now just special situations where for all primitive formulas p either $s[p] = \{0\}$ or $s[p] = \{1\}$.

Closed terms are evaluated at a given situation s exactly the same way as in the case of world states, that is,

1. $s(n) = n$;

---

7 See also Chapter 3 where we discuss this issue in the context of $\mathcal{KL}$.

2. $s(f(t_1, \ldots, t_k)) = s[f(n_1, \ldots, n_k)]$, where $n_i = s(t_i)$.

In order to obtain an evaluation of sentences, all we need to do is modify the rules for atomic sentences of the previous section and add rules to deal with quantification. (For simplicity, we will ignore K from now on.)

1. $e, s \models_T P(t_1, \ldots, t_k)$ iff $1 \in s[P(n_1, \ldots n_k)]$, where $n_i = s(t_i)$;
   $e, s \models_F P(t_1, \ldots, t_k)$ iff $0 \in s[P(n_1, \ldots n_k)]$, where $n_i = s(t_i)$;
2. $e, s \models_T \neg \alpha$ iff $e, s \models_F \alpha$;
   $e, s \models_F \neg \alpha$ iff $e, s \models_T \alpha$;
3. $e, s \models_T \alpha \vee \beta$ iff $e, s \models_T \alpha$ or $e, s \models_T \beta$;
   $e, s \models_F \alpha \vee \beta$ iff $e, s \models_F \alpha$ and $e, s \models_F \beta$;
4. $e, s \models_T \exists x \alpha$ iff for some $n \in N$, $e, s \models_T \alpha_n^x$;
   $e, s \models_F \exists x \alpha$ iff for all $n \in N$, $e, s \models_F \alpha_n^x$;
5. $e, s \models_T \boldsymbol{B}\phi$ iff for all $s' \in e$, $e, s' \models_T \phi$;
   $e, s \models_F \boldsymbol{B}\phi$ iff $e, s \not\models_T \boldsymbol{B}\phi$.

As before, we define truth as true-support in a (two-valued) world state and a set of situations, and validity as truth in all world states and sets of situations.

Theorem 12.2.1: *The problem of determining the validity of* $(\boldsymbol{B}\phi \supset \boldsymbol{B}\psi)$ *for objective ϕ and ψ is undecidable.*

Proof: Let Σ be the set of predicate symbols appearing in ϕ and ψ. Let $\phi' = \phi \wedge \bigwedge_{P \in \Sigma}(\forall \vec{z}[P(\vec{z}) \vee \neg P(\vec{z})])$ and $\psi' = \psi \vee \bigvee_{P \in \Sigma}(\exists \vec{x}[P(\vec{x}) \wedge \neg P(\vec{x})])$.

We now reduce regular first-order logical implication to belief implication by showing that $\models (\phi \supset \psi)$ iff $\models (\boldsymbol{B}\phi' \supset \boldsymbol{B}\psi')$. Hence, since first-order logical implication is undecidable, the undecidability of belief implication follows immediately.

To prove the if direction, let $\models (\boldsymbol{B}\phi' \supset \boldsymbol{B}\psi')$. Then $\models (\phi' \supset \psi')$ holds because $e \models (\boldsymbol{B}\phi' \supset \boldsymbol{B}\psi')$ for any singleton $e = \{w\}$ where w is an arbitrary world state. Let w be a world state such that $w \models \phi$. Since $\bigwedge_{P \in \Sigma}(\forall \vec{z} P(\vec{z}) \vee \neg P(\vec{z}))$ is valid, $w \models \phi'$ and, by assumption, $w \models \psi'$. Since $\bigvee_{P \in \Sigma}(\exists \vec{x} P(\vec{x}) \wedge \neg P(\vec{x}))$ is unsatisfiable, $w \models \psi$ follows.

To prove the only-if direction, let $\models (\phi \supset \psi)$ and let e be an epistemic state such that $e \models \boldsymbol{B}\phi'$. Let s be any situation in e. It suffices to show that $s \models \psi'$. If s is a world state, then $s \models \psi$ by assumption and, therefore, $s \models \psi'$. Now consider the case where s is a situation but not a world state. If $s \models \bigvee_{P \in \Sigma}(\exists \vec{x} P(\vec{x}) \wedge \neg P(\vec{x}))$, we are done. Otherwise, s is like a world state with respect to all predicate symbols in Σ. In this case, a simple induction proves that any world state w which agrees with s on all primitive formulas with predicate symbols from Σ agrees with s on all sentences with predicate symbols from Σ. In particular, it follows that $w \models \psi'$ and, therefore, $s \models \psi'$. ∎

Corollary 12.2.2: *Propositional tautological entailment is co-NP hard.*

Proof: Note that, in the propositional case, the above argument can be simplified as follows: $\models (\phi \supset \psi)$ iff $\models [\boldsymbol{B}(\phi \wedge \bigwedge_{p \in \Sigma}(p \vee \neg p)) \supset \boldsymbol{B}(\psi \vee \bigvee_{p \in \Sigma}(p \wedge \neg p))]$, hence reducing the problem of logical implication to the problem of tautological entailment in polynomial time. ∎

We now consider a variant of $\mathcal{BL}$ that avoids this undecidability problem.

Definition 12.2.3: Let ϕ be an objective formula. A variable x is said to be *existentially (universally) quantified* in ϕ iff x is bound in the scope of an *even (odd)* number of $\neg$-operators. In the following we will often write "existential" as short hand for "existentially quantified variable."

Definition 12.2.4: An objective formula ϕ is called *existential-free* if ϕ contains no existentials.

Definition 12.2.5: Let ϕ be an objective formula and x an existential in ϕ. A term t is said to be an *admissible* substitution for x with respect to ϕ iff every variable y in t is universally quantified in ϕ and x is bound within the scope of y.

If the context is clear, we often say *t is admissible for x* or *t is admissible*.

The main reason why admissible terms are defined this way is because the semantics given below is only defined for *sentences*. Thus we have to make sure that a term substituted for an existentially quantified variable within a sentence again yields a sentence. Notice that, in contrast to the more familiar idea of Skolemization, more than one function symbol may occur in an admissible term, these function symbols may already occur elsewhere in ϕ, and not every universally quantified variable in whose scope x is bound needs to occur in the term. For example, if

$$\phi = \forall x \forall y \exists z (P(x, y, z) \wedge Q(f(a, x))),$$

then $f(a, y)$, $g(y, x)$, x, and $^\#1$ are all admissible terms for z. On the other, neither z nor $f(x, u)$, where u is a variable different from x and y, are admissible.

Definition 12.2.6: Let ϕ be an objective formula with existentially quantified variables $x_1, \ldots, x_k$. Then $\phi^\#$ denotes ϕ with all $\exists x_i$ removed for all $1 \leq i \leq k$.

To illustrate the previous definitions, let $\phi = \exists x (\forall y [P(x, y) \vee \exists z Q(z)])$. Then $\phi^\#$ is simply $(\forall y [P(x, y) \vee Q(z)])$. Note that x and z, which are existentially bound in ϕ, occur free in $\phi^{\#}$. Now consider the constant a and the term $f(y)$, where y is the same variable as in ϕ and $\phi^{\#}$. Then a and $f(y)$ are admissible for x and z, respectively, and $(\phi^{\#})^{xz}_{a f(y)} = \forall y [P(a, y) \vee Q(f(y))]$. Note that for the substitution to work as intended, we need the assumption that no variable within ϕ is bound by an existentially quantifier more than once.

We now have the machinery in place, which allows us to replace the old semantic rule for explicit belief by a new one, which treats existential quantifiers in a much more restricted way.

5'. Let $\vec{x}$ be a sequence of the existentially quantified variables in ϕ.
$e, s \models_T B\phi$ iff for some admissible $\vec{t}$ and for all $s' \in e$, $e, s' \models_T (\phi^{\#})^{\vec{x}}_{\vec{t}}$;
$e, s \models_F B\phi$ iff $e, s \not\models_T B\phi$.

In the new logic, which we call $\mathcal{BL}^q$, truth, validity, and satisfiability are defined in a way completely analogous to the propositional case. Given a world state w, an epistemic state e, and a sentence α, we say that α is *true* at w and e if $e, w \models_T \alpha$ and *false* if $e, w \not\models_T \alpha$. α is valid ($\models \alpha$) if α is true in all world states w and epistemic states e.

12.2.1 Some properties

We do not have a complete proof theory for the logic $\mathcal{BL}^q$. As mentioned before, while having a proof provides a nice independent characterization of the valid sentences, it is not essential for our purposes. We should, however, find a syntactic characterization of belief implication, since our main objective is to show that explicit belief has a computational advantage over implicit belief. We will take this up in the next subsection. Here we outline the main similarities and differences between explicit and implicit belief.

We begin by observing that first-order explicit belief indeed eliminates existential generalization.

Theorem 12.2.7: *Let P be a unary predicate symbol and a and b distinct constants. Then*

$$\not\models B(P(a) \vee P(b)) \supset B\exists x P(x).$$

Proof: Let $e = \{s_1, s_2\}$, where $s_1 \models_T P(^\#1)$ and $s_2 \models_T P(^\#2)$ for distinct standard names $^\#1$ and $^\#2$. No other atoms are supported in s_1 or s_2. In s_1, let the constant a co-refer with $^\#1$ and all other primitive terms with $^\#2$. In s_2, let b co-refer with $^\#2$ and all other primitive terms with $^\#1$. Since $s_1 \models_T P(a)$ and $s_2 \models_T P(b)$, $e \models_T B(P(a) \vee P(b))$. Also, the admissible substitutions for x in $B\exists x P(x)$ are precisely the closed terms. It is easy to see that for every closed term t, either $s_1 \not\models_T P(t)$ or $s_2 \not\models_T P(t)$. Therefore, $e \models_T \neg B\exists x P(x)$. ∎

As pointed out in the introduction to this section, the reason existential generalization fails is the *constructive* interpretation of existential quantification. In other words, an existential is believed only if the agent is able to produce a witness (in the form of an admissible term). In fact, we immediately obtain

Theorem 12.2.8: *For any epistemic state e, if $e \models_T B\exists x\phi$ then $e \models_T B\phi_t^x$ for some t.*

Note, however, that $(B\exists x\phi \supset B\phi_t^x)$ is not valid for any t, in general, since the admissible term varies with e.

When modifying the semantics of a logical symbol, one has to be careful to examine possible side-effects on other operators and their properties. Since we have modified quantification, we start this investigation by reexamining quantifying-in. It turns out that except for a restriction in the Barcan formula, the story is the same as for implicit belief.

Theorem 12.2.9: *For any objective ϕ, any standard name n,*

1. $\models \exists x B\phi \supset B\exists x\phi$
2. $\models B\phi_n^x \supset \exists x B\phi$
3. $\models \forall x B\phi \supset B\forall x\phi$ *if ϕ is existential-free*
4. $\models B\forall x\phi \supset \forall x B\phi$

Proof:

1. Let $e \models_T \exists x B\phi$. Then $e \models_T B\phi_n^x$ for some $n \in N$ and, consequently, for all $s \in e$, $s \models_T [(\phi_n^x)^\#]_{t_1}^{x_1} \cdots {}_{t_k}^{x_k}$, where t_i is an admissible substitution for the existential x_i in ϕ_n^x. But n is also an admissible substitution for x, which implies $s \models_T (\phi^\#)_{n t_1}^{x x_1} \cdots {}_{t_k}^{x_k}$ and, therefore, $e \models_T B\exists x\phi$.

2. Follows immediately from the meaning of existential quantification.

3. Let $e \models_T \forall x B\phi$. Then, since ϕ is assumed to be existential-free, for all $n \in N$ and for all $s \in e$, $s \models_T \phi_n^x$. Therefore, for all $s \in e$ and for all $n \in N$, $s \models_T \phi_n^x$, which implies $e \models_T B\forall x\phi$.

4. Let $e \models_T B\forall x\phi$. Then there are admissible $t_1, \ldots, t_k$ for the existentials $x_1, \ldots, x_k$ of ϕ such that for all $s \in e$, $s \models_T \forall x(\phi^\#)_{t_1}^{x_1} \cdots {}_{t_k}^{x_k}$. This implies that for all $s \in e$ and for all $n \in N$, $s \models_T ((\phi^\#)_{t_1}^{x_1} \cdots {}_{t_k}^{x_k})_n^x$, which can be rewritten as $s \models_T [(\phi_n^x)^\#]_{t'_1}^{x_1} \cdots {}_{t'_k}^{x_k}$, where $t'_i = (t_i)_n^x$. Therefore, $e \models_T B\phi_n^x$ for all $n \in N$ because the $(t_i)_n^x$ are admissible. $e \models_T \forall x B\phi$ follows now immediately. ∎

The following example demonstrates that the proviso in (3) is indeed necessary.

Example 12.2.10: Let $n_1, n_2, \ldots$ be an arbitrary ordering of the standard names, $\phi = \exists y P(x, y)$, and $e = \{s\}$, where $s \models_T P(n_i, n_{i+1})$ and s has no true- or false-support for any other primitive formula. Also every primitive term t is assumed to co-refer with n_1.

Certainly $e \models_T \forall x \mathbf{B} \exists y P(x, y)$, since for all $n_i \in N$, $e \models_T \mathbf{B} P(n_i, n_{i+1})$. However, there is no admissible term t we could substitute for y such that $e \models_T \mathbf{B} \forall x P(x, t)$. There are three cases to consider. If $t = n_i$ for a fixed standard name n_i, then $s \not\models_T P(n_i, n_i)$. The same holds if $t = x$. Otherwise t is a term which contains at least one function symbol and $t^x_{n_i}$ co-refers with n_1 for any n_i and $s \not\models_T P(n_1, n_1)$. So no matter which t we try, $e \not\models_T \mathbf{B} \forall x P(x, t)$.

There is one more important question to consider, and that is the conversion into normal form of formulas. It turns out that all conversion rules from first order logic are valid in this logic except that we need provisos in the following three cases.

Lemma 12.2.11: *Let ϕ, ψ, and η be objective and, in addition, let ϕ be existential-free. Let ϕ' be like ϕ except that the names of bound variables are distinct from those in ϕ, ψ, and η.*

1. *$\models \mathbf{B}((\phi \vee \psi) \wedge (\phi' \vee \eta)) \supset \mathbf{B}(\phi \vee (\psi \wedge \eta))$.*
2. *$\models \mathbf{B}((\phi \wedge \psi) \vee (\phi' \wedge \eta)) \supset \mathbf{B}(\phi \wedge (\psi \vee \eta))$.*
3. *Let x not occur free in ϕ. Then*
   $\models \mathbf{B} \forall x(\psi \vee \phi) \supset \mathbf{B}((\forall x \psi) \vee \phi)$.

We leave the proof as an exercise.

The following examples show that the proviso is indeed necessary in case (1) and (3). (2) is left as an exercise.

Example 12.2.12: The following sentences are satisfiable:

1. $\mathbf{B}((\exists x P(x) \vee Q(c)) \wedge (\exists y P(y) \vee R(d))) \wedge \neg \mathbf{B}(\exists x P(x) \vee (Q(c) \wedge R(d)))$;
2. $\mathbf{B} \forall x(P(x) \vee \exists y Q(y)) \wedge \neg \mathbf{B}[(\forall x P(x)) \vee \exists y Q(y)]$.

Proof: 1. Let $e = \{s_1, s_2\}$ such that s_1 only supports the truth of $P(a)$ and $R(d)$, while s_2 only supports the truth of $P(b)$ and $Q(c)$, where a, b, c, d are distinct standard names. Assume that all primitive terms co-refer with standard names other than a, b, c, d in both situations. One can easily verify that $e \models_T \mathbf{B}((P(a) \vee Q(c)) \wedge (P(b) \vee R(d)))$, from which $e \models_T \mathbf{B}((\exists x P(x) \vee Q(c)) \wedge (\exists y P(y) \vee R(d)))$ follows because a is an admissible substitution for x and b for y. However, $e \models_F \mathbf{B}(\exists x P(x) \vee (Q(c) \wedge R(d)))$ because $s_1 \not\models_T Q(c)$, $s_2 \not\models_T R(d)$, and for every admissible term t, either $s_1 \not\models_T P(t)$ or $s_2 \not\models_T P(t)$.

2. Consider $e = \{s_1, s_2\}$, where

$s_1 \models_T P(^\#1), Q(f(^\#2)), P(^\#3), P(^\#4), \ldots$ and
$s_2 \models_T Q(f(^\#1)), P(^\#2), P(^\#3), P(^\#4), \ldots$

Assume further that $s_1[f(^\#1)] = n \neq m = s_2[f(^\#2)]$. Also, let no other primitive term co-refer with either n or m. Then $e \models_T \boldsymbol{B} \forall x (P(x) \lor Q(f(x)))$, from which $e \models_T \boldsymbol{B} \forall x (P(x) \lor \exists y Q(y))$ follows because $f(x)$ is an admissible substitution for y. However, we have that $s_1 \not\models_T P(^\#2)$ and $s_2 \not\models_T P(^\#1)$ and thus neither situation supports $\forall x P(x)$. Also, for any closed term t, either $s_1 \not\models_T Q(t)$ or $s_2 \not\models_T Q(t)$ and thus $e \not\models_T \boldsymbol{B}(\forall x P(x)) \lor (\exists y Q(y))$. ∎

One of the unfortunate consequences of these counterexamples is that an arbitrary sentence cannot in general be converted into an equivalent prenex normal form. For example, it is possible to believe explicitly that $\forall u \exists v \forall x \exists y [P(u, v) \lor Q(x, y)]$ yet fail to believe $[\forall u \exists v P(u, v)] \lor [\forall x \exists y Q(x, y)]$. One advantage of sentences in normal form is that they are often easier to deal with when it comes to defining syntactic derivations than arbitrary sentences. In fact, the Resolution method requires sentences to be in prenex conjunctive normal form. It turns out that not all is lost in this respect under our notion of explicit belief. In fact, the following section on computing belief implication shows that we are able to work with existential-free sentences alone, and normal form transformations work as usual on that class of sentences. For example:

Lemma 12.2.13: *If ϕ is an objective existential-free sentence and ϕ_{PCNF} is ϕ converted into prenex conjunctive normal form, then $\models B\phi \equiv B\phi_{PCNF}$.*

Proof: The proof is essentially the same as in classical logic. ∎

12.2.2 Deciding belief implication

In this subsection, we develop a complete syntactic characterization of belief implication from which a decision procedure follows. The main theorem of his section (Theorem 12.2.21) tells us, roughly, that the validity of sentences of the form $(B\phi \supset B\psi)$ can be determined by testing whether every clause of the CNF of ψ is subsumed by a clause of the CNF of ϕ for appropriate substitutions of terms for the variables in both sentences. Since we only need to test a finite number of substitutions, decidability obtains.

Since the technical details are fairly involved, those readers who would just like to get a rough sense of why decidability obtains are invited to skip right to the main theorem (page 217) and the discussion following it.

For those who want to read on, here is a very brief guide to the technical material which leads up to the main result: the idea is to gradually eliminate quantifiers from $(B\phi \supset B\psi)$

through Skolemization and other substitutions (Theorems 12.2.16, 12.2.17, and 12.2.19) so that, in the end, the problem reduces to testing for the subsumption of clauses (Theorem 12.2.20).

In the following, we will often use the following notation: Let α be a formula with free variables $x_1, \ldots, x_k$ and let $\vec{t} = \langle t_1, \ldots, t_k \rangle$ be a sequence of terms. We often write $\alpha[t_1, \ldots, t_k]$ or $\alpha[\vec{t}]$ instead of $\alpha_{t_1}^{x_1} \cdots {}_{t_k}^{x_k}$ if we do not care about which or how many variables are involved as long as all free variables are substituted by appropriate terms. (We also apply this convention when substituting the variables in a term.)

The first result is about Skolemizing the existentials of the left hand side of belief implications. To this end, we need the following technical definition and lemma.

Definition 12.2.14: Let $y_1, y_2, \ldots$ be an ordering of the variables. A <u>term map</u> τ is a function from function symbols into terms such that $\tau(f) = t$, where f is a k-ary function symbol and t is a term whose free variables are contained in $\{y_1, \ldots, y_k\}$. (Not all the y_i need to occur in t).

Next τ is extended to a mapping τ^* from terms into terms such that

$$\tau^*(t) = \begin{cases} n & \text{if } t \text{ is the standard name } n \\ x & \text{if } t \text{ is the variable } x \\ \tau(f)_{\tau^*(t_1)}^{y_1} \cdots {}_{\tau^*(t_k)}^{y_k} & \text{otherwise } [t = f(t_1, \ldots, t_k)] \end{cases}$$

We extend τ^* to apply to formulas in the obvious way: for any formula α, let $\tau^*(\alpha)$ be α with every term t in α replaced by $\tau^*(t)$. Let s be a situation. Then let s_τ be a situation obtained from s such that s and s_τ agree on all primitive formulas and $s_\tau[f(\vec{n})] = m$, where $s \models_T m = \tau(f)_{\vec{n}}^{\vec{y}}$.

Lemma 12.2.15:

1. Let s be a situation and τ a term map. Then for any closed term t,
   $s_\tau(t) = s(\tau^*(t))$.
2. Let τ be a term map and e a set of situations. Let $e_\tau = \{s_\tau \mid s \in e\}$.
   If τ^* is surjective, then for any sentence α,

   $e_\tau, s_\tau \models_T \alpha$ iff $e, s \models_T \tau^*(\alpha)$

   $e_\tau, s_\tau \models_F \alpha$ iff $e, s \models_F \tau^*(\alpha)$

Proof: The proof is by induction. Since it is not very illuminating, we omit it here. ∎

Theorem 12.2.16: Let ϕ and ψ be objective sentences. Let $\phi_{SK\exists}$ denote the sentence obtained from ϕ by replacing all occurrences of every existential x in ϕ by $f_x(U(x))$, where f_x is a function symbol occurring nowhere else in ϕ or ψ and $U(x)$ is a sequence of the universally quantified variables in whose scope x is bound. f_x is usually called a Skolem function. Then

$$\models (B\phi \supset B\psi) \text{ iff } \models (B\phi_{SK\exists} \supset B\psi)$$

Proof: Assume $\models (B\phi \supset B\psi)$ and let $e \models_\tau B\phi_{SK\exists}$. Then $e, s \models_\tau \phi_{SK\exists}$ for all $s \in e$, which can be rewritten as $e, s \models_\tau \phi^{\#}[f_{x_1}(U(x_1)), \ldots, f_{x_k}(U(x_k))]$. Since $f_{x_i}(U(x_i))$ is admissible for x_i, $e \models_\tau B\phi$ follows immediately and, by assumption, $e \models_\tau B\psi$.

Now assume $\models (B\phi_{SK\exists} \supset B\psi)$ and let $e \models_\tau B\phi$. In order to show that $e \models_\tau B\psi$, we first construct the following term map τ.

For every Skolem function f_x let $f_x^1, f_x^2, \ldots$ be an infinite sequence of new Skolem functions of the same arity as f_x. For convenience, we identify f_x with f_x^0. From $e \models_\tau B\phi$ we get $e, s \models_\tau \phi^{\#}[t_1, \ldots, t_k]$ for all $s \in e$, where the t_i are admissible terms substituted for the existentials in ϕ. Let $y_1, y_2, \ldots$ be an enumeration of the variables. Then define τ such that

$$\begin{aligned}
\tau(f_{x_i}) &= t_i{}_{y_1 \ldots y_n}^{z_1 \ldots z_n} & \text{for } 1 \leq i \leq k \text{ and } U(x_i) = \langle z_1, \ldots, z_n \rangle \\
\tau(f_{x_i}^j) &= f_{x_i}^{j-1}(y_1, \ldots, y_n) & \text{for } j > 0 \\
\tau(g) &= g(y_1, \ldots, y_m) & \text{where } g \text{ is any other } m\text{-ary function symbol}
\end{aligned}$$

Now we need to show that τ^* is surjective. We prove that every term t is an image of a term under τ^* by induction on the number of function symbols in t.

1. If $t = n$ or $t = x$ for a standard name n or variable x, $\tau^*(t) = t$ by definition of τ^*.
2. Let $t = f(\vec{n})$ be primitive.
   (a) If $f = f_x^i$, then, by construction of τ, $\tau^*(f_x^{i+1}(\vec{n})) = \tau(f_x^{i+1})[\tau^*(\vec{n})] = f_x^i(\vec{n})$.
   (b) Otherwise $\tau^*(f(\vec{n})) = f(\vec{n})$.
3. Let $t = f(t_1, \ldots, t_k)$ be a non-primitive term. By induction, there are variables $u_1, \ldots, u_k$ such that $t = f(\tau^*(u_1), \ldots, \tau^*(u_k))$.
   (a) If $f = f_x^i$, then $f(\tau^*(u_1), \ldots, \tau^*(u_k)) = \tau(f_x^{i+1})[\tau^*(u_1), \ldots, \tau^*(u_k)]$
       $= \tau^*(f_x^{i+1}(u_1, \ldots, u_k))$.
   (b) Otherwise $f(t_1, \ldots, t_k) = \tau^*(f(u_1, \ldots, u_k))$.

Given the above construction of the term map τ, let $e_\tau = \{s_\tau \mid s \in e\}$, where s_τ is defined as in Definition 12.2.14. The main proof of the if direction of the theorem proceeds now as follows. By assumption, $e \models_\tau B\phi$, that is, $e, s \models_\tau \phi^{\#}[t_1, \ldots, t_k]$ for all $s \in e$. By construction of τ it follows that $\tau^*(\phi_{SK\exists}) = \phi^{\#}[t_1, \ldots, t_k]$, since every Skolem function f_{x_i} in $\phi_{SK\exists}$ is mapped into t_i. Therefore, $e, s \models_\tau \tau^*(\phi_{SK\exists})$ for all $s \in e$ and, by Lemma 12.2.15, for all $s_\tau \in e_\tau$, $e_\tau, s_\tau \models_\tau \phi_{SK\exists}$. Hence $e_\tau \models_\tau B\phi_{SK\exists}$. Since $\models (B\phi_{SK\exists} \supset B\psi)$ by assumption,

$e_\tau \models_\text{T} B\psi$ and, by Lemma 12.2.15, $e \models_\text{T} \tau^*(B\psi)$. However, $\tau^*(B\psi) = B\psi$ because ψ contains none of the f_x^j. Therefore, $e \models_\text{T} B\psi$ and we are done. ∎

Being able to Skolemize the left hand side of belief implications is convenient, yet it does not shed any light on the computational advantage of explicit belief, since, after all, Skolemization works just as well in the case of classical logical implication.

The next theorem, however, indeed reveals a crucial difference. It says that, if an explicit belief ψ is implied at all by believing an existential-free sentence ϕ, then ψ is believed with all existentials replaced by appropriate terms. Note that this is not the case for logical implication. For example, $\models [(P(a) \vee P(b)) \supset \exists x P(x)]$, yet no term can be replaced for x that would allow the implication to go through.

Theorem 12.2.17: *Let ϕ and ψ be objective sentences with ϕ existential-free.*

$\models (B\phi \supset B\psi)$ *iff* $\models (B\phi \supset B\psi^\#[\vec{t}])$ *for some admissible* $\vec{t}$.

Proof: To prove the only-if direction, assume $\models (B\phi \supset B\psi)$. We show that we can choose an admissible $\vec{t}$ for the existentials in ψ independent of the set of situations in question. For that reason, let us consider $e_\text{max} = \{s \mid s \models_\text{T} \phi\}$. Since ϕ is existential-free, that is, $\phi = \phi^\#$, $e_\text{max} \models_\text{T} B\phi$. Then, by assumption, $e_\text{max} \models_\text{T} B\psi$, i.e., for some admissible $\vec{t}$ and for all $s \in e_\text{max}$, $s \models_\text{T} \psi^\#[\vec{t}]$. Now consider any set of situations e such that $e \models_\text{T} B\phi$. Since ϕ is existential-free, $s \models_\text{T} \phi$ for all $s \in e$ and, therefore, $e \subseteq e_\text{max}$. Thus for all $s \in e$, $s \models_\text{T} \psi^\#[\vec{t}]$, where $\vec{t}$ are the same admissible terms as in the case of e_max. Finally, since $\psi^\#[\vec{t}]$ is existential-free, $e \models_\text{T} B\psi^\#[\vec{t}]$.

Conversely, assume that $\models (B\phi \supset B\psi^\#[\vec{t}])$ for some admissible $\vec{t}$ and let $e \models_\text{T} B\phi$. Then $e \models_\text{T} B\psi^\#[\vec{t}]$, i.e., for all $s \in e$, $s \models_\text{T} \psi^\#[\vec{t}]$ (since $\psi^\#[\vec{t}]$ is existential-free), from which $e \models_\text{T} B\psi$ follows immediately. ∎

The main point of the previous theorem is that the problem of determining whether a belief implication with an arbitrary right hand side holds reduces to the problem of determining whether certain belief implications hold whose right hand side is existential-free.

To reduce the problem even further, let us assume that ψ in $(B\phi \supset B\psi)$ is in fact existential-free. By Lemma 12.2.13, ψ can now be transformed into prenex conjunctive normal form (ψ_PCNF) such that $\models (B\phi \supset B\psi)$ iff $\models (B\phi \supset B\psi_\text{PCNF})$. Next, using Theorem 12.2.9, we can successively move all the universal quantifiers in ψ_PCNF outside the B operator. For example, let $\forall x$ be the left most quantifier and let ψ'_PCNF be ψ_PCNF with $\forall x$ removed. Then $\models (B\phi \supset B\psi_\text{PCNF})$ iff $\models (B\phi \supset \forall x B\psi'_\text{PCNF})$.

With these transformations in mind, the following theorem tells us that universal quantifiers on the right hand side of belief implications can be removed altogether by replacing

Avoiding Logical Omniscience 215

them with standard names occurring nowhere else.

First, we need a lemma about renaming standard names consistently in situations and sentences.

Lemma 12.2.18: *Let $v : N \longrightarrow N$ be a surjection over the standard names. v is extended to apply to terms and formulas in the usual way. Let $\dot{N} \subseteq N$ such that v restricted to $\dot{N}$, which we write as $\dot{v}$, is a bijection from $\dot{N}$ into N. (Such a $\dot{N}$ always exists because v is assumed to be surjective.)*

For any situation s define a situation s_v such that $s_v[P(\vec{n})] = s[P(v(\vec{n}))]$ for every primitive formula $P(\vec{n})$ and $s_v[t] = \dot{v}^{-1}(s[(v(t))])$ for every primitive term t. Finally, given any set of situations e, let $e_v = \{s_v \mid s \in e\}$. Then for any set of situations e, situation s, and sentence α,

$$e_v, s_v \models_T \alpha \text{ iff } e, s \models_T v(\alpha)$$
$$e_v, s_v \models_F \alpha \text{ iff } e, s \models_F v(\alpha)$$

Proof: We omit the proof since it is neither very hard nor illuminating. ∎

Theorem 12.2.19: *Let n^* be a standard name not occurring in ϕ or ψ.*

$$\models (B\phi \supset \forall x B\psi) \text{ iff } \models (B\phi \supset B\psi^x_{n^*})$$

Proof: Assume $\models (B\phi \supset \forall x B\psi)$ and $e \models_T B\phi$ for some set of situations e. By assumption, $e \models_T \forall x B\psi$. Therefore, $e \models_T B\psi^x_n$ for all $n \in N$ and, in particular, $e \models_T B\psi^x_{n^*}$.

Conversely, assume $\models (B\phi \supset B\psi^x_{n^*})$ and $\not\models (B\phi \supset \forall x B\psi)$. Thus for some e, $e \models_T B\phi$ and $e \not\models_T B\psi^x_{n'}$ for some $n' \in N$. Let v be a surjection over the standard names such that $v(n) = n$ for all standard names n occurring in ϕ or ψ and $v(n^*) = n'$. Let e_v and $v(\phi)$ be defined as in Lemma 12.2.18. By the definition of v, $v(B\phi) = B\phi$ and therefore, $e \models_T v(B\phi)$. Thus, by Lemma 12.2.18, $e_v \models_T B\phi$ and, by assumption, $e_v \models_T B\psi^x_{n^*}$. Finally, by Lemma 12.2.18, $e \models_T v(B\psi^x_{n^*})$, which is the same as $e \models_T B\psi^x_{n'}$ contradicting our assumption that $e \not\models_T B\psi^x_{n'}$. ∎

Given this reduction, if we also assume that ϕ is existential-free, it can now be shown that belief implication reduces to testing for the subsumption of clauses:

Theorem 12.2.20: *Let $\phi = \forall \vec{x} \bigwedge \phi_i$ be an existential-free sentence in prenex conjunctive normal form. Let $\psi = \bigwedge \psi_j$ be a quantifier free sentence in conjunctive normal form. Then*

$$\models B\forall \vec{x} \bigwedge \phi_i \supset B \bigwedge \psi_j$$

iff for all ψ_j there is a clause ϕ_i and a substitution $\vec{t}$ of closed terms for $\vec{x}$ such that the literals contained in $\phi_i[\vec{t}]$ are a subset of the literals in ψ_j, denoted as

$$\phi_i[\vec{t}] \subseteq \psi_j$$

Proof: For the if-direction assume that for all ψ_j there are ϕ_i and closed terms $\vec{t}$ such that $\phi_i[\vec{t}] \subseteq \psi_j$. Let $e \models_T \boldsymbol{B}\forall \vec{x} \bigwedge \phi_i$. Then $e, s \models_T (\bigwedge \phi_i)[\vec{t}]$ for all $s \in e$ and for all closed terms $\vec{t}$. Since, by assumption, for every ψ_j there are ϕ_i and $\vec{t}$ such that $\phi_i[\vec{t}] \subseteq \psi_j$, it follows that $e, s \models_T \psi_j$ for all $s \in e$ and for all ψ_j and thus $e \models_T \boldsymbol{B}\psi$.

Conversely, assume that for some ψ_j and all ϕ_i and $\vec{t}$, $\phi_i[\vec{t}] \not\subseteq \psi_j$. It needs to be shown that $\not\models (\boldsymbol{B}\forall \vec{x} \bigwedge \phi_i \supset \boldsymbol{B}\psi_j)$, which immediately implies that $\not\models (\boldsymbol{B}\forall \vec{x} \bigwedge \phi_i \supset \boldsymbol{B} \bigwedge \psi_j)$. This is done by constructing a situation s such that $s \models_T \forall \vec{x} \bigwedge \phi_i$ and $s \not\models_T \psi_j$.

The idea is as follows. For every substitution $\vec{t}$ of closed terms for $\vec{x}$ and for every ϕ_i, choose a literal that is in $\phi_i[\vec{t}]$ and not in ψ_j. Let s have true-support only for these literals. Then $s \models_T \forall \vec{x} \bigwedge \phi_i$ and $s \not\models_T \psi_j$, from which $\not\models (\boldsymbol{B}\forall \vec{x} \bigwedge \phi_i \supset \boldsymbol{B} \bigwedge \psi_j)$ follows.

For the construction of an appropriate s it is useful to choose a co-reference relation with certain desirable properties. Let $n_1, n_2, \ldots$ be an ordering of the standard names such that $n_1, \ldots, n_k$ are precisely the standard names occurring in ϕ and ψ. Let $t_1, t_2, \ldots$ be an ordering of the primitive terms. Then we would like the following to hold:

1. No two primitive terms co-refer.

2. No primitive term co-refers with any standard name in $\{n_1, \ldots, n_k\}$.

3. For any closed term t there is a t' such that t and t' co-refer and none of the standard names in t' co-refers with any primitive term. Let us call such t' *minimal*.

4. Let t_1 and t_2 be closed terms such that none of the standard names occurring in t_1 or t_2 co-refers with a primitive term. If t_1 and t_2 co-refer, then $t_1 = t_2$.

The key idea is that we want to be able to convert any term into a canonical form, that is, a minimal term so that it becomes trivial to decide whether two terms co-refer (Property 4).

We can satisfy the above properties using the following assignment of the primitive terms t_i:

$$s[t_i] = n_{p^m},$$

where p is the $(i + k)$-th prime number, $m = max\{1, j_1, \ldots, j_l\}$, and $n_{j_1}, \ldots, n_{j_l}$ are the standard names occurring in t_i.

Note that this choice guarantees that if $s[t(n_1, \ldots, n_l)] = n$ then n comes later in the ordering of standard names than the n_i. Together with the fact that every natural number has a unique prime factorization, the proof that the above properties hold is not hard and we omit it here.

Now we are ready to define the true- and false-support of primitive formulas at s.

Avoiding Logical Omniscience   217

By assumption, for each substitution of terms $\vec{t}$ for $\vec{x}$ and each ϕ_i, there is a literal l in ϕ_i such that $l[\vec{t}]$ is in $\phi_i[\vec{t}]$ and not in ψ_j. For each minimal $\vec{t}$ and i, choose one such l and call it $\eta(\vec{t}, i)$. For each non-minimal $\vec{t}$ and i choose a minimal co-referring $\vec{t}'$ and let $\eta(\vec{t}, i) = \eta(\vec{t}', i)$. (Note that such minimal $\vec{t}'$ always exists according to Property 3 of the co-reference relation.) Now let

$$s \models_T P(\vec{n}) \quad \text{iff} \quad \text{for some } \vec{t} \text{ and } i, \eta(\vec{t}, i) = P(\vec{u}) \text{ and } s \models_T u_i[\vec{t}] = n_i.$$
$$s \models_F P(\vec{n}) \quad \text{iff} \quad \text{for some } \vec{t} \text{ and } i, \eta(\vec{t}, i) = \neg P(\vec{u}) \text{ and } s \models_T u_i[\vec{t}] = n_i.$$

Given such a situation s, we need to prove that $s \models_T \phi$ and $s \not\models_T \psi_j$.

To show $s \models_T \phi$, it suffices to establish that $s \models_T \phi_i[\vec{n}]$ for every i and standard names $\vec{n}$. Let $\eta(\vec{n}, i) = P(\vec{u})$, where $P(\vec{u})$ occurs in ϕ_i. (The case where $\eta(\vec{n}, i) = \neg P(\vec{u})$ is completely symmetric and is omitted.) From the definition of s it follows that $s \models_T P(\vec{u})[\vec{n}]$ and, therefore, $s \models_T \phi_i[\vec{n}]$.

To show $s \not\models_T \psi_j$, let us assume, to the contrary, that $s \models_T \psi_j$. Then $s \models_T P(\vec{u})$, where $P(\vec{u})$ is a literal in ψ_j. (Again, we omit the symmetric case of $\neg P(\vec{u})$.) By definition of s, there must be an i and a $\vec{t}$ such that $\eta(\vec{t}, i) = P(\vec{v})$ ($P(\vec{v}) \in \phi_i$) and $s \models_T u_i = v_i[\vec{t}]$. Also, there must be a minimal sequence of terms $\vec{t}'$ such that t_i' co-refers with t_i for all i and $P(\vec{v})[\vec{t}'] \notin \psi_j$. $v_i[\vec{t}']$ is minimal and co-refers with u_i. By Property 4 of the co-reference relation, $P(\vec{v})[\vec{t}'] = P(\vec{u})$, contradicting the assumption that $P(\vec{u}) \in \psi_j$. ∎

With Theorem 12.2.20, we are finally in a position to prove the main result of this section, which gives a complete syntactic characterization of belief implication.

Theorem 12.2.21: *Let ϕ and ψ be objective sentences. Let $\vec{z}$ be a sequence of the universally quantified variables in ϕ, $\vec{x}$ a sequence of the existentials in ψ, and $\vec{y}$ a sequence of the universally quantified variables in ψ. Let $\phi_{SK\exists}$ be ϕ with all existentials Skolemized and let $\forall \vec{z} \bigwedge \phi_i$ be $\phi_{SK\exists}$ in prenex conjunctive normal form. Let $\bigwedge \psi_j$ be the matrix of ψ in conjunctive normal form. Finally, let $\vec{n}^*$ be a substitution for $\vec{y}$ consisting of distinct standard names not occurring in ϕ or ψ.*

Then $\models (\mathbf{B}\phi \supset \mathbf{B}\psi)$ iff there are admissible substitutions $\vec{t}$ for $\vec{x}$ not containing any standard names of $\vec{n}^*$ such that for every ψ_j there are ϕ_i and closed terms $\vec{u}$ for $\vec{z}$ such that $\phi_i{}^{\vec{z}}_{\vec{u}} \subseteq (\psi_j{}^{\vec{x}}_{\vec{t}})^{\vec{y}}_{\vec{n}^*}$.

Proof: We begin by proving a slightly weaker result in that we allow the standard-name substitutions for $\vec{y}$ to vary, depending on the choice of admissible terms for $\vec{x}$.

$\models (\boldsymbol{B}\phi \supset \boldsymbol{B}\psi)$

iff $\models (\boldsymbol{B}\phi_{\text{SK}\exists} \supset \boldsymbol{B}\psi)$ Skolemization (Theorem 12.2.16)

iff $\models (\boldsymbol{B}\phi_{\text{SK}\exists} \supset (\boldsymbol{B}\psi^{\#})_{\vec{t}}^{\vec{x}})$ for admissible $\vec{t}$ substituting for existentials $\vec{x}$ (Theorem 12.2.17)

iff $\models (\boldsymbol{B}(\forall \vec{z} \bigwedge \phi_i) \supset \boldsymbol{B}(\forall \vec{y} \bigwedge \psi_j)_{\vec{t}}^{\vec{x}})$ conversion of existential-free formulas into PCNF (Lemma 12.2.13)

iff $\models (\boldsymbol{B}(\forall \vec{z} \bigwedge \phi_i) \supset (\forall \vec{y} \boldsymbol{B} \bigwedge \psi_j)_{\vec{t}}^{\vec{x}})$ The Barcan formula is valid for for existential-free sentences (Theorem 12.2.9)

iff $\models (\boldsymbol{B}(\forall \vec{z} \bigwedge \phi_i) \supset (\boldsymbol{B}(\bigwedge (\psi_j)_{\vec{t}}^{\vec{x}})_{\vec{n}'}^{\vec{y}}))$ where $\vec{n}'$ are distinct new standard names not occurring in ϕ, ψ, and $\vec{t}$ (Theorem 12.2.19)

iff for every $((\psi_j)_{\vec{t}}^{\vec{x}})_{\vec{n}'}^{\vec{y}}$ there are $\vec{u}$ and ϕ_i such that $(\phi_i)_{\vec{u}}^{\vec{z}} \subseteq ((\psi_j)_{\vec{t}}^{\vec{x}})_{\vec{n}'}^{\vec{y}}$ by Theorem 12.2.20.

The if direction of the theorem follows immediately from the above argument because $\vec{n}^*$ satisfies the restrictions imposed on $\vec{n}'$. To prove the only-if direction it suffices to prove that if $\models \boldsymbol{B}(\forall \vec{z} \bigwedge \phi_i) \supset (\boldsymbol{B}(\bigwedge \psi_j{}_{\vec{t}}^{\vec{x}})_{\vec{n}'}^{\vec{y}})$ for admissible $\vec{t}$ and $\vec{n}'$ not occurring in ϕ, ψ, or $\vec{t}$, then there are admissible $\vec{t}'$ such that $\models \boldsymbol{B}(\forall \vec{z} \bigwedge \phi_i) \supset (\boldsymbol{B}(\bigwedge \psi_j{}_{\vec{t}'}^{\vec{x}})_{\vec{n}^*}^{\vec{y}})$ and $\vec{t}'$ contains none of the $\vec{n}^*$.

Let $\models \boldsymbol{B}(\forall \vec{z} \bigwedge \phi_i) \supset (\boldsymbol{B}(\bigwedge \psi_j{}_{\vec{t}}^{\vec{x}})_{\vec{n}'}^{\vec{y}})$. The $\vec{t}'$ can now be constructed as follows: define a bijection ν over standard names that maps the i-th element of $\vec{n}'$ into the i-th element of $\vec{n}^*$ for all i. Note that this is possible since $\vec{n}^*$ and $\vec{n}'$ are both assumed to be sequences of *distinct* standard names. Also, let ν map all standard names occurring in $\bigwedge \phi_i$ and $\bigwedge \psi_j$ onto themselves. Then $\vec{t}'$ can be taken to be $\vec{t}$ with all standard names n replaced by $\nu(n)$. Notice that $\vec{t}'$ is admissible and it contains no occurrences of $\vec{n}^*$ because of the way ν is defined ($\vec{t}'$ could only contain a member of $\vec{n}^*$ if $\vec{t}$ contained a member of $\vec{n}'$, which has been ruled out by assumption). Finally, since a bijective ν is also surjective, we can apply Lemma 12.2.18 and $\models \boldsymbol{B}(\forall \vec{z} \bigwedge \phi_i) \supset (\boldsymbol{B}(\bigwedge \psi_j{}_{\vec{t}'}^{\vec{x}})_{\vec{n}^*}^{\vec{y}})$ follows. This completes the proof of the theorem. ∎

Given this theorem it is not hard to construct an algorithm that computes belief implication and is guaranteed to always halt. The algorithm makes use of the well-known fact that, if a set of logical expressions can be made identical (unified) by substituting some of the variables by other terms, then there is always a *most general unifier* from which all other unifiers can be obtained by further substitutions. Most importantly, most general unifiers are unique up to renaming of variables and they can be computed in linear time.

More formally, for every ϕ and ψ_j let Φ_{ij} be the set of most general substitutions such that $\phi_i \theta \subseteq \psi_j \theta$. Without loss of generality we assume that any two θ and θ' in Φ_{ij} share at most variables from $\vec{x}$ and $\vec{y}$. (Possible occurrences of $\vec{z}$ in a substitution can always be replaced by variables occurring nowhere else.) Each Φ_{ij} has cardinality at most $|\psi_j|^{|\phi_i|}$. Let $\Phi_j = \bigcup_i \Phi_{ij}$. Unifying substitutions for $\vec{x}$ that work for all ψ_j (namely admissible terms) exist iff there are θ_j from each Φ_j which unify and for which the following condition hold:

1. Each y from $\vec{y}$ is substituted by y (so that we can replace y by a standard name occurring nowhere else).

2. Each substitution of x is admissible, that is, contains at most those variables from $\vec{y}$ in whose scope x is bound in the original sentence.

So far we only know that the problem is decidable. We still have not answered the question whether the problem is computable with reasonable effort. It turns out that the algorithm has a worst case running time exponential in $|\bigwedge \phi_i| \times |\bigwedge \psi_j|$, where $\bigwedge \phi_i$ and $\bigwedge \psi_j$ are the matrices of the prenex conjunctive normal forms of ϕ and ψ, respectively. Furthermore, there cannot be a polynomial time algorithm unless P = NP because t-entailment, or belief implication for that matter, includes as a special case *theta-subsumption*,[8] which is known to be NP-complete. However, for the following restrictions our algorithm performs much better:

1. the knowledge base is in conjunctive normal form or convertible to conjunctive normal form without significantly increasing its size,

2. clauses in the conjunctive normal form version of the knowledge base are short (size limited by a small constant),

3. the knowledge base has many different predicates, and only a small number, logarithmic in the size of the knowledge base, of knowledge base clauses theta-subsumes a given clause, and

4. queries are small, logarithmic in the size of the knowledge base, even when converted to conjunctive normal form.

Under these conditions, the running time is on the order of $k \times k^{loglogk}$, where k is the size of the knowledge base. Thus computing belief implication has an acceptable running time as long as $loglogk$ is small, say, ≤ 3. Note also that classical logical implication remains undecidable under the above four restrictions.

---

8 Theta-subsumption is equivalent to the following form of belief-implication. Let ϕ and ψ be clauses with free variables $\vec{x}$ and $\vec{y}$, respectively. Then ϕ theta-subsumes ψ iff $\models [\boldsymbol{B}(\forall \vec{x} \phi) \supset \boldsymbol{B}(\exists \vec{y} \psi)]$.

12.3 Bibliographic notes

The propositional version of explicit belief was first introduced in [81]. Its first-order extension was presented in [65] and later in [70]. The logical omniscience problem was first discussed by Hintikka [50], who also proposed a solution, however, without considering issues of complexity [51]. A description of tautological entailment can be found in [2], and its four-valued semantics is due to Dunn [27]. In [33], an alternative account of the four-valued semantics using classical world states and a non-standard interpretation of negation is considered, an idea originally proposed in [128, 127]. Frisch [36] considers a 3-valued variant of situations (see also Exercise 3 below). These were also used in [85] to define a tractable form of abduction. The naive extension of the semantics to the first-order case, which we dismissed for computational reasons, relates to first-order tautological entailment [2] the same way as the propositional version relates to tautological entailment. The undecidability result (Theorem 12.2.1) was proved in [1], cast in terms of first-order tautological entailment. The notion of first-order belief implication, which we finally settled on, is closely related to a form of first-order entailment developed by Patel-Schneider [115], which he calls t-entailment. A variant of t-entailment is also used in [35]. Just like explicit belief its semantics is based on sets of four-valued situations. The main difference between the semantics of t-entailment and explicit belief is that disjunction receives a nonstandard interpretation under t-entailment while the rule for existential quantification remains the usual. Nevertheless, as shown in [70], both amount to the same thing in that $\models (B\phi \supset B\psi)$ iff ϕ t-entails ψ. In fact, the algorithm we presented above to determine belief implication is only a slight variant of one Patel-Schneider proposed for t-entailment. He points out the connection to theta-subsumption, and the analysis showing that belief implication or t-entailment has a reasonable complexity under certain assumptions is also due to Patel-Schneider. Theta-subsumption itself is described in [39]. While the benefits of using explicit belief over t-entailment are relatively minor when it comes to non-nested beliefs, the pay-off will become apparent in the next chapter, where we cover introspection together with *de re* and *de dicto* belief. Since t-entailment is defined as a meta-logical relation between sentences, it is far from obvious how it could be adapted to handle these extensions.

12.4 Where do we go from here?

Since we are not done yet with our exploration of limited reasoning, we defer a discussion of open issues to the next chapter.

12.5 Exercises

1. Show that $(BKB \supset K\phi)$ is valid iff KB logically implies ϕ for propositional objective KB and ϕ.
2. Show that $(KKB \supset B\phi)$ is not valid for any objective KB and ϕ.
3. Consider the following variant of propositional situations: a 3-valued situation s is a mapping which assigns each propositional variable a *non-empty* subset of $\{0, 1\}$. In other words, we rule out the case where a variable has neither true- nor false-support. It is easily seen that an agent whose epistemic state consists of 3-valued situations believes all tautologies. Show that Theorem 12.1.7 still holds.
   Hint: Modify the algorithm in the proof of that theorem.
4. Prove Lemma 12.2.11.
5. Give a counterexample for Lemma 12.2.11, part 2, if the proviso is dropped.
6. Show that the following sentence is not valid:
   $$B\forall u \exists v \forall x \exists y [P(u, v) \lor Q(x, y)] \supset B([\forall u \exists v P(u, v)] \lor [\forall x \exists y Q(x, y)]).$$
7. Consider the sentences $\phi = \forall x[P(f(x)) \lor Q(g(x))]$ and $\psi = \forall x \exists y(P(y) \lor Q(y))$. While $(\phi \supset \psi)$ is valid, $(B\phi \supset B\psi)$ is not, given our nonstandard interpretation of existentials within beliefs.

    (a) Find a sentence ψ' which is logically equivalent to ψ such that $(B\phi \supset B\psi')$ is valid after all.

    (b) Can you think of a systematic transformation of sentences containing existential quantifiers which would allow us to recover some of the existential generalizations implicit in a knowledge base as in the above example?

13 The logic $\mathcal{EOL}$

Now that we have a basic understanding of explicit belief in the first-order case, we want to extend $\mathcal{BL}^q$ to a variant of $\mathcal{OL}$ called $\mathcal{EOL}$, which uses explicit instead of implicit belief and otherwise includes nested beliefs with full introspection, a notion of explicitly only-knowing, and equality. There are at least two reasons for doing this: (a) to increase the expressive power of limited belief and (b) to be able to specify limited versions of **ASK** and **TELL** within the logic itself.

The language of $\mathcal{EOL}$ is essentially the same as $\mathcal{OL}$ except that K is replaced by B. We continue to use O but change its meaning to explicitly only-knowing. In a nutshell, $\mathcal{EOL}$ turns out to be very similar to $\mathcal{OL}$ as far as introspection is concerned and differs mainly in its deductive capabilities, which $\mathcal{EOL}$ inherits from $\mathcal{BL}^q$. For example, while

$$OTeacher(best\_friend(tina)) \supset B[\exists x Teacher(x) \land \neg BTeacher(x)]$$

is valid in $\mathcal{EOL}$,

$$O[Teacher(tina) \lor Teacher(tom)] \supset B[\exists x Teacher(x) \land \neg BTeacher(x)]$$

is not. The latter implication fails because, as in $\mathcal{BL}^q$, existential generalization from disjunctions is not a valid inference.

We will not give as detailed an analysis of $\mathcal{EOL}$ as we did for $\mathcal{OL}$, since this would take us too far afield. Instead, after introducing the semantics, we will briefly look at some of the key logical properties. We will then focus on a representation theorem similar to that for implicit belief and the specification of limited **ASK** and **TELL** routines.

13.1 Semantics

13.1.1 Equality

What should the meaning of = be in a logic based on a four-valued semantics? Certainly, if our aim is to reason efficiently about arbitrary equality expressions, we have little choice but to look for an appropriate nonstandard semantics. On the other hand, the classical interpretation of equality is arguably the most intuitive, which makes it hard to give up.[1] However, as we will see later the deciding factor for us not to tamper with = has been the fact that it allows for a representation theorem for explicit belief very similar to that of implicit belief (Chapter 7). The price we are willing to pay is that an unrestricted use of equality expressions results in undecidability. It remains an interesting open question

---

[1] Recall that = in classical logic is a *logical symbol* just like the logical connectives, and we have been careful so far to preserve their meaning as much as possible.

whether there are nonstandard interpretations of = which are both reasonably intuitive and attractive on computational grounds.

13.1.2 Nested explicit belief

Nested beliefs could be handled in a straight forward way if we were to confine ourselves to sentences without quantifying-in. For example, consider the following epistemic state

$$e = \{s \mid s \models_T Teach(father(tom), sara)\}.$$

Without any change to the semantics of non-nested **B** at all we immediately obtain beliefs about beliefs such as

1. $e \models_T \boldsymbol{BB}\exists x\,Teach(x, sara)$ or
2. $e \models_T \boldsymbol{B}\neg\boldsymbol{B}Teach(tom, sara)$.

To see that (1.) holds let $s \in e$. Then for every $s' \in e$, $s' \models_T Teach(father(tom), sara)$ $[= (\exists x\,Teach(x, sara))^{\#}[x/father(tom)]]$. Hence $e, s \models_T \boldsymbol{B}\exists x\,Teach(x, sara)$. Since this is true for all $s \in e$, we get $e \models_T \boldsymbol{BB}\exists x\,Teach(x, sara)$. Similarly, (2.) follows by negative introspection.

A closer look at the example also reveals why quantifying-in poses a problem. Intuitively, all that is known at e is that Tom's father is one of Sara's teachers. In particular, it is not known at e who Tom's father is. Hence, if we let

$$\gamma = \exists x[Teach(x, sara) \wedge \neg\boldsymbol{B}Teach(x, sara)],$$

we would expect $e \models_T \boldsymbol{B}\gamma$ to hold.

According to $\mathcal{BL}^q$ this means that we can find an admissible term t and substitute it for x so that $\gamma^{\#}[x/t]$ holds at every situation in e. Clearly, the only plausible candidate is $t = father(tom)$. However, this yields

$$\gamma^{\#}[x/t] = Teach(father(tom), sara) \wedge \neg\boldsymbol{B}Teach(father(tom), sara)],$$

which seems patently false because it means that e both believes and does not believe that Tom's father teaches Sara. A moment's thought reveals that it is simply a mistake to substitute t for the occurrence of x within the scope of **B** in γ. In fact, this is no different from $\mathcal{KL}$, where we saw that the axiom of specialization ($\forall x\alpha \supset \alpha[x/t]$) fails for essentially the same reason. Free occurrences of x within the scope of a modal operator may only be replaced by standard names. For implicit belief this does not pose a problem for the semantics since there variables are never substituted by anything but standard names.

For explicit belief it does, but the problem can be fixed without giving up on the idea of substituting existentials by admissible terms. To see what is involved, consider the sentence γ again. Not knowing who Tom's father is means that it is not known of the individual (standard name) co-referring with *father(tom)* that he is a teacher. The idea is to still substitute both occurrences of x by *father(tom)* but to somehow mark the second

The logic $\mathcal{EOL}$ 225

occurrence to remember that it gets its referent from the first occurrence of *father(tom)* outside the **B**. That way, the referent is fixed and does not vary across situations in *e*. This "marking" is achieved by introducing a new one-place function symbol $^\triangle$, which we write in postfix notation. Terms substituting for existentials inside a modal operator are then marked by appending $^\triangle$ to them. In our example, we obtain

$$\gamma' = \textit{Teach}(\textit{father}(\textit{tom}), \textit{sara}) \wedge \neg \textbf{\textit{B}}\textit{Teach}(\textit{father}(\textit{tom})^\triangle, \textit{sara}).$$

Now let *s* be any situation in *e* with $s[\textit{father}(\textit{tom})] = n$. To determine the meaning of ***B****Teach*(*father*(*tom*)$^\triangle$, *sara*) at *e* and *s* we first replace *father*(*tom*)$^\triangle$ by *n* and then evaluate ***B****Teach*(*n*, *sara*) in the usual way. Hence, for all *s*,

$$e, s \models_{\textrm{T}} \textbf{\textit{B}}\textit{Teach}(\textit{father}(\textit{tom}), \textit{sara})$$

yet $e, s \not\models_{\textrm{T}} \textbf{\textit{B}}\textit{Teach}(\textit{father}(\textit{tom})^\triangle, \textit{sara})$, which gives us the desired distinction between knowing that and knowing who in the case of explicit belief. Choosing $^\triangle$ to be a function symbol rather than a special marker is merely a convenience, which saves us the trouble of having to redefine the language.

For the most part, we are interested only in properties of formulas without $^\triangle$, that is, we care about $^\triangle$ only as far as it is used for the interpretation of belief. To distinguish formulas with occurrences of $^\triangle$ from those without, we call the latter *ordinary*.

The following somewhat technical definitions are preparations for our new, and final, formulation of explicit belief.

Definition 13.1.1: A logical connective, term, or formula contained in a formula α occurs at the *objective level* of α if it does not occur within the scope of a modal operator.

The notions of *existentially quantified variables, admissible terms, existential-free formulas*, and $\alpha^{\sharp}$, which we defined for objective formulas in the previous chapter, all carry over to non-objective formulas by considering only variables at the objective level of a formula.

Example 13.1.2: Consider $\alpha = \forall x_1 \exists x_2 P(x_1, x_2) \wedge \textbf{\textit{B}}\exists x_3 Q(x_1, x_2, x_3)$. Then we get $\alpha^{\sharp} = \forall x_1 P(x_1, x_2) \wedge \textbf{\textit{B}}\exists x_3 Q(x_1, x_2, x_3)$. Note that only the existential quantifiers at the objective level are removed. Everything inside a modal operator is left as is.

Definition 13.1.3: Let α be a formula with free variables $\vec{x} = \langle x_1, \ldots, x_k \rangle$. ($\alpha$ may contain other free variables as well.) Let $\vec{t} = \langle t_1, \ldots, t_k \rangle$ be a sequence of terms. $\alpha[\![\vec{x}/\vec{t}]\!]$ is α with every occurrence of x_i at the objective level replaced by t_i and every occurrence of x_i inside the scope of a modal operator replaced by $t_i^{\triangle}$.

Example 13.1.4: Let $\alpha = \forall x_1 P(x_1, x_2) \wedge \boldsymbol{B}\exists x_3 Q(x_1, x_2, x_3)$. Then $\alpha[\![x_2/f(x_1)]\!] = \forall x_1 P(x_1, f(x_1)) \wedge \boldsymbol{B}\exists x_3 Q(x_1, f(x_1)^\triangle, x_3)$.

Note the difference with $\alpha^{x_2}_{f(x_1)}$ which is $\forall x_1 P(x_1, f(x_1)) \wedge \boldsymbol{B}\exists x_3 Q(x_1, f(x_1), x_3)$.

Definition 13.1.5: Let α be a sentence, t a closed term, and s a situation. Then α^s is obtained from α by replacing every occurrence of $t^\triangle$ by the standard name $s(t)$.

Example 13.1.6: Let $\alpha = P(a^\triangle) \wedge \boldsymbol{B}(Q(f(a)^\triangle) \vee R(a^\triangle))$ and s a situation with $s[a] = {}^{\#}1$ and $s[f({}^{\#}1)] = {}^{\#}2$. Then $\alpha^s = P({}^{\#}1) \wedge \boldsymbol{B}(Q({}^{\#}2) \vee R({}^{\#}1))$

We then obtain the following semantic rules for the operator $\boldsymbol{B}$.

$e, s \models_T \boldsymbol{B}\alpha$ iff there are admissible $\vec{t}$ such that for all $s' \in e$, $e, s' \models_T \alpha^{s\sharp}[\![\vec{x}/\vec{t}]\!]$

$e, s \models_F \boldsymbol{B}\alpha$ iff $e, s \not\models_T \boldsymbol{B}\alpha$

Note the difference with our original semantic definition of $\boldsymbol{B}$ for $\mathcal{BL}^q$. Instead of simply substituting admissible terms for the existentials in α we need to first replace terms marked by $\triangle$ by their co-referring standard names and make sure that admissible terms which are substituted within the scope of a modal operator are marked by $\triangle$.

13.1.3 Explicitly only-knowing

To complete the semantics of $\mathcal{EOL}$ we still need to define the meaning of explicitly only-knowing. For existential-free sentences the definition of implicitly only-knowing carries over in a straightforward way, that is, a set of situations e (explicitly) only-knows a sentence α iff e is the largest set that believes α. The story is not quite as simple in the case of non-existential-free sentences. To see why consider $\alpha = \exists x P(x)$. What should it mean for e to only-know α? Since a necessary requirement is that e believes α, there must be some closed term a such that $e \models_T \boldsymbol{B}P(a)$. It may be tempting to let $e = \{s \mid s \models_T P(a)\}$ for some closed term a. But that seems too strong. For example, to say that all the KB believes is that someone is a teacher conveys a lot less information than all the KB believes is that Tom's father is a teacher.

Perhaps the easiest way around this problem is to require the terms that are used when only-knowing an existentially quantified sentence to convey no information about the world. In other words, the terms should behave like Skolem functions or internal identifiers. For that reason, we introduce a countably infinite set of new function symbols of every arity $\mathcal{F}_{SK}$. Members of $\mathcal{F}_{SK}$ are called *sk-functions*. Then we say that $e \models_T \boldsymbol{O}\exists x P(x)$ just in case $e = \{s \mid s \models_T P(a)\}$ for some constant $a \in \mathcal{F}_{SK}$. The idea that function symbols in $\mathcal{F}_{SK}$ do not convey information about the world can be dealt with pragmatically when us-

The logic $\mathcal{EOL}$

ing the logic to query and update a KB (see Section 13.6 below). In particular, queries and new information to be added to the KB should be restricted to sentences without function symbols from $\mathcal{F}_{SK}$. Those in $\mathcal{F}_{SK}$ are reserved for internal use by the KB.

Definition 13.1.7: Let α be a sentence with existentials $\vec{x} = \langle x_1, \ldots, x_k \rangle$ at the objective level. Let $U(x_i)$ be a sequence of the universally quantified variables in whose scope x_i is bound. Let $f_i \in \mathcal{F}_{SK}$ ($1 \leq i \leq k$) be pairwise distinct function symbols of arity $|U(x_i)|$ occurring nowhere else in α. Then $f_i(U(x_i))$ is called an <u>sk-term</u> (for x_i) and $\vec{t} = \langle f_1(U(x_1)), \ldots, f_k(U(x_k)) \rangle$ is called an sk-term substitution of $\vec{x}$ in α.

Note that every sk-term is also an *admissible* substitution.

Let α be a sentence, where $\vec{x}$ is a sequence of the existentially quantified variables at the objective level.

$e, s \models_T O\alpha$ iff there is an sk-term substitution $\vec{t}$ for $\vec{x}$ such that for all s',
$$s' \in e \text{ iff } e, s' \models_T \alpha^{s\sharp}[\![\vec{x}/\vec{t}_{SK}]\!]$$
$e, s \models_F O\alpha$ iff $e, s \not\models_T O\alpha$

As before let us call a formula of $\mathcal{EOL}$ *basic* if it does not mention O. Then just like in the case of world states and implicit beliefs there are sets of situations which agree on all basic explicit beliefs but disagree on what they only-know. In fact, this can be demonstrated using the same example as in Chapter 6 and replacing "world states" by "situations." We could even solve this problem the same way as before by introducing maximal sets of situations in a way completely analogous to implicit belief. However, for simplicity and since none of the following results depend on whether or not we use maximal sets, we will ignore this issue altogether and allow arbitrary sets of situations instead.

Hence, the definitions of truth, validity, satisfiability, and logical implication for $\mathcal{EOL}$ are precisely the same as for the non-nested version of the logic introduced in the previous chapter.

13.2 Some properties of $\mathcal{EOL}$

$\mathcal{EOL}$ shares many of its properties with $\mathcal{OL}$. This is particularly true of introspection, properties of subjective sentences, and quantifying-in. Indeed, the differences between the two logics can all be traced to the weak deductive component, which $\mathcal{EOL}$ inherits from $\mathcal{BL}^q$. We will not give a complete account of all the logical properties of the logic but concentrate on key features different from $\mathcal{BL}^q$. In particular, as in the case of $\mathcal{OL}$, we do not have a complete axiomatization.

Here we are only interested in ordinary sentences, that is, we consider $^\triangle$ only when introduced by the interpretation of B or O.[2] In the following, α and β are arbitrary ordinary sentences and ρ and σ are ordinary subjective sentences. To start with, introspection in $\mathcal{EOL}$ works as expected, that is, we obtain

$\models (B\alpha \supset BB\alpha)$.

$\models (\neg B\alpha \supset B\neg B\alpha)$.

As in the case of implicit belief this leads to the accuracy of all and the completeness of existential-free subjective beliefs.

$\models (\neg B\alpha \supset (B\sigma \supset \sigma))$.

$\models (\sigma \supset B\sigma)$ provided σ is existential-free.

To see that the proviso is indeed necessary, consider the following example.

Example 13.2.1: Let e be any nonempty set of situations such that for all $s \in e$, $s \models_T P(n, n)$ for all standard names n and $s \models_T (t = {}^{\#}1)$ for all primitive terms t. Then clearly $e \models_T \forall x \exists y BP(x, y)$, yet there is no term t such that $e \models_T \forall x BP(x, t^\triangle)$ and, hence, $e \not\models_T B\forall x \exists y BP(x, y)$.

While objective beliefs are certainly not closed under *modus ponens*, we retain logical omniscience as far as believing subjective sentences is concerned, that is,

$\models (B(\rho \supset \sigma) \wedge B\rho \supset B\sigma)$.

This should not surprise since subjective sentences are two-valued because $e \models_F B\alpha$ iff $e \not\models_T B\alpha$ for all e and α. The proof is not quite as straightforward as it may seem, however, and is left as an exercise.

Furthermore, the properties regarding quantifying-in with nested beliefs are precisely the same as in the non-nested case of $\mathcal{BL}^q$ (see Theorem 12.2.9 on page 209).

Finally, and again not surprisingly, we obtain the expected connection between explicitly only-knowing and believing.

$\models O\alpha \supset B\alpha$.

We end this section with an interesting observation regarding the basic explicit beliefs an agent may hold. Let us call a set of basic sentences Γ an *explicit basic belief set* if there is an e such that $\Gamma = \{\alpha \mid \alpha \text{ is basic and } e \models B\alpha\}$. Recall that in $\mathcal{OL}$ basic belief sets or stable sets, for that matter, are not determined by their objective subsets (Theorem 9.3). Interestingly, explicit basic beliefs set are much better behaved in this respect.

---

[2] This is not to say that non-ordinary sentences have no special properties. For example, $(BP(a^\triangle) \supset \exists x BP(x))$ for any closed term t is valid. It is just that these properties are of no concern to our main purpose: defining a limited KR service.

Theorem 13.2.2: *The explicit basic belief sets are uniquely determined by their objective subsets.*

The proof is technical, tedious, and not very illuminating. So rather than going through those details, we look at an example which illustrates why the theorem holds for explicit belief, in contrast to implicit belief. Consider the sentence

$$\lambda = \exists x P(x) \wedge \neg BP(x)$$

and let λ_K be λ with B replaced by K. Recall that in order to conclude that the above theorem fails for implicit belief, we used λ_K and the fact that there are two epistemic states which agree on all objective beliefs yet disagree on $K\lambda_K$ (see the proof of Theorem 4.6.2). It turns out that this property no longer holds in the case of explicit belief.

Theorem 13.2.3: *Let e_1 and e_2 be sets of situations which agree on all objective beliefs. Then $e_1 \models_T B\lambda$ iff $e_2 \models_T B\lambda$.*

Proof: Since the two directions are completely symmetric, we only consider the only-if direction. Let $e_1 \models_T B\lambda$. Then for some closed term t, $e_1 \models_T B[P(t) \wedge \neg BP(t^\triangle)]$. Since e_1 and e_2 agree on their objective beliefs by assumption, we have $e_1 \models_T BP(t)$ and $e_2 \models_T BP(t)$. We also have that $e_1 \models_T B(t \neq n)$ for every n such that $e_1 \models_T BP(n)$ Again, the same holds for e_2 and indeed for precisely the same known P's. Then for every situation $s \in e_2$, $s \models_T P(t) \wedge s = m$ where $s \not\models_T BP(m)$. But then $e_2 \models_T B\lambda$. ∎

13.3 Representing the explicitly believed instances of a formula

The main goal of this chapter is to develop limited versions of **ASK** and **TELL** using $\mathcal{EOL}$ as the underlying logic. More precisely, we assume from now on that we are given an ordinary and objective KB. As in the case of $\mathcal{KL}$ and $\mathcal{OL}$, we are concerned with reducing arbitrary queries to objective ones. Together with the complexity results of the previous chapter, this will often result in tractable or decidable methods for query evaluation under explicit belief. While the language allows queries which mention O, we will restrict our attention to *basic* queries only, mainly to simplify the presentation.

The results in the following two sections closely mirror those in Chapter 7 on the Representation Theorem for implicit beliefs. Indeed, the method to represent the known instances of a formula and the reduction of arbitrary beliefs to objective ones are the same as before, except that K is replaced by B. We assume for now that KB's are not only objective but also existential-free. At the end of this section, we will show that this can be done without loss of generality, essentially for the same reason why Skolemization is not a limitation in the case of classical logic.

We begin with the definition of RES for explicit belief, denoted as RES_B, which describes the known instances of an objective formula for a given KB.

Let ϕ be an objective formula and KB be a finite set of objective existential-free sentences. Suppose that $n_1, \ldots, n_k$, are all the names in ϕ or in KB, and that n' is some name that does not appear in ϕ or in KB. Then $\text{RES}_B[\![\phi, \text{KB}]\!]$ is defined by:

1. If ϕ has no free variables, then $\text{RES}_B[\![\phi, \text{KB}]\!]$ is
   $\forall z(z = z)$, if $\Re[\![\text{KB}]\!] \models B\phi$, and
   $\neg \forall z(z = z)$, otherwise.

2. If x is a free variable in ϕ, then $\text{RES}_B[\![\phi, \text{KB}]\!]$ is
   $[((x = n_1) \wedge \text{RES}_B[\![\phi^x_{n_1}, \text{KB}]\!]) \vee \ldots$
   $((x = n_k) \wedge \text{RES}_B[\![\phi^x_{n_k}, \text{KB}]\!]) \vee$
   $((x \neq n_1) \wedge \ldots \wedge (x \neq n_k) \wedge \text{RES}_B[\![\phi^x_{n'}, \text{KB}]\!^{n'}_x])]$.

As in the case of implicit belief, we can make this definition completely determinate by choosing n' and x to be the first (in lexicographic order) standard name and variable that satisfy their respective criterion.

Note that RES_B is identical to its counterpart for implicit belief (see Definition 7.2.1 on page 115) except that K is replaced by B. This is possible precisely because we have chosen the same two-valued semantics of = for both explicit and implicit belief.

Lemma 13.3.1: *Let $*$ be a bijection from standard names to standard names. For a given situation s we define s^* the same way as we did for world states (see Theorem 2.8.5 on page 32). First recall that for any term t or wff α, we let t^* or α^* indicate the expression resulting from simultaneously replacing in t or α every name by its mapping under $*$. Then for every primitive formula α, $s^*[\alpha] = s[\alpha^*]$ and for every primitive term t, $s^*[t]$ is $(s[t^*])^{*-1}$. For any set of situations e let $e^* = \{s^* \mid s \in e\}$. Then for any basic sentence α,*

$$e^*, s^* \models_T \alpha \quad \text{iff} \quad e, s \models_T \alpha^*;$$
$$e^*, s^* \models_F \alpha \quad \text{iff} \quad e, s \models_F \alpha^*.$$

Proof: The lemma is proved by induction on the structure of terms and basic sentences similar to the proof of Theorem 2.8.5. ∎

Lemma 13.3.2: *Let ϕ be an objective formula with free variables $x_1, \ldots, x_k$ and let $e = \Re[\![\text{KB}]\!]$. Then for any $s \in e$ and any substitutions of closed terms $t_1, \ldots, t_k$ for $x_1, \ldots, x_k$,*

$e, s \models_T \text{RES}_B[\![\phi, \text{KB}]\!][t_1, \ldots, t_k]$ iff $e, s \models_T (B\phi)[\![t_1, \ldots, t_k]\!]$

(Note the use of $[\![\cdot]\!]$ on the right hand side, making sure that $t_i{}^\triangle$ is substituted for x_i.)

Proof: By induction on the number of free variables in ϕ.
Let ϕ have no free variables. Then $e, s \models_T \text{RES}_B[\![\phi, \text{KB}]\!]$ iff $\text{RES}_B[\![\phi, \text{KB}]\!] = \forall x(x = x)$ iff $e \models_T B\phi$.

Assume the lemma holds for k variables. Let ϕ have $k + 1$ variables. Let t be a closed term intended as a substitution for the variable x with $s[t] = n$. Suppose n occurs in KB or ϕ. Then $e, s \models_T \text{RES}_B[\![\phi, \text{KB}]\!][t_1, \ldots, t_k, t]$ iff $e, s \models_T \text{RES}_B[\![\phi_n^x, \text{KB}]\!][t_1, \ldots, t_k]$ (all other candidates are not satisfied because $s[t] = n$.) iff $e, s \models_T (B\phi_n^x)[\![t_1, \ldots, t_k]\!]$ (by induction) iff $e, s \models_T (B\phi)[\![t_1, \ldots, t_k, t]\!]$.

Now suppose n does not occur in KB and ϕ. Let $*$ be a bijection over the standard names that is the identity function except that it swaps n and n', where n' is the name not occurring in ϕ and KB used in the definition of RES_B. Let t' and t'_i be such that $t'^* = t$ and $t'^*_i = t_i$. Then

$e, s \models_T \text{RES}_B[\![\phi, \text{KB}]\!][t_1, \ldots, t_k, t]$
iff $e, s \models_T (\text{RES}_B[\![\phi, \text{KB}]\!][t'_1, \ldots, t'_k, t'])^*$
iff $e, s \models_T (\text{RES}_B[\![\phi_{n'}^x, \text{KB}]\!]_x^{n'}[t'_1, \ldots, t'_k, t'])^*$ by the defn. of RES_B
iff $e^*, s^* \models_T \text{RES}_B[\![\phi_{n'}^x, \text{KB}]\!]_x^{n'}[t'_1, \ldots, t'_k, t']$ (by Lemma 13.3.1)
iff $e^*, s^* \models_T \text{RES}_B[\![\phi_{n'}^x, \text{KB}]\!][t'_1, \ldots, t'_k]$ (since $s^*(t') = s(t)^{*-1} = n^{*-1} = n^* = n'$)
iff $e^*, s^* \models_T (B\phi_{n'}^x)[\![t'_1, \ldots, t'_k]\!]$ (by induction)
iff $e, s \models_T ((B\phi_{n'}^x)[\![t'_1, \ldots, t'_k]\!])^*$ (by Lemma 13.3.1)
iff $e, s \models_T (B\phi)[\![t_1, \ldots, t_k, t]\!]$. ∎

13.4 Reducing arbitrary sentences to objective terms

As in the previous section, let us assume for now that KB is an existential-free and objective sentence. This means, in particular, that the set of situations which only-knows KB is the unique set $\Re[\![\text{KB}]\!] = \{s \mid s \models_T \text{KB}\}$.

First, we define $\|\alpha\|_{\text{KB}}$ as in Chapter 7 on page 117 except that we let

$\|B\alpha\|_{\text{KB}} = \text{RES}_B[\![\|\alpha\|_{\text{KB}}, \text{KB}]\!],$

where KB is objective and α an arbitrary objective sentence.

Lemma 13.4.1: $\|\alpha^{\sharp}\|_{\text{KB}} = \|\alpha\|_{\text{KB}}{}^{\sharp}$.

Proof: A simple induction on the structure of α. Note that $\alpha^{\sharp}$ only removes existential quantifiers at the objective level and that $\|\cdot\|_{KB}$ does not alter the objective level of a formula. ∎

Lemma 13.4.2: *Let* $e = \Re[\![KB]\!]$, *let* α *be a basic formula with free variables* $\vec{x} = \langle x_1, \ldots, x_k \rangle$, *and* $\vec{t} = \langle t_1, \ldots, t_k \rangle$ *a sequence of closed terms.*

$$e, s \models_T \alpha[\![\vec{x}/\vec{t}]\!] \quad \text{iff} \quad e, s \models_T \|\alpha\|_{KB}[\vec{x}/\vec{t}]$$
$$e, s \models_F \alpha[\![\vec{x}/\vec{t}]\!] \quad \text{iff} \quad e, s \models_F \|\alpha\|_{KB}[\vec{x}/\vec{t}]$$

Proof: By induction on the structure of α.
Let $\vec{t}^\triangle = \langle t_1^\triangle, \ldots, t_k^\triangle \rangle$. If α is objective, the lemma holds trivially because $\alpha[\![\vec{t}]\!] = \|\alpha\|_{KB}[\vec{t}]$. Given the definition of $\|\alpha\|_{KB}$, the cases for $\neg$, $\vee$, and $\forall$ follow easily by induction. Finally, we get that

	$e, s \models_T (\boldsymbol{B}\alpha)[\![\vec{t}]\!]$	
iff	$e, s \models_T \boldsymbol{B}(\alpha[\vec{t}^\triangle])$	by definition of $[\![\vec{t}]\!]$
iff	$e, s' \models_T \alpha[\vec{t}^\triangle]^{s\sharp}[\![\vec{t}']\!]$	for some admissible $\vec{t}'$ and for all $s' \in e$
iff	$e, s' \models_T \alpha[\vec{n}]^\sharp[\![\vec{t}']\!]$	for $s[\vec{t}] = \vec{n}$
iff	$e, s' \models_T \alpha^\sharp[\![\vec{n}, \vec{t}']\!]$	
iff	$e, s' \models_T \|\alpha^\sharp\|_{KB}[\vec{n}, \vec{t}']$	by induction
iff	$e, s' \models_T \|\alpha\|_{KB}^\sharp[\vec{n}, \vec{t}']$	by Lemma 13.4.1
iff	$e, s' \models_T \|\alpha\|_{KB}[\vec{t}^\triangle]^{s\sharp}[\vec{t}']$	
iff	$e, s \models_T \boldsymbol{B}(\|\alpha\|_{KB}[\vec{t}^\triangle])$	
iff	$e, s \models_T (\boldsymbol{B}\|\alpha\|_{KB})[\![\vec{t}]\!]$	
iff	$e, s \models_T \text{RES}_{\boldsymbol{B}}[\![\|\alpha\|_{KB}, KB]\!][\vec{t}]$	by Lemma 13.3.2
iff	$e, s \models_T \|\boldsymbol{B}\alpha\|_{KB}[\vec{t}]$	by definition of $\|\cdot\|_{KB}$ ∎

Theorem 13.4.3: *Let KB be an objective existential-free sentence and* α *a basic sentence. Then* $\models (OKB \supset (\boldsymbol{B}\alpha \equiv \boldsymbol{B}\|\alpha\|_{KB}))$.

Proof: Let $e \models_T OKB$, that is, $e = \{s \mid s \models_T KB\}$. Then $e \models_T \boldsymbol{B}\alpha$ iff $e \models_T \|\boldsymbol{B}\alpha\|_{KB}$ (by Lemma 13.4.2) iff $e \models_T \text{RES}_{\boldsymbol{B}}[\![\|\alpha\|, KB]\!]$ (by the definition of $\|\cdot\|_{KB}$) iff $e \models_T \boldsymbol{B}\|\alpha\|$ (by Lemma 13.3.2). ∎

Up to now we have assumed that KB is not only objective but also existential-free. We now show how to handle arbitrary objective KB's.

Lemma 13.4.4: *Let* $^\diamond$ *be a bijection from function symbols to function symbols such that* f *and* $f^\diamond$ *are of the same arity for all* f *and* $^\diamond$ *is the identity function everywhere except*

The logic $\mathcal{EOL}$ 233

for $f \in \mathcal{F}_{SK}$.

For any situation s, we define

$$s^\diamond[f(\vec{n})] = s[f^\diamond(\vec{n})] \quad \text{for all primitive terms } f(\vec{n}).$$
$$s^\diamond[p] = s[p] \quad \text{for all primitive formulas } p.$$

In other words, $s^\diamond$ is exactly like s except that f is replaced by $f^\diamond$ for all $f \in \mathcal{F}_{SK}$. For any set of situations e, let $e^\diamond = \{s^\diamond \mid s \in e\}$. For any formula α (term t), let $\alpha^\diamond$ ($t^\diamond$) denote α (t) with every function symbol f replaced by $f^\diamond$. If α is basic, then

$$e^\diamond, s^\diamond \models_T \alpha \quad \text{iff} \quad e, s \models_T \alpha^\diamond$$
$$e^\diamond, s^\diamond \models_F \alpha \quad \text{iff} \quad e, s \models_F \alpha^\diamond$$

Proof: The proof is a simple induction on the structure of α. ∎

Theorem 13.4.5: *Let KB be an objective sentence and α an arbitrary basic sentence. Let* $\text{KB}' = \text{KB}^\sharp[\vec{t}]$ *for arbitrary sk-terms $\vec{t}$ not occurring in KB or α. Then*

1. $\models (O\text{KB} \supset B\alpha)$ *iff* $\models (O\text{KB}' \supset B\alpha)$.
2. $\models (O\text{KB} \supset \neg B\alpha)$ *iff* $\models (O\text{KB}' \supset \neg B\alpha)$.

Proof: Let e and e' be two sets of situations such that $e \models_T O\text{KB}$ and $e' \models_T O\text{KB}'$. Then $e' = \{s \mid s \models_T \text{KB}'\}$ and $e = \{s \mid s \models_T \text{KB}^*\}$ for some $\text{KB}^* = \text{KB}^\sharp[\vec{u}]$, where $\vec{u}$ is an sk-term substitution of the existentials in KB.

Define a bijection $^\diamond$ such that $\vec{t}^\diamond = \vec{u}$ and all $f \in \mathcal{F}_{SK}$ that occur in KB or α are mapped onto themselves. Let $e'^\diamond = \{s^\diamond \mid s \in e'\}$ as defined in Lemma 13.4.4.

First, we show that $e'^\diamond = e$. By assumption, for all situations s, $s \in e'$ iff $s \models_T \text{KB}'$. By the definition of $^\diamond$, $\text{KB}' = \text{KB}^{*\diamond}$. Therefore, by Lemma 13.4.4, for all s, $s \in e'^\diamond$ iff $s \models_T \text{KB}^*$, that is, $e'^\diamond = e$.

Now let α be basic and without sk-functions. By Lemma 13.4.4, we get that $e' \models_T B\alpha$ iff $e'^\diamond \models_T B\alpha$ because $\alpha^\diamond = \alpha$. Also, $e = e'^\diamond$. Thus $e \models_T B\alpha$ iff $e' \models_T B\alpha$.

Since this holds for arbitrary e with $e \models_T O\text{KB}$, the theorem follows. ∎

Corollary 13.4.6: *Let KB and α be as above. Then exactly one of $(O\text{KB} \supset B\alpha)$ or $(O\text{KB} \supset \neg B\alpha)$ is valid.*

Proof: Let KB' be as in the theorem. Since KB' is existential-free, there is a unique epistemic state e such that $e \models_T O\text{KB}'$. Hence exactly one of $(O\text{KB}' \supset B\alpha)$ or $(O\text{KB}' \supset \neg B\alpha)$ is valid. The result then follows from the theorem. ∎

Corollary 13.4.7: $\models (O\text{KB} \supset (B\alpha \equiv B\|\alpha\|_{\text{KB}'}))$.

Proof: Let $e \models_\text{T} O\text{KB}$ and assume, without loss of generality, that $e \models_\text{T} B\alpha \wedge \neg B\|\alpha\|_{\text{KB}'}$. By the previous corollary, $\models (O\text{KB} \supset B\alpha)$ and hence $\models (O\text{KB}' \supset B\alpha)$. We also have $\models (O\text{KB} \supset \neg B\|\alpha\|_{\text{KB}'})$ because $\|\alpha\|_{\text{KB}'}$ obviously does not mention any $f \in \mathcal{F}_{\text{sk}}$ introduced by KB'. Therefore, $\models (O\text{KB}' \supset \neg B\|\alpha\|_{\text{KB}'})$, contradicting Theorem 13.4.3. ∎

Hence, to apply Theorem 13.4.3 to arbitrary objective KB's, all we need to do is replace the existentials in KB by sk-terms occurring nowhere else.

13.5 Decidability results

Here we list some complexity results we have obtained for determining the explicit beliefs of an objective knowledge base. First we consider the propositional case with nested beliefs (Theorem 13.5.2), and then the first-order case without quantifying-in or equality (Theorem 13.5.3). Those results follow easily by induction on the depth of nested beliefs given the corresponding results of the previous chapter for the non-nested case. The proofs are omitted here.

Adding quantifying-in on the part of the query is more complicated. Since the proof is overly technical and not very illuminating, we will spare the reader the details.

We begin by defining conjunctive normal form for formulas with nested beliefs.

Definition 13.5.1: A basic quantifier-free formula α is in conjunctive normal form, if α is a conjunction of disjunctions, where every disjunct is either a literal or of the form $B\beta$ or $\neg B\beta$ such that β itself is in conjunctive normal form.

Theorem 13.5.2: *Let KB be a finite set of quantifier-free objective sentences and α a basic quantifier-free sentences without equality in conjunctive normal form.*
Then the validity of $(O\text{KB} \supset B\alpha)$ can be determined in $O(nm)$ time, where $n = |\text{KB}|$ and $m = |\alpha|$.

Theorem 13.5.3: *Let KB be a finite set of ordinary objective sentences without equality. Let α be an ordinary basic sentence without equality or quantifying-in, that is, no subformula $B\beta$ of α contains free variables. Then the validity of $(O\text{KB} \supset B\alpha)$ is decidable.*

While decidability is fairly easy to establish if we do not allow equality or quantifying-in, these cases are perhaps also the least interesting. Recall that it was quantifying-in that made $\mathcal{KL}$ particularly useful as an interaction language. For example, one can tell the KB that it knows all of Tina's students using the sentence $\forall x\, Teach(tina, x) \supset \mathbf{K}Teach(tina, x)$. Furthermore, being able to express equalities or inequalities as in $\exists x\, Teach(tina, x) \wedge x \neq sara$ is very useful as well. It would be nice to handle examples like these under explicit belief as well. As we will see in a moment, this can be done, but there is a price to pay. In particular, we have to restrict the syntactic form of both KB and α if we want to preserve decidability. It turns out that it takes very little to become undecidable even if we allow no other predicate symbol apart from $=$.

Definition 13.5.4: Let Ψ be the set of sentences constructed from at most two unary function symbols f_1 and f_2, $=$, and one leading universal quantifier.

An example of the kinds of sentences contained in Ψ is
$$\forall x(f_1(x) = f_2(x)) \vee (f_1(f_1(x)) = x).$$
Note that neither constants nor standard names are allowed and that there is exactly one variable, which is universally quantified, that is, the sentences in Ψ are existential-free.

Let Γ be the set of equality axioms introduced in Chapter 2 (page 31) and let us denote classical logical implication by $\models^{\text{FO}}$.[3] Then:

Theorem 13.5.5: *[Gurevich] The problem of determining whether $\Gamma \models^{\text{FO}} \phi$ holds is undecidable for sentences ϕ in Ψ.*

Recall that in Chapter 2, we considered an infinite set of axioms Δ (see page 30) which rule out interpretations with finite domains. As we will see in a moment, it is not hard to show that the validity problem for a special class of belief implications reduces to the question whether $\Gamma \cup \Delta \models^{\text{FO}} \phi$ holds for sentences in Ψ. Fortunately, the restrictions on Ψ make it possible to ignore Δ so that the above theorem can be applied.

Lemma 13.5.6: *For any sentence $\phi \in \Psi$, $\Gamma \cup \Delta \models^{\text{FO}} \phi$ iff $\Gamma \models^{\text{FO}} \phi$.*

Proof: The lemma follows easily from the following property: for any Tarskian interpretation $I = \langle D, \Phi \rangle$ with a finite domain D there is an interpretation $I' = \langle D', \Phi' \rangle$ with a countably infinite domain such that I and I' agree on all sentences in Ψ. I' can be constructed as follows: for any $D = \{d_1, \ldots, d_n\}$, let $D' = \{d_1^1, \ldots, d_n^1, d_1^2, \ldots, d_n^2, \ldots\}$, $\Phi'(f_1)(d_i^k) = d_j^k$ where $\Phi(f_i)(d_i) = d_j$, and similarly for f_2. We leave the details of the proof that I and I' agree on all sentences of Ψ[4] to the reader. ∎

---

3 Note that $\models^{\text{FO}}$ treats $=$ as an ordinary predicate.
4 Note that this does not hold for arbitrary objective sentences. For example, $\exists x \forall y(x = y)$ is satisfiable only

Theorem 13.5.7: *Let ϕ and ψ be arbitrary objective sentences. Then the validity problem for sentences of the form $(B\phi \supset B\psi)$ is undecidable.*

Proof: It suffices to consider the case where $\psi \in \Psi$ and $\phi = \text{TRUE}$. $(B\text{TRUE} \supset B\psi)$ is valid iff $B\psi$ is valid iff for all situations s, $s \models_T \psi$ (since ψ is existential-free) iff for all world states w, $w \models \psi$ (because $=$ is the only predicate, which has a classical two-valued semantics) iff ψ is valid in $\mathcal{L}$ iff $\Gamma \cup \Delta \models^{\text{FO}} \psi$ by Theorem 2.8.4 iff $\Gamma \models^{\text{FO}} \psi$ by the lemma. By Theorem 13.5.5, there is no decision procedure for this problem and, hence, there cannot be a decision procedure which determines the validity of belief implications $(B\phi \supset B\psi)$ for arbitrary objective ϕ and ψ. ∎

In the following theorem, we restrict both equality and quantifying-in in order to obtain decidability. The restrictions may seem awkward and somewhat arbitrary, and perhaps they are. Nevertheless, we will see later that the theorem allows us to handle fairly sophisticated queries and be guaranteed an answer after a finite amount of processing.

Theorem 13.5.8: *Let* $\text{KB} = \text{KB}' \cup E$ *be a finite set of ordinary objective sentences such that*

1. *the only sentences in* KB' *containing equality are of the form*

$$\forall \vec{x} \pm P(\vec{t}) \supset [(\vec{x} = \vec{n}_1) \vee \ldots \vee (\vec{x} = \vec{n}_k)] \text{ and}^5$$

2. *E is a set of ground equality literals.*

Let α be an ordinary basic sentence with the following restrictions:

1. *Equality atoms $t_1 = t_2$ are such that t_i is either a variable or a ground term.*

2. *Any free variable x in a subformula $B\gamma$ in α is either a universally quantified variable in α or does not appear in the scope of a universal quantifier. In other words, only universally quantified variables or existentials without leading universals may participate in quantifying-in.*

Then the validity of $(O\text{KB} \supset B\alpha)$ is decidable.

Examples of a knowledge base and queries which fall under the constraints of the theorem are discussed in Section 13.6.4.

---

when the domain is a singleton set. For that reason it is important that the sentences in Ψ must not contain constants or existential quantifiers.

5 We could replace $\supset$ by $\equiv$, but we chose not to because it is simpler and we can always fake the if direction by explicitly listing the instances of P without using equality.

The logic $\mathcal{EOL}$

13.6 ASK and TELL

In this section, we adapt definitions of **ASK** and **TELL** to deal with explicit belief, both at the knowledge and symbol level.

13.6.1 ASK

Similar to implicit belief **ASK** takes two arguments, a sentence (query) α and an epistemic state represented by a set of situations e. As before, we assume for simplicity that we are only asking the KB whether or not α is *believed*. This way the answer will always be *yes* or *no*. We already discussed in Chapter 5 how the usual four answers to a query, *yes*, *no*, *unknown*, and *inconsistent*, can be constructed by asking questions for both α and $\neg\alpha$.

Definition 13.6.1:

$$\text{ASK}[\alpha, e] = \begin{cases} yes & \text{if } e \models_T B\alpha \\ no & \text{otherwise.} \end{cases}$$

For epistemic states representable by an objective sentence KB we immediately get a specification of **ASK** within $\mathcal{EOL}$ thanks to the definition of O. We restrict KB to be existential-free. As mentioned earlier, this is no more serious than Skolemization in classical logic, since we can always find sk-term substitutions for the existentials in KB.

Theorem 13.6.2: *Let KB be an objective and existential-free sentence and let α be basic.*

$$\text{ASK}[\alpha, \Re[\![KB]\!]] = \begin{cases} yes & \text{if } \models (O\text{KB} \supset B\alpha) \\ no & \text{otherwise.} \end{cases}$$

Proof: It suffices to show that $\Re[\![KB]\!] \models_T B\alpha$ iff $\models (O\text{KB} \supset B\alpha)$. To prove the if-direction, let $\models (O\text{KB} \supset B\alpha)$. If $\Re[\![KB]\!]$ is empty, then $\Re[\![KB]\!] \models_T B\alpha$ holds vacuously. Otherwise, since $\Re[\![KB]\!] \models_T O\text{KB}$, $\Re[\![KB]\!] \models_T B\alpha$ follows by assumption.

Conversely, let $\Re[\![KB]\!] \models_T B\alpha$. Since KB is objective and existential-free, by Theorem 13.4.5, exactly one of $\models (O\text{KB} \supset B\alpha)$ and $\models (O\text{KB} \supset \neg B\alpha)$ holds. Since, by assumption, $\Re[\![KB]\!] \models_T B\alpha$, $\models (O\text{KB} \supset B\alpha)$ follows. ∎

We remark that open queries can be handled as in the case of implicit belief (see Section 5.8.2 and 7.6).

13.6.2 TELL

Given an epistemic state e, we can make sense of TELLing it an existential-free sentence in a way completely analogous to implicit belief:

$$\textbf{TELL}[\alpha, e] = e \cap \{s \mid e, s \models_T \alpha\}. \tag{13.1}$$

In other words, we remove those situations from e which do not support the truth of α.

If α is not existential-free, then the story gets somewhat more complicated, which is best explained by an example. Suppose $\alpha = \exists x P(x)$. If we use the above definition of **TELL**, then for most e, the resulting epistemic state would not even believe α because there is no appropriate admissible term for x. It seems natural then to first replace α by $\alpha' = P(a)$ for some term a which is not "used" anywhere else and pass α' to **TELL**. The problem is that e does not necessarily tell us which terms are already in use. There are at least two ways of solving this problem. One is to keep track explicitly of the terms that have been introduced so far by **TELL**. Another is to restrict ourselves to epistemic states which are representable by an objective KB. In that case, **TELL** only needs to look at KB to determine the terms introduced so far. Although this approach is a bit counter to the idea of a purely knowledge level definition of **TELL**, we have opted for it mainly because it is conceptually simpler (no explicit bookkeeping is required) and the representable epistemic states are the ones we care about most.

Definition 13.6.3: A set of situations e is (finitely) *explicitly representable* iff there is an objective existential-free sentence KB such that $e = \Re[\![KB]\!]$.

Next, we need a way to uniquely pick new sk-terms relative to a given sentence. To make this process determinate, let us assume that the function symbols are ordered lexicographically.

Definition 13.6.4: Let α and β be arbitrary sentences and let $\vec{x} = \langle x_1, \ldots, x_k \rangle$ be the existentials at the objective level of β. Then let $\vec{t}_\alpha = \langle f_1(U(x_1)), \ldots, f_k(U(x_k)) \rangle$ be an sk-term substitution of $\vec{x}$ in α such that the f_i are the least function symbols in $\mathcal{F}_{SK}$ which do not appear in α.

Definition 13.6.5: Let e be explicitly representable with $e \models_T OKB$ for some objective existential-free KB. Let α be a basic sentence with existentials $\vec{x}$ at the objective level. Then

$$\textbf{TELL}[\alpha, e] = e \cap \{s \mid e, s \models_T \alpha^\sharp [\![\vec{x}/\vec{t}_{KB}]\!]\}$$

Note that the definition reduces to the usual one (13.1) when α is existential-free.

Since basic beliefs reduce to objective beliefs, it is not hard to see that we obtain a representation theorem for **TELL** similar to the one for implicit belief. Moreover, **TELL** preserves representability in that the resulting objective sentence is also existential-free.

Theorem 13.6.6: *Let KB be an objective, and existential-free sentence and α a basic sentence with existentials $\vec{x}$. Then* **TELL**$[\alpha, \Re[\![KB]\!]]$ *is explicitly representable and*

$$\text{TELL}[\alpha, \Re[\![KB]\!]] = \Re[\![KB \wedge \|\alpha\|_{KB}{}^{\#}[\vec{x}/\vec{t}_{sk}]]\!].$$

Proof: Suppose that **TELL**$[\alpha, \Re[\![KB]\!]] = e$. In other words we have that $e = \Re[\![KB]\!] \cap \{s \mid \Re[\![KB]\!], s \models_T \alpha^{\#}[\![\vec{x}/\vec{t}_{sk}]\!]\}$. By Lemma 13.4.2, we then get that $e = \Re[\![KB]\!] \cap \{s \mid s \models_T \|\alpha\|_{KB}{}^{\#}[\vec{x}/\vec{t}_{sk}]\}$. Thus $e = \{s \mid s \models_T KB \wedge \|\alpha\|_{KB}{}^{\#}[\vec{x}/\vec{t}_{sk}]\}$. ∎

13.6.3 Decidability

Given the decidability results from Section 13.5, it is clear that these can be applied to **ASK** and **TELL** as well. In particular, **ASK** and **TELL** are decidable operations provided the KB and the query or new information satisfy the syntactic restrictions of Theorem 13.5.8. An unpleasant side-effect of the restrictions on the form of the knowledge base is that the result of a **TELL** may take us beyond those KB's that we can handle. While this is not as pretty a result as one might hope for, we nevertheless obtain a nontrivial decidable query mechanism, as the following examples demonstrate.

13.6.4 Examples of ASK

In order to better compare the differences and similarities between **ASK** under explicit and implicit belief, we refer to the same knowledge base (Figure 13.1) and example queries as in Section 5.7.

Recall that all proper names are taken to be standard names. Since KB contains an existential, we first replace it by an arbitrary sk-term a. Let KB' be the resulting KB and let $e = \Re[\![KB']\!]$. We consider queries of the form **ASK**$[\alpha, e]$ and denote the answers by TRUE, FALSE, and UNKNOWN. As before, TRUE means that α was answered *yes* and $\neg \alpha$ *no*, and similar for the others. (While explicit belief is able to handle inconsistencies in a nontrivial way, there will not be any in the examples.)

The first thing to note is that there are very few differences compared to the answers under implicit belief. More precisely, we have that Queries 1.–7. and 9.–11. of Section 5.7 yield the same answers when ***K*** is replaced by ***B***. This has to do mainly with the simplicity of the knowledge base, the fact that = has its usual two-valued meaning, and the fact that most queries do not require reasoning by cases apart from equality reasoning. Indeed, as

- *Teach(tom, sam)*
- *Teach(tina, sue)* ∧ [*Teach(tom, sue)* ∨ *Teach(ted, sue)*]
- ∃*xTeach(x, sara)*
- ∀*x*[*Teach(x, sandy)* ≡ (*x* = *ted*)]
- ∀*x*∀*y*[*Teach(x, y)* ⊃ (*y* = *sam*) ∨ (*y* = *sue*) ∨ (*y* = *sara*) ∨ (*y* = *sandy*)]

Figure 13.1: The Example KB

we will see, differences do arise whenever cases need to be considered with respect to the disjunction *Teach(tom, sue)* ∨ *Teach(ted, sue)*.

In the following, we consider one example (2.) where implicit and explicit belief agree and then go through those where there actually is a difference. The answers under implicit belief are shown in parentheses.

2. *Teach(tom, sandy)*   FALSE (FALSE)
   Since = is two-valued, it follows from the KB that Tom does not teach Sandy. Note, however, that the KB does not rule out situations with true support for teachers of Sandy other than Ted. It is just that these like all other situations must also have false support for those.

8. ∃*x*[*Teach(x, sue)* ∧ ¬***B****Teach(x, sue)*]   UNKNOWN (TRUE)
   Here the answer is UNKNOWN because the KB only knows about the existence of one teacher of Sue, namely Tina. The KB does not know whether there are any other and, hence, any unknown ones because existential generalization fails in the case of *Teach(tom, sue)* ∨ *Teach(ted, sue)*.

12. ∃*y*(*y* ≠ *sam*) ∧ ¬***Bif***[∀*xTeach(x, y)* ⊃ ***B****Teach(x, y)*]   TRUE (FALSE)
    where ***Bif*** α is an abbreviation for ***B***α ∨ ***B***¬α.
    Recall that the question asks whether there is someone other than Sam for whom we do not know whether he or she has any unknown teachers. In contrast to implicit belief, the answer is yes because under explicit belief we are no longer sure whether or not we are missing any of Sue's teachers (see Question 8).

13. ∃*y****B***∃*x*[*Teach(x, y)* ∧ ∃*z*[(*y* ≠ *z*) ∧ ***B****Teach(x, z)*]]   FALSE (TRUE)
    The question is whether there is an individual *y* of whom it is known that one of her teachers *x* is known to teach somebody else *z*. Previously, the answer was yes because of Sue (replacing *y*) and Sue only. Now, under explicit belief, the answer is no and, again, because of Sue. Cases other than Sue fail for the same reasons as under implicit belief. The case for Sue now fails as well because the only one believed to teach Sue is Tina, and Tina is not known to teach anyone else. In other words, we obtain that
    ⊨ ***O***KB ⊃ ¬***B***∃*x*[*Teach(x, sue)* ∧ ∃*z*[(*sue* ≠ *z*) ∧ ***B****Teach(x, z)*]].

This then ends our extended excursion into the realm of explicit belief and decidable reasoning. In the next chapter, we turn to a quite different topic, that is, we look at the logical underpinnings of knowledge bases able to model dynamic worlds.

13.7 Bibliographic notes

The contents of this chapter is based in part on [72]. Some of the issues which we only sketch here are treated in more detail there. The undecidability result by Gurevich (Theorem 13.5.5) can be found in [6]. In [67] a generalization of Theorem 13.2.2, which states that, in contrast to $\mathcal{OL}$, basic belief sets in $\mathcal{EOL}$ are uniquely determined by their objective subsets, is considered. It is shown that the theorem holds even if we use two-valued world states (but stick to the non-standard interpretation of existentials otherwise). There is surprisingly little other work on introspective reasoning under limited belief in the literature. Konolige [56, 57] seems to be the first to address this issue. However, rather than proposing an actual instance of a computationally attractive reasoner, he presents a general framework in which one can be formalized. Elgot-Drapkin and Perlis [30, 29] attack the issue of reasoning under resource limitations including introspection by explicitly modeling proof-steps. In [56], Konolige questions whether quantifying-in is compatible with nice computational behaviour. His argument is that, while introspection in monadic predicate calculus is decidable, it becomes undecidable if we add quantifying-in. In our work, we are able to at least partially absolve quantifying-in from being the computational villain. In [77] $\mathcal{EOL}$ is strengthened by reintroducing certain forms of classical reasoning while retaining decidability. The idea is, roughly, to use a typed (or sorted) language and to give types, which are special unary predicates, an ordinary two-valued interpretation. The information about types like "all dogs are mammals" is kept separate from the KB, which mentions types only as restrictions of variables. Using results from [37], it is shown that under certain restrictions on the information about types, reasoning remains decidable. In particular, the decision procedure developed in the previous chapter can be used without change except that unification is replaced by sorted unification.

13.8 Where do we go from here?

Following up on the extension of $\mathcal{EOL}$ just outlined above, there seems to be plenty of room for improvement. In particular, the restrictions on the types considered so far are extremely severe. Essentially, there is little else allowed other than sentences like the dogs-are-mammals example. Most likely, these restrictions can be loosened without giving up on decidability. Apart from extending $\mathcal{EOL}$ along these lines, there seems to be plenty of room to improve on the decidability results we have obtained so far. In particular, the restrictions of Theorem 13.5.8 are most likely too conservative.

Another area of further investigation concerns equality. Recall that the main reason for keeping its original two-valued interpretation was its usefulness in obtaining a representation theorem (Theorem 13.4.3) similar to that of $\mathcal{KL}$. The downside, of course, is that we can afford to include only very limited equality statements in the KB if we want to guarantee decidability. Various questions come to mind. For example, is it possible to use a weaker form of equality which still allows some form of representation theorem but has better computational properties when it comes to reasoning about equality itself? Or, how about keeping the original form of equality and adding a second, weaker one to the logic? What should the relationship be between the two? We tried for a while to find answers ourselves, but without success except to notice that these are tough research issues.

Finally, $\mathcal{EOL}$ should allow us to consider forms of limited autoepistemic reasoning. Consider, for example, the default

$$\delta = \forall x [\boldsymbol{B}Bird(x) \land \neg \boldsymbol{B} \neg Fly(x) \supset Fly(x)]$$

and KB $= \{Bird(tweety)\}$. Similar to implicit belief, it is not hard to show that Tweety indeed flies assuming all that is known is KB and δ, that is, $\boldsymbol{O}(\text{KB} \land \delta) \supset \boldsymbol{B}Fly(tweety)$ is valid.[6] An intriguing question is what kind of default reasoning $\mathcal{EOL}$ gives rise to and, perhaps more importantly, to what extent we obtain decidability. So far, however, the issue of limited autoepistemic reasoning remains largely unexplored.

13.9 Exercises

1. Prove that for all ordinary subjective sentences ρ and σ, $\models (\boldsymbol{B}(\rho \supset \sigma) \land \boldsymbol{B}\rho \supset \boldsymbol{B}\sigma)$.
   Hint: First prove the following: for any α, e and admissible $\vec{t}$, if $e, s \models_T \alpha^{\#}[\![\vec{x}/\vec{t}]\!]$ then $e, s \models_T \alpha$.

2. Show that, similar to implicit belief, $\textbf{ASK}[\forall x P(x) \supset \neg \boldsymbol{B}P(x), e_0] = yes$, where e_0 is the set of all situations.

3. Let e be the epistemic state of the example KB.
   Show that $\textbf{ASK}[\exists x \boldsymbol{B}Teach(x, sara), e] = no$.

4. What is $\textbf{ASK}[\exists x [\neg \exists y (\boldsymbol{B}Teach(x, y)) \land Teach(x, sue)], e]$?

5. Describe the result of $\textbf{TELL}[\alpha, e]$, where
   (a) $\alpha = \forall x. \boldsymbol{B}Teach(x, sue) \supset Teach(x, sara)$.
   (b) $\alpha = \forall x. Teach(x, sue) \supset \boldsymbol{B}Teach(x, sue)$.

---

6 Note the use of $\boldsymbol{B}Bird(x)$ instead of just $Bird(x)$ within the default. In fact, if we used the latter, the default would not go through. (See also Exercise 8).

(c) $\alpha = \exists x.Teach(x, sue) \supset \neg \mathbf{B}Teach(x, sue)$.

6. Find a sentence which, when **TELL**ing it to the example KB, results in a KB not covered by Theorem 13.5.8.

7. Recall that Query 8 gave the answer UNKNOWN. Rephrase the query in a way which would be equivalent under implicit belief but not under explicit belief and where the answer comes out TRUE.

8. Consider the default
$$\delta = \forall x[\mathbf{B}Bird(x) \wedge \neg \mathbf{B}\neg Fly(x) \supset Fly(x)]$$
and KB = $\{Bird(tweety)\}$. Show that $\models (\mathbf{O}(\text{KB} \wedge \delta) \supset \mathbf{B}Fly(tweety))$. Now consider δ with $\mathbf{B}Bird(x)$ replaced by $Bird(x)$. Show that the default conclusion no longer goes through. What can be inferred instead?

14 Knowledge and Action

As the final topic of this book, we explore the possibility of an *active* knowledge-based system, that is, a system that is capable of performing actions that cause conditions in the world to change. A typical example of such a system is a robot. When a robot performs the action of moving, it causes its location (and that of any object it is carrying) to change. Obviously such actions also affect what a robot *knows* about the world: after moving, the robot should know that its current location is no longer what it was. Here we have an example of a system acquiring new beliefs not as result of a **TELL** operation (at least directly), but as a result of performing an action. We will also see that there are sensing actions, whose effect is not to change the world, but only to change what the system knows about the world. For example, a robot might perform the action of listening carefully thereby acquiring beliefs about whether or not someone is approaching.

To be able to describe how knowledge depends on action, we first need to be clear about actions themselves, independent of knowledge. So we begin by reviewing a popular way of representing actions and their effects, using the language of the situation calculus. We then present in Section 14.2 a new language called $\mathcal{AOL}$ which integrates the situation calculus with $\mathcal{OL}$. This gives us a language for talking about action, knowing, and only-knowing. In Section 14.3, we discuss the general principles that allow us to determine what is known after performing an action (including a sensing action) in terms of what was true before. Ultimately, what is known after a sequence of actions will reduce to some function of what was known initially. We illustrate the properties of $\mathcal{AOL}$ with a simple robotic example in Section 14.4. Finally, in Section 14.5, we present axioms for $\mathcal{AOL}$ in second-order logic.

14.1 A theory of action

To describe the behaviour of actions, that is, under what conditions an action can be successfully performed, and what conditions will change or not change as the result of performing an action, we use a language called the *situation calculus*.[1] The situation calculus is a dialect of the predicate calculus specifically designed for representing dynamically changing worlds. It is similar in many ways to our first-order language $\mathcal{L}$ from Chapter 2, but with some second-order features, and with two distinguished sorts of terms:

actions: Any change to the world is assumed to be the result of some named *action*. An action is denoted by an action term, which can be a constant, a variable, or a function

---

[1] This section reviews an existing approach to representing actions that was not developed by us. See the bibliographic notes for all references.

symbol with arguments. For example, put(x, y) might stand for the action of putting some object x on object y.

situations: In the situation calculus, a possible world history is called a *situation*.[2] A situation is denoted by a situation term,[3] which can be either a variable, the constant S_0, or a term of the form $do(a, s)$, where a is an action term, and s is a situation term. The constant S_0 stands for the initial situation in which no actions have yet occurred. The term $do(a, s)$ stands for the situation that results from performing the action a in situation s. For example, $do(\text{put}(x, y), S_0)$ stands for a situation where a single action was done, putting object x on object y.

Notice that in the situation calculus, both actions and situations are denoted by first-order terms. For example,

$$do(\text{put}(A, B), do(\text{put}(B, C), S_0))$$

is a term that denotes a situation where exactly two actions have occurred: object B was placed on object C, and then object A was placed on object B.

Relations whose truth values vary from situation to situation are called relational *fluents* and are denoted by predicate symbols taking a situation term as their last argument. For example, $Broken(x, s)$ might be a relational fluent meaning that object x is broken in situation s. Functions whose denotations vary from situation to situation are called functional fluents, and are denoted by function symbols taking a situation term as their last argument. For example, $position(x, s)$ might be a functional fluent denoting the position of object x in situation s.

Finally, we use the distinguished predicate $Poss(a, s)$ to state that action a is possible to execute in situation s, and $SF(a, s)$ to state that action a returns a binary sensing value of 1 in situation s. This completes the syntactic description of the language.[4]

14.1.1 Action preconditions

To describe when an action is possible in some application domain, we need to write an axiom[5] that characterizes $Poss$ for that action. For example, in a blocks world, we might

---

[2] The reader should be aware that the meaning of situations here is completely different from their meaning in the previous two chapters on limited reasoning. The two uses of the same expression evolved independently. Since there is no overlap, we did not see the need to invent a new name.

[3] In the literature, the syntactic terms themselves are often also called situations. This is less of a problem than it may seem since, as we will see, axioms force there to be exactly as many situations as there are situation terms, not unlike our treatment of standard names.

[4] To deal with knowledge, we will introduce additional distinguished predicates and functions below.

[5] Traditionally, researchers using the situation calculus have preferred to talk about logical theories rather than KB's and so use the term *axiom* instead of *basic belief*.

have:[6]

$$Poss(\mathsf{drop}(x), s) \equiv Holding(x, s)$$

which says that an object x can be dropped provided that it is currently being held (the agent in both cases being implicit). In general, we have *precondition axioms* of the following form:

Definition 14.1.1: Precondition axioms are axioms of the form

$$Poss(A(\vec{x}), s) \equiv \phi_A(\vec{x}, s).$$

Here ϕ_A is a formula with at most $\vec{x}$ and s free that specifies the necessary and sufficient properties of the situation s under which A can be executed in s.

It is typical in situation calculus applications to focus on the *legal* situations, which are those situations that result from performing actions whose preconditions are satisfied. So, for example,

$$do(\mathsf{put}(A, B), do(\mathsf{put}(B, C), S_0))$$

would be a legal situation iff

$$Poss(\mathsf{put}(B, C), S_0) \wedge Poss(\mathsf{put}(A, B), do(\mathsf{put}(B, C), S_0))$$

was true. Precondition axioms serve to place constraints on what constitutes a legal situation.

14.1.2 Action sensing

We assume that in addition to actions like $\mathsf{drop}(x)$ that change the state of the world, there are actions whose main purpose is to provide information about some property of the world. For simplicity, we assume that every action returns a (binary) sensing value on completion. By convention, actions not involved with sensing always return the value 1. For example, we might have the action of inspecting an object (to see whether it is broken) and an axiom

$$SF(\mathsf{inspect}(x), s) \equiv Broken(x, s)$$

saying that the $\mathsf{inspect}(x)$ action will return a value of 1 iff $Broken(x)$ holds. In general, we have *sensed-fluent axioms* of the following form:

Definition 14.1.2: Sensed-fluent axioms are axioms of the form

$$SF(A(\vec{x}), s) \equiv \phi_A(\vec{x}, s).$$

---

[6] In situation calculus formulas, free variables are considered to be implicitly universally quantified. We will continue using this convention in this section.

Here ϕ_A is a formula with at most $\vec{x}$ and s free that specifies the conditions under which action A returns a sensing value of 1.

For ordinary actions A that do not do sensing, we can use TRUE as the ϕ_A. Sensed-fluent axioms are to SF what precondition axioms are to $Poss$, and will only be used later when we consider knowledge in $\mathcal{AOL}$.

14.1.3 Action effects

How an action causes the world to change can be described by *effect axioms*. These are the causal laws which state which fluents will hold or not hold as a result of performing the action. For example, dropping a fragile object causes it to be broken:

$$Fragile(x, s) \supset Broken(x, do(\mathsf{drop}(x), s)). \tag{14.1}$$

Exploding a bomb next to an object causes it to be broken:

$$NextTo(b, x, s) \supset Broken(x, do(\mathsf{explode}(b), s)). \tag{14.2}$$

Repairing an object causes it to be not broken:

$$\neg Broken(x, do(\mathsf{repair}(x), s)). \tag{14.3}$$

However, axiomatizing a dynamic world requires more than these effect axioms. So-called *frame axioms* are also necessary. These specify the action *invariants* of the domain, namely, those fluents that are not changed by a given action. For example, an object's colour is not affected by dropping something:

$$colour(y, do(\mathsf{drop}(x), s)) = colour(y, s).$$

One way of describing how the fluent $Broken$ is unaffected is:

$$\neg Broken(y, s) \land (y \neq x \lor \neg Fragile(y, s)) \supset \neg Broken(y, do(\mathsf{drop}(x), s)).$$

14.1.4 The frame problem and a solution

The problem with such frame axioms is that there are in general a great many of them. Relatively few actions will affect the truth value of a given fluent; all other actions leave the fluent unchanged. For example, an object's colour is not changed by picking things up, opening a door, going for a walk, electing a new prime minister of Canada, and so on. This is problematic for the axiomatizer who must think of all these axioms; it is also problematic for any automated reasoning process as it must reason efficiently in the presence of so many frame axioms.

Suppose that the person responsible for axiomatizing an application domain has specified all of the causal laws for the world being axiomatized. More precisely, suppose they have succeeded in writing down *all* the effect axioms, that is, for each fluent F and each

Knowledge and Action

action A that can cause F's truth value to change, they have written axioms of the form

$$R(\vec{x}, s) \supset (\neg) F(\vec{x}, do(A, s)),$$

where R is a first-order formula specifying the contextual conditions under which the action A will have its specified effect on F. A reasonable solution to the frame problem would be a systematic procedure for generating, from these effect axioms, all the frame axioms. If possible, we would also want a *parsimonious* representation for these frame axioms (because in their simplest form, there are too many of them).

We now consider a simple solution to the frame problem, best illustrated with an example. Suppose that (14.1), (14.2), and (14.3) are all the effect axioms for the fluent *Broken*, that is, they describe all the ways that any action can change the truth value of *Broken*. We can rewrite (14.1) and (14.2) in the logically equivalent form:

$$a = \mathsf{drop}(r, x) \wedge \mathit{Fragile}(x, s) \ \vee \ \exists b.\, a = \mathsf{explode}(b) \wedge \mathit{NextTo}(b, x, s)$$
$$\supset \mathit{Broken}(x, do(a, s)).$$

Similarly, consider the negative effect axiom for *Broken*, (14.3); this can be rewritten as:

$$a = \mathsf{repair}(x) \supset \neg \mathit{Broken}(x, do(a, s)).$$

In general, we can assume that the effect axioms for a fluent F have been written in the forms:

$$\gamma_F^+(\vec{x}, a, s) \supset F(\vec{x}, do(a, s)), \tag{14.4}$$
$$\gamma_F^-(\vec{x}, a, s) \supset \neg F(\vec{x}, do(a, s)). \tag{14.5}$$

Here $\gamma_F^+(\vec{x}, a, s)$ is a formula describing the conditions under which doing the action a in situation s causes the fluent F to become true in the successor situation $do(a, s)$; similarly $\gamma_F^-(\vec{x}, a, s)$ describes the conditions under which performing a in s causes F to become false in the successor situation. The solution to the frame problem rests on a *completeness assumption*, which is that the causal axioms (14.4) and (14.5) characterize all the conditions under which action a can lead to a fluent $F(\vec{x})$ becoming true (respectively, false) in the successor situation. In other words, axioms (14.4) and (14.5) describe all the causal laws affecting the truth values of the fluent F. Therefore, if the truth value of $F(\vec{x})$ changes from *false* to *true* as a result of doing a, then $\gamma_F^+(\vec{x}, a, s)$ must be *true* and similarly for a change from *true* to *false*. It is possible to automatically derive a <u>successor-state axiom</u> from the causal axioms (14.4) and (14.5) and this completeness assumption.

Definition 14.1.3: Successor-state axioms are axioms of the following form

$$F(\vec{x}, do(a, s)) \equiv \gamma_F^+(\vec{x}, a, s) \vee (F(\vec{x}, s) \wedge \neg \gamma_F^-(\vec{x}, a, s))$$

for relational fluents F, and of the form

$$f(\vec{x}, do(a, s)) = z \equiv \gamma_f^+(z, \vec{x}, a, s) \vee (f(\vec{x}, s) = z \wedge \neg \exists z' \gamma_f^-(\vec{x}, z', a, s)),$$

for functional fluents f. These axioms characterize the state of a fluent F (or the value of a functional fluent f) in the successor situation given properties of the current state and the action executed.

Notice that a successor state axiom universally quantifies over actions a. In fact, the ability to universally quantify over actions is one of the keys to obtaining a parsimonious solution to the frame problem.

Returning to our example about breaking things, we obtain the following successor-state axiom:

$$Broken(x, do(a, s))$$
$$\equiv \quad a = \mathsf{drop}(x) \land Fragile(x, s)$$
$$\lor \ \exists b. \, a = \mathsf{explode}(b) \land NextTo(b, x, s)$$
$$\lor \ Broken(x, s) \land a \neq \mathsf{repair}(x).$$

In Section 14.4 below, we will see an example of a successor state axiom for a functional fluent.

14.1.5 Basic action theories

It is important to note that the above solution to the frame problem presupposes that there are no *state constraints*, that is, axioms that relate the state of two fluents across all situations. For example, in the blocks world, $\forall s. On(x, y, s) \supset \neg On(y, x, s)$ is such a law. Such constraints can implicitly contain effect axioms (so-called indirect effects), in which case the above completeness assumption will not be true.[7] For our purposes, we assume that whatever needs to be said about a dynamic world can be formulated using axioms of the following form:

- axioms describing the initial situation—what is true initially, before any actions have occurred;[8]
- action-precondition axioms as above, one for each action type $A(\vec{x})$;
- sensed-fluent axioms as above, one for each action type $A(\vec{x})$;
- successor-state axioms as above, one for each fluent.

In addition, we need to be able to conclude that an action like drop is in fact distinct from an action like inspect or any other action:

- unique-name axioms for the primitive actions: for any pair of action types $A(\vec{x})$ and

---

[7] The approach discussed in this section can be extended to deal with some forms of state constraints, by compiling their effects into the successor-state axioms.
[8] This is any finite set of sentences that mention only the situation term S_0.

$B(\vec{y})$, we have
$$A(\vec{x}) \neq B(\vec{y}) \wedge [A(\vec{x}) = A(\vec{x}') \supset \vec{x} = \vec{x}'].$$

Finally, we need a collection of axioms that characterize the set of situations:

- a set of domain-independent foundational axioms for the situation calculus. These are used to state that the situations are all and only those reachable from S_0 by performing a finite sequence of actions.

A set of axioms of this form is called a *basic action theory*. In Section 14.5, we will examine a generalization of such a theory to handle knowing and only-knowing.

14.2 The logic $\mathcal{AOL}$

$\mathcal{AOL}$ can be thought of as an amalgamation of the situation calculus and $\mathcal{OL}$. We begin by proposing a semantics that extends that of $\mathcal{OL}$ by adding actions and applying it to a slightly extended language of the situation calculus to account for knowledge and standard names. Later, we will also provide a set of axioms which are sound and complete for the semantics.

The language of $\mathcal{AOL}$ starts with the situation calculus as introduced in Section 14.1. The first change required in $\mathcal{AOL}$ concerns ordinary objects. We use the same objects as $\mathcal{OL}$, that is, standard names and add them to the language. In order to facilitate the axiomatization later on, we also include a special unary function *succ* from standard names to standard names. Similar to standard names themselves *succ* is given a fixed interpretation such that $succ(^\#i) = {^\#}(i+1)$ for all i.

To deal with knowledge in $\mathcal{AOL}$, the biggest change from the situation calculus is that we imagine that in addition to S_0 and its successors, there are a number of other initial and non-initial situations considered as possible epistemic alternatives. To state what is known in S_0 and any of its successors, we require the special predicate K_0. Taking S_0 to be the situation counterpart to the given world state w in $\mathcal{OL}$, K_0 is a 1-place predicate over situations that is the counterpart to the given epistemic state e in $\mathcal{OL}$. In other words, $K_0(s)$ is intended to hold if s is a situation considered by the agent in S_0 to be possible. How knowledge changes when performing an action a in situation s is governed by $SF(a, s)$ and $Poss(a, s)$, and will be discussed later in Section 14.3.

To be compatible with $\mathcal{OL}$, we assume that for every n-ary predicate or function symbol of $\mathcal{OL}$ there is a corresponding fluent of arity n+1 with the same name. For example, if $colour(x, y)$ is in $\mathcal{OL}$, then $colour(x, y, s)$ is in $\mathcal{AOL}$. The set of action types is assumed to be finite as is normally done in the situation calculus. For simplicity, we also assume here that there are no other non-fluent predicates or functions besides *succ*.

So, to review, $\mathcal{AOL}$ is a dialect of second-order logic with the following predicate and function symbols playing a special role: S_0, *do*, *Poss*, *SF*, K_0, *succ* and the standard names (here viewed as constants). Note that, in contrast to $\mathcal{OL}$, $\mathcal{AOL}$ does not use modal operators. We will see below that knowing and only-knowing can be defined elegantly using second-order formulas that refer to the predicate K_0.

Semantics

Recall that in the semantics of $\mathcal{OL}$ (and, for that matter, of $\mathcal{L}$ and $\mathcal{KL}$), there are world states corresponding to all possible interpretations of the predicate and function symbols (over the domain of standard names). Different applications, of course, will use different subsets as part of the given e, but the complement of e is still relevant because of only knowing. We need the same in $\mathcal{AOL}$ with respect to K_0, but more: we need to allow for all possible interpretations of the predicate and function symbols *after all possible sequences of actions*. This is to ensure that it is possible to know the initial value of a term or formula without also necessarily knowing its value in successor situations (although in some applications this generality will not be required). Thus instead of defining a world state as any function from primitive expressions to suitable values as in $\mathcal{OL}$, we define a world state in $\mathcal{AOL}$ as any function from primitive expressions and sequences of actions to these values.

More precisely, the primitive formulas of $\mathcal{AOL}$ are predicate applications (including *Poss* and *SF* but excluding K_0) whose arguments are primitive terms, but with the situation argument if any suppressed.[9] The primitive terms are either constants (excluding S_0 and standard names) or function applications (excluding *do* and *succ*) whose arguments are either standard names or primitive action terms, again with any situation argument suppressed. Let Act^* be the set of all sequences of primitive action terms including the empty sequence ϵ. If $\vec{A}$ is such a sequence and A a primitive action term, then $\vec{A} \cdot A$ denotes the sequence $\vec{A}$ followed by A.

World states in $\mathcal{AOL}$ extend those of $\mathcal{OL}$ to deal with the dynamics of actions. Formally, let α be a primitive formula, t a primitive term, and $\vec{A}$ a sequence of primitive actions. Then

- $w[\alpha, \vec{A}]$ is either 0 or 1 and
- $w[t, \vec{A}]$ is a standard name.

Let $\mathcal{W}$ denote the set of all world states in this new sense. An <u>action model</u> M is a pair $\langle e, w \rangle$, where $w \in \mathcal{W}$ and $e \subseteq \mathcal{W}$.

As in $\mathcal{OL}$, w is taken to specify the actual world state, and e specifies the epistemic state as those world states an agent has not yet ruled out as being the actual one. Seman-

---
9 We assume for simplicity that a situation argument only shows up for fluents and in the final argument position.

Knowledge and Action

tically, a situation is then a pair, $(w, \vec{A})$, consisting of a world state and a sequence of actions that have happened so far; as we will see below, a fluent p is considered true in this situation if $w[p, \vec{A}] = 1$. A pair (w, ϵ) is called an *initial situation*.

Before we begin, we need to address a minor complication as far as the interpretation of variables is concerned. Recall that, in $\mathcal{OL}$, the interpretation of variables is handled substitutionally, which is possible because the standard names are part of the language. Of course, we could do the same here for object variables and even for action variables, since primitive actions are part of the language as well.[10] However, the same does not work for situations since they are outside the language, and must be so because there are uncountably many of them. Rather than dealing with two accounts of variable interpretations, we have chosen to interpret all variables non-substitutionally, which is more uniform and also more convenient later on when we compare action models with arbitrary first-order structures.

A variable map v maps object, action, and situation variables into standard names, primitive actions, and situations, respectively. In addition, v assigns relations and functions of the appropriate type[11] to second-order relational and functional variables. For a given v, v_t^x denotes the variable map which is like v except that x is mapped into t.

The meaning of terms

We write $|\cdot|_{M,v}$ for the denotation of terms with respect to an action model $M = \langle e, w \rangle$ and a variable map v. Then

$$
\begin{aligned}
|n|_{M,v} &= n, \text{ where } n \text{ is a standard name;} \\
|succ(t)|_{M,v} &= {}^{\#}(i + 1), \text{ where } |t|_{M,v} = {}^{\#}i; \\
|f(\vec{t}, t_s)|_{M,v} &= w'[f(|\vec{t}|_{M,v}), \vec{A}], \text{ where } f(\vec{t}, t_s) \\
& \quad \text{ is a functional fluent and } |t_s|_{M,v} = (w', \vec{A}); \\
|A(\vec{t})|_{M,v} &= A(|\vec{t}|_{M,v}), \text{ where } A(\vec{t}) \text{ is an action term;} \\
|S_0|_{M,v} &= (w, \epsilon); \\
|do(t_a, t_s)|_{M,v} &= (w', \vec{A} \cdot A), \text{ where } |t_s|_{M,v} = (w', \vec{A}), \text{ and } |t_a|_{M,v} = A; \\
|X(\vec{t})|_{M,v} &= v(X)(|\vec{t}|_{M,v}), \text{ where } X \text{ is a function variable;} \\
|x|_{M,v} &= v(x), \text{ where } x \text{ is any variable, including} \\
& \quad \text{ predicate and function variables.}
\end{aligned}
$$

Observe that in a model $M = \langle e, w \rangle$, the only way to refer to a situation that does not use the given world state w is to use a situation variable.

---

10 In a sense, primitive actions act like standard names for actions.
11 The type determines the arity and the sort of each argument of the relations/functions the variable ranges over. Since, in our examples, the type will always be obvious from the context, we leave this information implicit.

The meaning of formulas

We write $M, \nu \models \alpha$ to mean α is true in action model M under variable map ν:

$M, \nu \models F(\vec{t}, t_s)$	iff	$w'[F(|\vec{t}|_{M,\nu}), \vec{A}] = 1$, where $F(\vec{t}, t_s)$ is a relational fluent and $|t_s|_{M,\nu} = (w', \vec{A})$;
$M, \nu \models X(\vec{t})$	iff	$|\vec{t}|_{M,\nu} \in \nu(X)$, where X is a relational variable;
$M, \nu \models Poss(t_a, t_s)$	iff	$w'[Poss(|t_a|_{M,\nu}), \vec{A}] = 1$, where $|t_s|_{M,\nu} = (w', \vec{A})$;
$M, \nu \models SF(t_a, t_s)$	iff	$w'[SF(|t_a|_{M,\nu}), \vec{A}] = 1$, where $|t_s|_{M,\nu} = (w', \vec{A})$;
$M, \nu \models K_0(t_s)$	iff	$|t_s|_{M,\nu} = (w', \epsilon)$ and $w' \in e$;
$M, \nu \models t_1 = t_2$	iff	$|t_1|_{M,\nu} = |t_2|_{M,\nu}$;
$M, \nu \models \neg\alpha$	iff	$M, \nu \not\models \alpha$;
$M, \nu \models \alpha \vee \beta$	iff	$M, \nu \models \alpha$ or $M, \nu \models \beta$;
$M, \nu \models \forall x.\alpha$	iff	$M, \nu^x_o \models \alpha$ for all o of the appropriate sort (object, action, situation, relation).

If α is a sentence, we write $M \models \alpha$ instead of $M, \nu \models \alpha$. Validity is defined in the usual way as truth in all action models, that is, a sentence α is valid in $\mathcal{AOL}$ iff for all action models $M = \langle e, w \rangle$, $M \models \alpha$.

Quantifying over situations

Before looking at how to characterize knowledge in $\mathcal{AOL}$, it is worth noting that care is required in using $\mathcal{AOL}$ to state properties of the world. Whereas in the situation calculus, we would write formulas like

$$\forall x, s. \neg Broken(x, do(\text{repair}(x), s))$$

to state that an object x would not be broken after repairing it, in $\mathcal{AOL}$, the universal quantifier over situations cannot be used so directly. In fact, it is not hard to see that the formula above is unsatisfiable: choose any w such that $w[Broken(n), \langle \text{repair}(n) \rangle] = 1$; then for any M and ν, we have that

$$M, \nu^s_{(w,\epsilon)} \models Broken(n, do(\text{repair}(n), s))$$

and so

$$M, \nu \models \neg \forall x, s. \neg Broken(x, do(\text{repair}(x), s)).$$

The problem here is that the universal quantifier over situations ranges not only over S_0 and its successors, as in the situation calculus, but also over *all* initial situations and their successors. This set of situations is determined in advance by the logic itself, just as the set of world states was determined in advance in $\mathcal{L}$, $\mathcal{KL}$, and $\mathcal{OL}$. So a formula like $\exists s. Broken(n, do(\text{repair}(n), s))$ is *valid* in $\mathcal{AOL}$, because we know that the corresponding world state must exist.

So how do we capture this situation calculus axiom in $\mathcal{AOL}$? The answer is to understand this axiom as a constraint on S_0 and its successors: while the logic of $\mathcal{AOL}$ guarantees that there are situations where repairing an object leaves it broken, we can assert that for S_0 and its successors, this is not the case.

To talk formally about the successors of a situation, we use the following abbreviation:

$$s \preceq s' \doteq \forall R[\ldots \supset R(s, s')]$$

with the ellipsis standing for the conjunction of

$\forall s_1.\ R(s_1, s_1)$
$\forall a, s_1.\ R(s_1, do(a, s_1))$
$\forall s_1, s_2, s_3.\ R(s_1, s_2) \wedge R(s_2, s_3) \supset R(s_1, s_3)$.

By using second-order quantification in this way, we guarantee that $s \preceq s'$ iff there is a finite (possibly empty) sequence of actions going from s to s'.[12]

So the correct way to formulate the effect axiom in $\mathcal{AOL}$ is as follows:

$\forall x, s.\ S_0 \preceq s \supset \neg Broken(x, do(\mathsf{repair}(x), s))$.

This states that the real world (as determined by the given w) is such that repairing an object always does the right thing. It leaves open how this might work in other possible worlds.

To talk formally about other possible worlds, it is convenient to introduce another abbreviation:

$$Init(s') \doteq \neg \exists a, s.s' = do(a, s).$$

Clearly $Init(S_0)$ is valid, and it is not too hard to show that every situation is a successor of some initial one:

Theorem 14.2.1: *The following is a valid sentence of $\mathcal{AOL}$:*

$\forall s \exists s'\ Init(s') \wedge s' \preceq s$.

Proof: Left as an exercise. ∎

14.3 Knowledge after action

While the given w in an action model is intended to represent the state of the world, the given e is intended to represent the epistemic state. Note that in our specification of the meaning of $\mathcal{AOL}$ formulas above, the only place where we use the e is in the interpretation of K_0. This means that knowing and only-knowing in $\mathcal{AOL}$ are defined in terms of K_0. We

---

[12] No first-order formula would work here, because of the transitive closure involved.

will end up presenting abbreviations *Knows*(ϕ, *s*) and *OKnows*(ϕ, *s*) for formulas of $\mathcal{AOL}$ that state that ϕ is known or all that is known in situation *s*. The ϕ in these abbreviations is a formula of $\mathcal{AOL}$ containing a special situation constant *now*. The definition of the abbreviations below will show how to replace this term by an ordinary situation term.

14.3.1 An informal characterization

To show how this works, we begin by characterizing what is known initially before any actions have been performed. Then we will show how to characterize what is known in other situations.

Initial knowledge

Roughly speaking, a formula ϕ will be considered to be known in S_0 if it comes out true at all the situations in K_0; a formula is considered to be all that is known in S_0 if, in addition, any initial situation where the formula comes out true is in K_0. This is the same intuition as in $\mathcal{OL}$.

Consider, for example, the formula *Broken*(*n*, *s*) which says that object *n* is broken in situation *s*. To say that in S_0 it is known that *n* is currently broken is to say that

$$\forall s. K_0(s) \supset Broken(n, s)$$

is true. So *Knows*(*Broken*(*n*, *now*), S_0) stands for this formula, and *Knows*(ϕ, S_0) will be an abbreviation for the formula

$$\forall s. K_0(s) \supset \phi_s^{now}.$$

Similarly, *OKnows*(ϕ, S_0) will be an abbreviation for the formula

$$\forall s. Init(s) \supset [K_0(s) \equiv \phi_s^{now}].$$

Clearly, the K_0 predicate is being used to specify what situations the agent considers possible initially. The question we now want to address is what situations the agent should consider possible in situations other than S_0, that is, after performing an action. We will be looking for something like

$$\forall s. \ldots \supset \phi_s^{now}$$

where the ellipsis stands for the situations the agent considers possible after performing the action.

Ordinary actions

Let us take the simplest case first, that of performing a simple action such as drop(*x*), dropping an object *x*. We would like to say that after doing this action, the agent does not learn anything new about the world other than the fact that the action was just performed successfully (and anything that follows from that). Since K_0 represents the situations believed to be possible initially, the new situations believed to be possible should be the successors of those in K_0 after a successful drop action. That is to say, the agent now considers s possible if $s = do(\text{drop}(x), s')$ for some s' in K_0 for which $Poss(\text{drop}(x), s')$ holds (the action completes successfully). In general, for any ordinary action a like $\text{drop}(x)$, $Knows(\phi, do(a, S_0))$ will stand for the formula

$$\forall s. \exists s'(s = do(a, s') \land K_0(s') \land Poss(a, s')) \supset \phi_s^{now}.$$

It is not hard to see that $Knows(Broken(n), do(\text{drop}(n), S_0))$ holds, meaning that one way to find out whether or not an object is broken is to drop it. This is usually not an advisable plan, however. Instead, we would like to use sensing to find out the state of the object.

Sensing actions

As we mentioned in Section 14.1.2, we use the special predicate SF to characterize the situations where an action returns a sensing value of 1. The effect sensing has on knowledge is that after an action like inspect, the agent should know whether or not the object was broken. Thus, the new situations believed to be possible should be the successors of those in K_0 after a successful inspect action where $Broken(x)$ has the same status as it does in S_0. That is to say, the agent now considers s possible if $s = do(\text{inspect}(x), s')$ for some s' in K_0 for which $Poss(\text{inspect}(x), s')$ holds and for which

$$Broken(x, s') \equiv Broken(x, S_0)$$

is true. More generally, we can use the sensed-fluent axioms from Section 14.1.2, and $Knows(\phi, do(a, S_0))$ will stand for the formula

$$\forall s. \exists s'(s = do(a, s') \land K_0(s') \land Poss(a, s') \land [SF(a, s') \equiv SF(a, S_0)]) \supset \phi_s^{now}.$$

Observe that with this abbreviation, if we assume that $Broken(x, S_0)$ is true, we will indeed get that $Knows(Broken(n), do(\text{inspect}(n), S_0))$ is also true. Also note that this abbreviation generalizes the previous one and works properly for actions like drop as well.

14.3.2 Knowing in $\mathcal{AOL}$

We are now ready to generalize this story to deal with arbitrary action sequences. Here is the idea: knowledge in a situation t will mean truth in all situations considered possible in t. The situations considered possible in t will be the least set satisfying the following:

- the elements of K_0 are considered possible in S_0 and in any other initial situation,
- a situation $do(a, s_1)$ is considered possible in $do(a, s_2)$ provided that s_1 was considered possible in s_2, and both $Poss(a, s_1)$ and $[SF(a, s_1) \equiv SF(a, s_2)]$ hold.

We can then use second-order logic to do the minimization, which gives the following:

Definition 14.3.1: Let $K(t_1, t_2)$ be an abbreviation for $\forall P[\ldots \supset P(t_1, t_2)]$ where the ellipsis is the conjunction of

$$K_0(s_1) \wedge Init(s_2) \supset P(s_1, s_2)$$
$$P(s_1, s_2) \wedge Poss(a, s_1) \wedge [SF(a, s_1) \equiv SF(a, s_2)] \supset P(do(a, s_1), do(a, s_2)).$$

Then $Knows(\phi, t)$ is an abbreviation for the formula

$$\forall s. K(s, t) \supset \phi_s^{now}.$$

So knowledge at arbitrary situations is not a primitive in $\mathcal{AOL}$ but is in fact definable as a certain second-order formula. As a key property of K we obtain:

Theorem 14.3.2: *The following is a valid sentence of $\mathcal{AOL}$:*

$$K(s', do(a, s)) \equiv \exists s''. s' = do(a, s'') \wedge$$
$$K(s'', s) \wedge Poss(a, s'') \wedge [SF(a, s'') \equiv SF(a, s)]$$

Proof: Let us call the ellipsis in the definition of K ell_K. To prove the if direction, let $M, \nu \models s' = do(a, s'') \wedge K(s'', s) \wedge Poss(a, s'') \wedge [SF(a, s'') \equiv SF(a, s)]$ and let $\nu(P)$ be an arbitrary binary relation over situations such that $M, \nu \models ell_K$. Since $M, \nu \models K(s'', s)$, we have that $M, \nu \models P(s'', s)$. Together with the second part of ell_K we thus obtain $M, \nu \models P(do(a, s''), do(a, s))$. Since this holds for all P satisfying ell_K and since $M, \nu \models s' = do(a, s'')$ by assumption, $M, \nu \models K(s', do(a, s))$ follows.

To show the converse, we need the fact that there is a unique minimal set of pairs of situations R_{min} which satisfies ell_K. R_{min} can be constructed as follows. Let

$$R_0 = \{((w', \epsilon), (w, \epsilon)) \mid w' \in e, w \in \mathcal{W}\}$$
$$R_{i+1} = \{((w', \vec{b} \cdot A), (w, \vec{b} \cdot A)) \mid ((w', \vec{b}), (w, \vec{b})) \in R_i,$$
$$w'[Poss(A), \vec{b}] = 1, w'[SF(A), \vec{b}] = w[SF(A), \vec{b}]\}$$

Then let $R_{min} = \bigcup R_i$. It is not hard to show that R_{min} satisfies ell_K and that any P which satisfies ell_K must be a superset of R_{min}. Hence R_{min} is unique and minimal.

Let $M, \nu \models K(s', do(a, s))$ with $\nu(s) = (w, \vec{b})$ for some world state w and action sequence $\vec{b}$. Thus $|do(a, s)|_{M,\nu} = (w, \vec{b} \cdot A)$ where $A = \nu(a)$. By the minimality of R_{min}, $(\nu(s'), |do(a, s)|_{M,\nu}) \in R_{min}$ and, therefore, $(\nu(s'), |do(a, s)|_{M,\nu}) \in R_{i+1}$ for some $i \geq 0$, that is, $\nu(s') = (w', \vec{b} \cdot A)$ for some w'. Now let $\nu(s'') = (w', \vec{b})$. Then clearly, $M, \nu \models s' = do(a, s'')$. Since $((w', \vec{b}), (w, \vec{b})) \in R_i$, we obtain $M, \nu \models K(s'', s)$. Also, given the definition of R_{i+1} we have that $M, \nu \models Poss(a, s'') \wedge [SF(a, s'') \equiv SF(a, s)]$. Taken together we obtain $M, \nu \models \exists s''. s' = do(a, s'') \wedge K(s'', s) \wedge Poss(a, s'') \wedge [SF(a, s'') \equiv SF(a, s)]$, which completes the proof. ∎

Knowledge and Action

Observe that this theorem is in the form of a successor-state axiom for what we can take to be a fluent K. In other words, with the above definition, we obtain as a consequence a solution to the frame problem for knowledge.

The theorem is helpful in establishing various properties of knowledge. For example, positive and negative introspection obtain, that is, both

$Knows(\phi, s) \supset Knows(Knows(\phi, now), s)$ and
$\neg Knows(\phi, s) \supset Knows(\neg Knows(\phi, now), s)$

are valid. Also an agent, after performing an action, cannot help but believe that the action was possible, even if it was actually impossible. Formally,

$Knows([\forall s''.now = do(a, s'') \supset Poss(a, s'')], do(a, s))$

is valid in $\mathcal{AOL}$. We leave the proof of these properties as exercises.

14.3.3 Only-knowing in $\mathcal{AOL}$

We can define only-knowing in $\mathcal{AOL}$ using a similar technique. The idea is that ϕ will be all that is known in situation s if it is known, and every situation s' of the right sort in which ϕ_s^{now} is true will be considered by the agent to be possible. In this case, a situation s' is the right sort for s if it consisted of the same sequence of actions as s. More precisely, we have

Definition 14.3.3: Let $t \simeq t'$ be an abbreviations for $\forall R[\ldots \supset R(t, t')]$ where the ellipsis stands for the conjunction of

$\forall s_1, s_2.\ Init(s_1) \land Init(s_2) \supset R(s_1, s_2)$
$\forall a, s_1, s_2.\ R(s_1, s_2) \supset R(do(a, s_1), do(a, s_2))$.

Then we let $OKnows(\phi, t)$ be an abbreviation for the formula

$\forall s.\ s \simeq t \supset (K(s, t) \equiv \phi_s^{now})$

where K is as above.

It is not hard to show that $Knows$ above satisfies

$\forall s.\ s \simeq t \supset (K(s, t) \supset \phi_s^{now})$

and so the difference between knowing and only-knowing is that an "if" becomes an "iff," exactly as we had in Chapter 8.

14.4 Using $\mathcal{AOL}$

We now turn to an example showing how knowing and only-knowing can be combined with actions in $\mathcal{AOL}$.

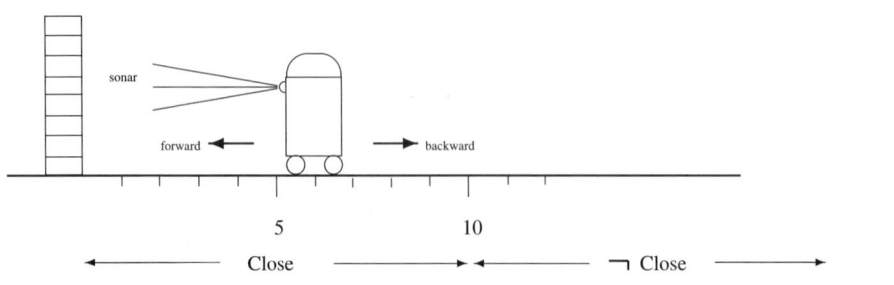

Figure 14.1: A one-dimensional robot world

Imagine a robot that lives in a 1-dimensional world, and that can move towards or away from a fixed wall. The robot also has a sonar sensor that tells it when it gets close to the wall, say, less than 10 units away. See Figure 14.1. So we might imagine three actions, forward and backward which move the robot one unit towards and away from the wall, and a sonar sensing action. We have a single fluent, *wall-distance(s)*, which gives the actual distance from the robot to the wall in situation s. We can use the formula *Close(s)* as an abbreviation for *wall-distance(s)* < 10.

We begin by writing precondition axioms for the three actions, parameterized by some initial situation s (as discussed in Section 14.2).

$\forall s'. s \preceq s' \supset Poss(\text{forward}, s') \equiv \textit{wall-distance}(s') > 0.$[13]

$\forall s'. s \preceq s' \supset Poss(\text{backward}, s') \equiv \text{TRUE}.$

$\forall s'. s \preceq s' \supset Poss(\text{sonar}, s') \equiv \text{TRUE}.$

Next, we define sensed fluent axioms for the three actions:

$\forall s'. s \preceq s' \supset SF(\text{forward}, s') \equiv \text{TRUE}.$

$\forall s'. s \preceq s' \supset SF(\text{backward}, s') \equiv \text{TRUE}.$

$\forall s'. s \preceq s' \supset SF(\text{sonar}, s') \equiv \textit{Close}(s').$

Finally, we define a successor state axiom for our single fluent *wall-distance*:

$\forall s'. s \preceq s' \supset \forall a. \textit{wall-distance}(do(a, s')) = z \equiv$
$\quad a = \text{forward} \wedge z = \textit{wall-distance}(s') - 1$
$\quad \vee\ a = \text{backward} \wedge z = \textit{wall-distance}(s') + 1$
$\quad \vee\ z = \textit{wall-distance}(s') \wedge a \neq \text{forward} \wedge a \neq \text{backward}.$

We let ALL(s) denote the conjunction of these formulas.

---

[13] Here and in the successor state axiom below, we freely use simple arithmetic involving $<$, $+$, and $-$, which can easily be defined in second-order terms with the standard names acting as natural numbers. We leave out the details here.

Now we are ready to consider some specifics having to do with what is true initially by building an initial action theory Γ_0. Assume the robot is located initially 6 units away from the wall in S_0. For simplicity, we also assume that all of the axioms above are true in S_0, that they are known in S_0, and they are all that is known in S_0. So let Γ_0 be the following:

$$ALL(S_0) \wedge \textit{wall-distance}(S_0) = 6 \wedge OKnows(ALL(now), S_0),$$

and we get

Theorem 14.4.1: *The following are entailments of* Γ_0:

1. $Close(S_0) \wedge \neg Knows(Close(now), S_0)$
   the robot is close to the wall initially, but does not know it;

2. $\neg Knows(Close(now), do(\textsf{forward}, S_0))$
   after moving forward, the robot still does not know it is close;

3. $Knows(Close(now), do(\textsf{sonar}, S_0))$
   after reading the sonar, the robot knows it is close to the wall;

4. $Knows(Close(now), do(\textsf{forward}, do(\textsf{sonar}, S_0)))$
   after reading the sonar and then moving forward, the robot continues to know that it is close;

5. $\neg Knows(Close(now), do(\textsf{backward}, do(\textsf{sonar}, S_0)))$
   after reading the sonar and then moving backward, the robot no longer knows that it is close to the wall;

6. $Knows(Close(now), do(\textsf{sonar}, do(\textsf{backward}, S_0)))$
   after moving backward and then reading the sonar, the robot does know that it is close;

7. $Knows(Close(now), do(\textsf{backward}, do(\textsf{forward}, do(\textsf{sonar}, S_0))))$
   after reading the sonar, moving forward, and then backward, the robot knows that it is still close;

8. $Knows(Close(do(\textsf{forward}, now)), do(\textsf{sonar}, S_0))$
   after reading the sonar, the robot knows that it will remain close after moving forward;

9. $\neg Knows(Knows(Close(now), do(\textsf{sonar}, now)), S_0)$
   the robot does not know initially that it will know that it is close after reading the sonar;

10. $Knows(KnowsIf(Close(now), do(\textsf{sonar}, now)), S_0)$
    the robot does know initially that after reading the sonar, it will know whether or not it is close to the wall, where $KnowsIf(\phi, t)$ is an abbreviation for the formula $(Knows(\phi, t) \vee Knows(\neg \phi, t))$;

11. $Knows(\neg Knows(Close(now), do(\textsf{backward}, now)), S_0)$
    the robot knows initially that it will not know that it is close after moving backwards.

Proof: We prove item 4 and leave the rest as exercises.

To establish that *Knows*(*Close*(*now*), *do*(forward, *do*(sonar, S_0))) follows from Γ_0, we must show that for any s such that $K(s, do(\text{forward}, do(\text{sonar}, S_0)))$ we have that *Close*(s)) holds. By Theorem 14.3.2, any such s is equal to *do*(forward, *do*(sonar, s')) for some s' such that $K(s', S_0)$ and $[SF(sonar, s') \equiv SF(sonar, S_0)]$ hold. By Γ_0, we get that *Close*(S_0) holds and, therefore, *Close*(s') holds also. Given the successor-state axiom for *wall-distance*, we know that it is unaffected by sonar, and decreased by forward. *Close*(s) thus follows. ∎

14.5 An axiomatization of $\mathcal{AOL}$

In this section, we will develop an axiomatization that is complete for $\mathcal{AOL}$, that is, any valid sentence of $\mathcal{AOL}$ follows from the axioms, where "follows" refers to classical logical implication in second-order logic.[14] Unlike $\mathcal{OL}$, the completeness result becomes possible because we are no longer restricted to first-order formulas. For convenience we assume that equality is built into the logic so that no special axioms are needed for it.

When axiomatizing $\mathcal{AOL}$, the main problem we will have to face is to find a characterization which ensures that there are enough initial situations, where enough means precisely as many situations as there are in the semantics of $\mathcal{AOL}$. It turns out that the required axiom is nontrivial even in the case where there are only finitely many fluents. Here we will in fact only consider the finite case in detail and merely hint at how to extend the axiomatization to deal with infinitely many fluents.

The first three axioms tell us that the set of objects is isomorphic to the set $\mathcal{N}$ of standard object names. Indeed the formulation is a variant of the usual second-order definition of the natural numbers, that is, the following axioms do no more than give us domain closure and unique-name axioms for objects.

F1: $succ(^\#i) = {^\#(i+1)}$ for all standard names $^\#i$.
F2: $^\#i \neq {^\#j}$ for any distinct standard names $^\#i$ and $^\#j$.
F3: $\forall P.([P(^\#1) \land \forall x, y. P(x) \land succ(x) = y] \supset P(y)] \supset \forall x. P(x)$.

Next we need to say that the actions consist precisely of the primitive actions and that they are all distinct. This can be done in the usual way.

F4: Domain closure and unique names axioms for actions.

---

14 Saying that we have an axiomatization must be taken with a grain of salt, since it is well known that the valid sentences of second-order logic cannot be fully captured by axioms and inference rules alone as in first-order logic, that is, we must still appeal to the semantic notion of second-order logical implication to obtain what follows from the axioms.

Knowledge and Action

Neither *SF* nor *Poss* need special axioms since their meaning is left completely user-defined. The only foundational axiom concerning K_0 is one saying that it only applies to initial situations:

F5: $Init(S_0) \land \forall s.K_0(s) \supset Init(s)$.

Finally we have the job of characterizing the set of situations. We first want to say that any non-initial situation is the result of applying *do* to an initial situation and that situations have unique names.

F6: $\forall s \exists s' \, Init(s') \land s' \preceq s$.
F7: $\forall a_1, a_2, s_1, s_2. \, do(a_1, s_1) = do(a_2, s_2) \supset (a_1 = a_2 \land s_1 = s_2)$.

Note the use $\preceq$, which we introduced earlier as an abbreviation for a second-order sentence, in Axiom F6. It has the effect of minimizing the set of situations, that is, it makes sure that there are no situations other than those which can be reached from some initial situation by a finite sequence of actions.

Then the only remaining job is to characterize the set of initial situations. Looking back at the semantics of $\mathcal{AOL}$, recall that for a correct interpretation of only-knowing, we had to insist that there be an initial situation corresponding to any conceivable outcome of the fluents initially and after any sequence of actions. Given the power of second-order logic, it is possible to precisely capture this property axiomatically. Since the axioms themselves are somewhat complex, we begin with a simplified version, which, while not quite strong enough, demonstrates the key principles. We need two axioms, one which says that there are enough initial situations for every conceivable valuation of the fluents and another one which stipulates that there are no more initial situations than actually required by the axiom. We begin with the following definition:

Suppose that our language contains relational fluents $F_1, \ldots, F_n$ and $f_1, \ldots, f_m$.[15]
Let $FV(P_1, \ldots, P_{n+2}, h_1, \ldots, h_m, s)$ stand for

$\forall \vec{x}_1, \ldots \vec{x}_n, \vec{y}_1, \ldots \vec{y}_m, a.$
$[F_1(\vec{x}_1, s) \equiv P_1(\vec{x}_1) \land \ldots \land F_n(\vec{x}_n, s) \equiv P_n(\vec{x}_n) \land$
$f_1(\vec{y}_1, s) = h_1(\vec{y}_1) \land \ldots \land f_m(\vec{y}_m, s) = h_m(\vec{y}_m) \land$
$Poss(a, s) \equiv P_{n+1}(a) \land SF(a, s) \equiv P_{n+2}(a)].$

If we let $\vec{P}$ stand for the sequence $P_1, \ldots, P_{n+2}, h_1, \ldots, h_m$, then the first version of the axioms looks as follows:

---

[15] Recall that our language has only finitely many relational and functional fluents. In fact, it is precisely this axiom that depends on this assumption.

$\forall \vec{P} \exists s. Init(s) \wedge FV(\vec{P}, s).$
$\forall s_1, s_2. Init(s_1) \wedge Init(s_2) \wedge s_1 \neq s_2 \supset \exists \vec{P}. FV(\vec{P}, s_1) \wedge \neg FV(\vec{P}, s_2).$

Note the role of $FV(\vec{P}, s)$ here: it makes sure that the fluents of situation s take on the values as chosen by $\vec{P}$.

Quantifying over $\vec{P}$ is intended to give us all the initial situations we need. Unfortunately, it does not. To see why consider the case of two initial situations s_1 and s_2 which agree on all fluents but where after doing an action a, the two resulting situations $do(a, s_1)$ and $do(a, s_2)$ differ. While we need situations such as these, the axioms and, in particular, the first one does not guarantee their existence. The axioms are satisfied by models where no two initial situations are identical.

Before fixing the axioms, let us be clearer about how many initial situations we need. We need initial situations not only for all conceivable valuations of the fluents before any actions have occurred but for all conceivable valuations of the fluents after any number of actions have occurred.

To understand the solution, let us assume for a moment that we had a sort for action sequences as part of the language. (We will see below how we can actually encode action sequences in $\mathcal{AOL}$.) The main idea then is to add to each P_i and h_j an extra argument which is an action sequence. This will allow us to make the desired distinctions.

Let $FV(\vec{P}, \vec{a}, s)$ stand for the following expression:

$\forall \vec{x}_1, ...\vec{x}_n, \vec{y}_1, ...\vec{y}_m, a.$
$\quad [F_1(\vec{x}_1, do(\vec{a}, s)) \equiv P_1(\vec{x}_1, \vec{a}) \wedge ... \wedge F_n(\vec{x}_n, do(\vec{a}, s)) \equiv P_n(\vec{x}_n, \vec{a}) \wedge$
$\quad f_1(\vec{y}_1, do(\vec{a}, s)) = h_1(\vec{y}_1, \vec{a}) \wedge ... \wedge f_m(\vec{y}_m, do(\vec{a}, s)) = h_m(\vec{y}_m, \vec{a}) \wedge$
$\quad Poss(a, do(\vec{a}, s)) \equiv P_{n+1}(a, \vec{a}) \wedge SF(a, do(\vec{a}, s)) \equiv P_{n+2}(a, \vec{a})].$

The axioms now become

$\forall \vec{P} \exists s. Init(s) \wedge \forall \vec{a}. FV(\vec{P}, \vec{a}, s).$
$\forall s_1, s_2. Init(s_1) \wedge Init(s_2) \wedge s_1 \neq s_2 \supset \exists \vec{P}, \vec{a}. FV(\vec{P}, \vec{a}, s_1) \wedge \neg FV(\vec{P}, \vec{a}, s_2).$

Of course, since action sequences are not part of our language, we cannot use these axioms. Fortunately, there is a way to refer to action sequences in $\mathcal{AOL}$ without having to represent them explicitly.

Given the definition of $\simeq$, we can use situations t where $S_0 \preceq t$ as a canonical way of talking about sequences of actions.

$FV(\vec{P}, t, s)$ can then can be expressed as

$\forall \vec{x}_1, ...\vec{x}_n, \vec{y}_1, ...\vec{y}_m, a, u. s \preceq u \wedge u \simeq t \supset$
$\quad [F_1(\vec{x}_1, u) \equiv P_1(\vec{x}_1, t) \wedge ... \wedge F_n(\vec{x}_n, u) \equiv P_n(\vec{x}_n, t) \wedge$
$\quad f_1(\vec{y}_1, u) = h_1(\vec{y}_1, t) \wedge ... \wedge f_m(\vec{y}_m, u) = h_m(\vec{y}_m, t) \wedge$

Knowledge and Action

$$Poss(a,u) \equiv P_{n+1}(a,t) \wedge SF(a,u) \equiv P_{n+2}(a,t)].$$

Then the final form of the axioms is:

F8: $\forall \vec{P} \exists s. Init(s) \wedge \forall t. S_0 \preceq t \supset FV(\vec{P},t,s)$.
F9: $\forall s_1, s_2. Init(s_1) \wedge Init(s_2) \wedge s_1 \neq s_2 \supset$
$\exists \vec{P}, t. S_0 \preceq t \wedge FV(\vec{P},t,s_1) \wedge \neg FV(\vec{P},t,s_2)$.

These are all the axioms we need, and we will refer to them collectively as AX from now on, and we let $AX \models \alpha$ stand for "α is logically implied by AX" in ordinary (second-order) logic.

Soundness and completeness

In this section we want to establish that action models and AX capture the same logic, that is, that the valid sentences of $\mathcal{AOL}$ are precisely those that follow from AX.

It is not hard to show that the axioms are indeed sound with respect to the semantics of $\mathcal{AOL}$.

Theorem 14.5.1: *The axioms* AX *are valid in* $\mathcal{AOL}$.

Proof: The soundness of F1–F7 is easy to establish. In fact, the soundness of F6 was already shown in Theorem 14.2.1. Here we only consider F8 and F9 informally.

F8 stipulates that for any choice of the P_i and h_j we can find an appropriate initial situation s such that the values for F_i, f_j, $Poss$ and SF agree with those of P_i and h_j in all situations reachable from s. Since world states in $\mathcal{AOL}$ cover all possible valuations of the primitive terms and formulas over all possible action sequences, we can clearly find an appropriate world w where s is mapped into (w, ϵ) such that the above holds.

F9 says that for any two distinct initial situations s_1 and s_2 there exists an action sequence $\vec{A}$ such that $do(\vec{A}, s_1)$ and $do(\vec{A}, s_2)$ differ in the value of at least one of the fluents or $Poss$ or SF. Since world states in $\mathcal{AOL}$ are defined extensionally, there are no duplicates, and hence F9 holds. ∎

In order to prove completeness, we actually establish a stronger result, namely that the action models are, in a sense, the only models of AX. More precisely, we show that for any arbitrary Tarskian model $\mathcal{I}$ of AX there is an action model M such that $\mathcal{I}$ and M agree on all sentences. Here $\mathcal{I}$ is understood as a structure $\langle D, \Phi \rangle$, where D is the (sorted) domain of discourse and Φ assigns functions and relations to the corresponding symbols of the language. $\mathcal{I}$ together with a variable map ν defines the truth of a formula α ($\mathcal{I}, \nu \models \alpha$) in the usual way. The key property is that the objects, actions, and situations of an arbitrary model of AX are isomorphic to the objects, actions, and situations, respectively, of action models.

Lemma 14.5.2: Let $\mathcal{I} = \langle D, \Phi \rangle$ be a model of AX. Then there is an isomorphism * between D and the objects, actions, and situations of action models.

Proof: Let D be partitioned into sets D_o, D_a, and D_s, which correspond to the domain elements of sort object, action, and situation, respectively. Since $\mathcal{I}$ satisfies F1–F3, we have that for every $d \in D_o$ there is a unique standard name n_d such that $\mathcal{I}, v_d^x \models x = n_d$. Hence we let $d^* = n_d$, which clearly results in an isomorphism on objects of both models. Similarly, since $\mathcal{I}$ satisfies the domain closure and unique names axiom for actions, we have that for every $d_a \in D_a$ there are a unique action symbol A and objects $d_1, \ldots, d_k \in D_o$ such that $d_a = \Phi(A)(d_1, \ldots, d_k)$. Hence we obtain an isomorphism on actions with $d_a^* = [\Phi(A)(d_1, \ldots, d_k)]^* = A(d_1^*, \ldots, d_k^*)$.

We are left to show that the situations in both models are isomorphic as well. We begin by constructing an appropriate mapping * for situations and then show that it is indeed an isomorphism. Let $\sigma \in D_s$ be an arbitrary situation. By Axiom F6, σ must be reachable from some initial situation σ', that is, $\mathcal{I}, v_{\sigma'/\sigma}^{s'/s} \models Init(s') \wedge s' \preceq s \wedge s = do(\vec{a}, s')$ for some $\vec{a} \in Act^*$. Then let $\sigma^* = (w, \vec{a})$ where w is a world state such that for all $\vec{b} \in Act^*$,

$w[F(d_1^*, \ldots, d_k^*), \vec{b}] = 1$ iff $(d_1, \ldots, d_k, \sigma_1) \in \Phi(F)$ with
$\mathcal{I}, v_{\sigma'/\sigma_1}^{s'/s_1} \models (s_1 = do(\vec{b}, s'))$ for all relational fluents F including *Poss* and *SF*;

$w[f(d_1^*, \ldots, d_k^*), \vec{b}] = d^*$ iff $\Phi(f)(d_1, \ldots, d_k, \sigma_1) = d$ with
$\mathcal{I}, v_{\sigma'/\sigma_1}^{s'/s_1} \models (s_1 = do(\vec{b}, s'))$ for all functional fluents f.

Such a world state w clearly exists and is unique.

We begin by showing that * is an isomorphism over initial situations. It follows immediately from the construction that * maps initial situations from D_s into initial situations of action models. To show that * is an injection assume, to the contrary, that there are distinct initial σ_1 and σ_2 such that $\sigma_1^* = \sigma_2^* = (w, \epsilon)$. Let us call the situations reachable from a given initial situation a situation tree. By the construction of *, we have that the situation trees for σ_1 and σ_2 agree at every situation. However, this conflicts with Axiom F9 which says that any two distinct initial situations must differ in the value of some fluent after some number of actions have occurred. Hence our initial assumption that $\sigma_1^* = \sigma_2^*$ must be false.

To show that * is a surjection assume, to the contrary, that there is some (w, ϵ) which is not an image under *. Thus for every initial situation σ' of $\mathcal{I}$ there is a situation σ reachable from σ' by doing $\vec{a}$ such that σ and $(w, \vec{a})$ differ in at least one fluent value. It is easily seen that this conflicts with Axiom F8. (Simply choose $\vec{P}$ such that $(\vec{n}, do(\vec{a}, S_0)) \in P_i$ iff

Knowledge and Action   267

$w[F_i(\vec{n}), \vec{a}] = 1$.) Hence $*$ is a surjection.

We are left to show that $*$ is an isomorphism for all situations. To show that $*$ is an injection let σ and σ' be arbitrary situations with $\sigma^* = \sigma'^* = (w, \vec{a})$. By F6, σ and σ' originate from initial situation σ_0 and σ'_0. Given our construction of $*$, $\sigma_0^* = \sigma'^*_0 = (w, \epsilon)$. Hence, by assumption, $\sigma_0 = \sigma'_0$ and, therefore, $\sigma = \sigma'$. To show that $*$ is surjective, let $(w, \vec{a})$ be any $\mathcal{AOL}$-situation. Since we know already that $*$ is an isomorphism over the initial situations, there is some σ_0 such that $\sigma_0^* = (w, \epsilon)$. Then let σ be such that $\mathcal{I}, v_{\sigma_0\sigma}^{s_0 s} \models s = do(\vec{a}, s_0)$. Then $\sigma^* = (w, \vec{a})$. ∎

Lemma 14.5.3: *Let $\mathcal{I} = \langle D, \Phi \rangle$ be a model of* AX *and let $*$ be the isomorphism of the previous lemma extended to apply to arbitrary n-ary relations R and functions f over D in the following way:*

- $(d_1, \ldots, d_n) \in R^*$ *iff* $(d'_1, \ldots, d'_n) \in R$ *with* $d_i = d'^*_i$.
- $f^*(d_1, \ldots, d_n) = d$ *iff* $f(d'_1, \ldots, d'_n) = d'$ *with* $d_i = d'^*_i$ *and* $d' = d'^*$.

(Note that f^* is totally defined since $*$ is an isomorphism.)

Then there is an action model $M = \langle e, w_0 \rangle$ such that for every term t, variable map v, and formula α,

1. $|t|^*_{\mathcal{I}, v} = |t|_{M, v^*}$.
2. $\mathcal{I}, v \models \alpha$ *iff* $M, v^* \models \alpha$, *where* $v^*(v) = v(v)^*$ *for any variable v.*

Proof: We define the action model corresponding to $\mathcal{I}$ as $M = \langle e, w_0 \rangle$, where

$(w_0, \epsilon) = |S_0|^*_{\mathcal{I}, v}$ and $e = \{w \mid \exists \sigma \text{ s.t. } \mathcal{I}, v_\sigma^s \models K_0(s) \text{ and } \sigma^* = (w, \epsilon)\}$.

The proof proceeds by induction on the structure of terms and formulas, respectively.

For the base case for terms, we consider variables and constants of the various types. If t is a variable v, then $|v|^*_{\mathcal{I}, v} = v(v)^* = v^*(v) = |v|_{M, v^*}$. (Note that $v(v)^*$ is well defined for predicate and function variables given the way we extended $*$ to apply to predicates and functions.) If t is an action constant or a standard name, then $|t|^*_{\mathcal{I}, v} = \Phi(t)^* = t = |t|_{M, v^*}$. Finally, if $t = S_0$, then $|S_0|^*_{\mathcal{I}, v} = (w_0, \epsilon) = |S_0|_{M, v^*}$.

Suppose the lemma holds for terms of length less than m and let t have length m. If $t = succ(t')$, then $|succ(t')|^*_{\mathcal{I}, v} = succ(|t'|^*_{\mathcal{I}, v})$ and $succ(|t'|^*_{\mathcal{I}, v}) = succ(|t'|_{M, v^*})$ by induction. Hence $succ(|t'|_{M, v^*}) = |succ(t')|_{M, v^*}$. If $t = A(\vec{t})$, then $|A(\vec{t})|^*_{\mathcal{I}, v} = A(|\vec{t}|^*_{\mathcal{I}, v})$ and $A(|\vec{t}|^*_{\mathcal{I}, v}) = A(|\vec{t}|_{M, v^*})$ by induction. Hence $A(|\vec{t}|_{M, v^*}) = |A(\vec{t})|_{M, v^*}$. Let $t = X(\vec{t})$, where X is a function variable. Then $|X(\vec{t})|^*_{\mathcal{I}, v} = v(X)^*(|\vec{t}|^*_{\mathcal{I}, v})$ and $v(X)^*(|\vec{t}|^*_{\mathcal{I}, v}) = v^*(X)(|\vec{t}|_{M, v^*})$ by induction. Hence $v^*(X)(|\vec{t}|_{M, v^*}) = |X(\vec{t})|_{M, v^*}$.

If $t = f(t_1, \ldots, t_k, t_s)$ ($f \neq do$), then $|f(t_1, \ldots, t_k, t_s)|^*_{\mathcal{I}, v} = \Phi(f)(d_1, \ldots, d_k, \sigma)^*$ with $|t_i|_{\mathcal{I}, v} = d_i$ and $|t_s|_{\mathcal{I}, v} = \sigma$. Then $\Phi(f)(d_1, \ldots, d_k, \sigma)^* = w[f(d_1^*, \ldots, d_k^*), \vec{b}]$

with $\sigma^* = (w, \vec{b})$. By induction, $|t_s|^*_{\mathcal{I},v} = \sigma^* = |t_s|_{M,v^*}$ and $|t_i|^*_{\mathcal{I},v} = d_i^* = |d_i|_{M,v^*}$. Thus $w[f(d_1^*, \ldots, d_k^*), \vec{b}] = w[f(|t_1|^*_{M,v^*}, \ldots, |t_k|^*_{M,v^*}), \vec{b}]$ (with $|t_s|_{M,v^*} = (w, \vec{b})$) = $|f(t_1, \ldots, t_k, t_s)|_{M,v^*}$.

Let $t = do(t_a, t_s)$ and $|do(t_a, t_s)|_{\mathcal{I},v} = \sigma$. Then $\mathcal{I}, v_{\sigma_0 \sigma}^{s_0 s} \models \mathit{Init}(s_0) \land s_0 \preceq s \land s = do(\vec{a} \cdot A, s_0)$ for some $\vec{a} \in \mathit{Act}^*$ and primitive action A with $\mathcal{I}, v \models A = t_a$. By construction, $\sigma^* = (w, \vec{a} \cdot A)$, $|t_s|^*_{\mathcal{I},v} = (w, \vec{a})$, and $|t_a|^*_{\mathcal{I},v} = A$. By induction, $|t_s|^*_{\mathcal{I},v} = |t_s|_{M,v^*}$ and $|t_a|^*_{\mathcal{I},v} = |t_a|_{M,v^*}$. Hence $|do(t_a, t_s)|^*_{\mathcal{I},v} = (w, \vec{a} \cdot A)$, where $(w, \vec{a}) = |t_s|_{M,v^*}$ and $A = |t_a|_{M,v^*}$. This is precisely the definition of $|do(t_a, t_s)|_{M,v^*}$.

To prove the second part of the lemma, only the base case is of interest. The induction step then follows trivially. Let F be either a relational fluent, Poss, or SF. Then $\mathcal{I}, v \models F(t_1, \ldots, t_k, t_s)$ iff $(|t_1|_{\mathcal{I},v}, \ldots, |t_k|_{\mathcal{I},v}, |t_s|_{\mathcal{I},v}) \in \Phi(F)$ iff (by construction) $w[F(|t_1|^*_{\mathcal{I},v}, \ldots, |t_k|^*_{\mathcal{I},v}), \vec{b}] = 1$ for $|t_s|^*_{\mathcal{I},v} = (w, \vec{b})$ iff (by the first part of the lemma) $w[F(|t_1|_{M,v^*}, \ldots, |t_k|_{M,v^*}), \vec{b}] = 1$ for $|t_s|_{M,v^*} = (w, \vec{b})$ iff $M, v^* \models F(t_1, \ldots, t_k, t_s)$.

Let X be a predicate variable. Then $\mathcal{I}, v \models X(\vec{t})$ iff $|\vec{t}|_{\mathcal{I},v} \in v(X)$ iff $|\vec{t}|^*_{\mathcal{I},v} \in v(X)^*$ iff $|\vec{t}|_{M,v^*} \in v^*(X)$ by Part 1 and the definition of v^*.

Finally, $\mathcal{I}, v \models K_0(t_s)$ iff $|t_s|_{\mathcal{I},v} \in \Phi(K_0)$ and $|t_s|_{\mathcal{I},v}$ is an initial situation iff $|t_s|^*_{\mathcal{I},v} \in e$ and $|t_s|^*_{\mathcal{I},v} (w, \epsilon)$ for some w (by construction) iff $|t_s|_{M,v^*} \in e$ and $|t_s|_{M,v^*} = (w, \epsilon)$. iff $M, v^* \models K_0(t_s)$. ∎

Theorem 14.5.4: *For any α, if α is valid in $\mathcal{AOL}$, then* $\mathrm{AX} \models \alpha$.

Proof: Let α be valid and let $\mathcal{I} \models \mathrm{AX}$ for an arbitrary Tarskian model $\mathcal{I}$. Let $M = \langle e, w \rangle$ be the model corresponding to $\mathcal{I}$ as constructed in the proof of Lemma 14.5.3. Since $M \models \alpha$, we immediately get $\mathcal{I} \models \alpha$ by the lemma. ∎

The axiomatization relies on the fact that there are only finitely many fluents in the language. In particular, Axioms F8 and F9 depend on this assumption. While there does not seem to be any direct way of extending the axioms to deal with an infinite language, it is possible to translate the infinite language into a finite one in a way that preserves validity. In other words, there is a mapping tr from infinite $\mathcal{AOL}$ into finite $\mathcal{AOL}$ such that for every sentence α in infinite $\mathcal{AOL}$, α is valid iff $\mathit{tr}(\alpha)$ is valid iff $\mathrm{AX} \models \mathit{tr}(\alpha)$, where AX are the appropriate axioms for finite $\mathcal{AOL}$. We will not go into the details here, which are rather tedious, except to note that the translation involves turning the predicate and function symbols as well as sequences of terms of the infinite language into terms of the finite language. For example, let $\ulcorner P \urcorner$ and $\ulcorner (t_1, \ldots t_k) \urcorner$ denote the encodings of the predicate symbol P and its arguments, respectively. Then $\mathit{tr}(P(t_1, \ldots t_k)) = \mathit{Holds}(\ulcorner P \urcorner, \ulcorner (t_1, \ldots t_k) \urcorner)$.

14.6 Bibliographic notes

This chapter is based on material that appeared in [75]. The situation calculus was first introduced by McCarthy [102]. The variant we use here, which views situations as histories rather than snapshots of the world as in [104], is summarized in [87] and discussed in much more detail in [123]. Successor state axioms were first introduced by Reiter [122] as a solution to the frame problem, which was recognized early as a fundamental problem in reasoning about dynamical domains [104]. Among other things, successor state axioms were a key to the development of the action language Golog [88], which allows the formulation of high-level plans using constructs such as while-loops and recursive procedures, and which has been applied in various domains including controlling the actions of a mobile robot [13]. The interaction between knowledge and action was first investigated by Moore [111]. Based on this work, Scherl and Levesque later proposed the successor state axiom for the K-relation [131], which we saw is valid in $\mathcal{AOL}$ (see Theorem 14.3.2). Finally, $\mathcal{AOL}$ itself has been studied further in [76]. In particular, a version of **ASK** appropriate for dynamic knowledge bases is introduced, and it is shown that queries of a certain form can be evaluated in a way completely analogous to $\mathcal{OL}$. Furthermore, based on earlier work by Lin and Reiter [92], it is shown how an $\mathcal{AOL}$ knowledge base can be updated as a result of an action.

14.7 Where do we go from here?

In our discussion of $\mathcal{AOL}$ we have gone very little beyond outlining the basic framework. More results have been obtained or are currently under investigation, but including them would go beyond the scope of this book, and many more issues remain to be resolved. Here we only briefly address three areas, the first of which concerns work which overlaps with the finishing of this book.

The fact that $\mathcal{AOL}$ has a complete axiomatization provides an interesting new angle for tackling the open question of a complete axiom system for $\mathcal{OL}$ (see Chapter 10). To see why, note that, if we restrict $\mathcal{AOL}$ to sentences about what is true and what is known about the initial situation, that is, if we ignore actions altogether, then we really have $\mathcal{OL}$ in disguise. In fact, there is a simple translation from a sentence α in $\mathcal{OL}$ into a sentence $\alpha[s]$ in $\mathcal{AOL}$, where s is a situation. Roughly, the translation amounts to adding the situation as an extra argument to every predicate and function and to translate the modal operators as follows:

$$(\boldsymbol{K}\alpha)[s] = \forall s'.K_0(s') \supset \alpha[s']$$
$$(\boldsymbol{O}\alpha)[s] = \forall s'.Init(s') \supset (K_0(s') \equiv \alpha[s'])$$

Note that the translations of **K** and **O** correspond to our characterization of *Knows* and *OKnows*, respectively, when restricted to initial situations (Section 14.3.1). With that it can be shown that α is valid in $\mathcal{OL}$ iff $\alpha[S_0]$ follows from the axioms of $\mathcal{AOL}$. Since, as we observed in Chapter 10, there cannot be a recursive first-order axiomatization of $\mathcal{OL}$, providing second-order axioms may in fact be the best we can hope for.

The work on updating $\mathcal{AOL}$ knowledge bases after actions mentioned at the end of the previous section requires severe restrictions on the form of successor state axioms. The problem is, as Lin and Reiter have shown, that the result of a first-order knowledge base may not always be first-order representable (although it is always second-order representable). The kinds of successor state axioms considered so far for which update remains first-order representable are quite simple and it seems worthwhile to look for more powerful classes.

As in the case of only-knowing under limited belief, $\mathcal{AOL}$ opens new avenues for studying default reasoning, this time taking into account dynamic worlds. Recall that default assumptions as discussed in Chapter 9 can be defeated in light of new information. We never went into details about how one can come to only-know a sentence that includes defaults and some new information.[16] One possibility is that the new information is the result of an action. For example, suppose the robot assumes that the coffee machine is turned on unless known otherwise. If it then fails to obtain coffee it may use its sensors to check the status light of the coffee machine to either verify or defeat the default assumption. In a dynamic world, defaults may also be applied to actions themselves. For instance, a robot may believe that, unless known otherwise, it is able to lift a cup from the table in front of it. Here "otherwise" could refer to the fact that the robot's arm is indeed broken. All these are interesting issues, and we believe that $\mathcal{AOL}$ provides a good basis to formalize and better understand them.

14.8 Exercises

1. Show that $\forall s \exists s'. \mathit{Init}(s') \wedge s' \preceq s$ is valid.
2. Show that the second-order definition of $K(t_1, t_2)$ guarantees that *Knows* satisfies positive and negative introspection:
   - $\mathit{Knows}(\phi, s) \supset \mathit{Knows}(\mathit{Knows}(\phi, \mathit{now}), s);$
   - $\neg \mathit{Knows}(\phi, s) \supset \mathit{Knows}(\neg \mathit{Knows}(\phi, \mathit{now}), s).$

---

16 Intuitively, we assumed that the system is somehow "told." However, it is far from obvious how to formalize this idea since, for example, we cannot just use **TELL**, which always produces a unique result, yet for nondeterminate sentences there can be more than one way to only-know the resulting KB.

3. Show that an agent always believes that the action she just performed was possible. In other words, show that the following sentence is valid:

$$Knows([\forall s''.now = do(a, s'') \supset Poss(a, s'')], do(a, s))$$

4. Prove the rest of Theorem 14.4.1.

Epilogue

What is the logic of knowledge bases? In what way is having a knowledge base different from believing the sentences it contains? As we have seen in this book, having a knowledge base means that these sentences are all that is known. This not only implies believing certain sentences, it also implies not believing others. By introspection then, it also implies believing that these others are not believed. So having a knowledge base means knowing what the knowledge base does *not* tell you about the world.

It is often said, quite correctly, that what distinguishes (some of) Artificial Intelligence from (much of) Computer Science is its concern with incomplete knowledge. Instead of modeling some domain or application directly, we model and reason with *beliefs* about that application. But what exactly is the difference between a symbolic model of an electrical circuit (say) and a symbolic model of beliefs about that circuit? The answer is that in very many cases, there is no difference. But in some cases, and especially when the beliefs about the circuit are incomplete, the two models can be very different in structure. Moreover, reasoning from a model of beliefs about a circuit can be very different from reasoning from a model of the circuit itself.

It is hard to see at first blush why handling incomplete knowledge would be considered desirable. Knowing more is surely better than knowing less and, as we have seen, reasoning correctly with incomplete knowledge can be very demanding computationally. So why not give up knowledge (incomplete or otherwise) and reason directly with models of the domain, as many in the "model-checking" school advocate?

The answer, we feel, is simple: in a natural setting, such as the world that people live in, incomplete knowledge is a fact of life. We learn about the world incrementally over time, and still need to make do. To assume otherwise, is just to imagine an artificially limited domain. Arguably, much of the research in Knowledge Representation and Reasoning is an attempt to deal with this troublesome fact head-on.

And this is where the logic of knowledge bases comes in. If having a knowledge base means knowing all you know, then having an incomplete knowledge base means knowing where that knowledge is incomplete. Armed with this meta-knowledge, an agent is in a position to do something better than giving up when it does not know something: it can apply a default, perform sensing, ask a question, and so on. This is a very powerful idea and, in our opinion, the main justification for an agent reasoning about its own knowledge. Of course at this stage of the game, it still remains to be seen just how this idea can be fully realized in AI systems.

References

[1] A. R. Anderson and N. D. Belnap, Jr., First Degree Entailments. *Math. Annalen*, **149**, 302–319, 1963.

[2] A. R. Anderson and N. D. Belnap, Jr., *Entailment, The Logic of Relevance and Necessity*. Princeton University Press, 1975.

[3] P. Atzeni, S. Ceri, S. Paraboschi, and R. Torlone,MI *Database Systems: Concepts, Languages, and Architectures*. McGraw-Hill Publishing Co., Berkshire, 2000.

[4] J. Barwise and J. Perry, *Situations and Attitudes*. MIT Press, Cambridge, 1983.

[5] S. Ben-David and Y. Gafni, All we believe fails in impossible worlds. Manuscript, Department of Computer Science, Technion, 1989.

[6] E. Börger and E. Grädel and Y. Gurevich, *The Classical Decision Problem*. Springer-Verlag, 1997.

[7] R. Brachman, R. Fikes and H. Levesque, KRYPTON: A functional approach to knowledge representation. *IEEE Computer*, **16**(10), 67–73, 1983. Reprinted in [9].

[8] R. Brachman and H. Levesque, Competence in knowledge representation. *Proceedings of the National Conference of the American Association for Artificial Intelligence (AAAI-82)*, AAAI Press/MIT Press, Cambridge, 189–192, 1982.

[9] R. Brachman and H. Levesque (eds.), *Readings in Knowledge Representation*. Morgan Kaufmann Publishers, Inc., San Francisco, 1985.

[10] R. Brachman and H. Levesque, *Knowledge Representation*. in preparation.

[11] R. Brachman and J. Schmolze, An overview of the KL-ONE knowledge representation system. *Cognitive Science*, **9**(2), 171–216, 1985.

[12] M. Brand and R. Harnish (eds.), *The Representation of Knowledge and Belief*. The University of Arizona Press, Tucson, 1986.

[13] W. Burgard, A. B. Cremers, D. Fox, D. Hähnel, G. Lakemeyer, D. Schulz, W. Steiner, and S. Thrun, The interactive museum tour-guide robot. *Proceedings of the National Conference on Artificial Intelligence (AAAI-98)*, AAAI Press/MIT Press, Cambridge, 11–18, 1998.

[14] B. Chellas, *Modal Logic: An Introduction*. Cambridge University Press, Cambridge, 1980.

[15] J. Chen, The logic of only knowing as a unified framework for non-monotonic reasoning. *Fundamenta Informaticae*, **21**, 205–220, 1994.

[16] J. Chen, The generalized logic of only knowing that covers the notion of epistemic specifications. *Journal of Logic and Computation*, **7**(2), 159-174, 1997.

[17] C. Cherniak, *Minimal Rationality*. MIT Press, Cambridge, 1986.

[18] B. Cohen, and G. Murphy, Models of concepts. *Cognitive Science*, **8**(1), 27–59, 1984.

[19] A. Cohn, L. Schubert and S. Shapiro (eds.), *Proceedings of the 6th International Conference on Principles of Knowledge Representation and Reasoning, KR-98*, Morgan Kaufmann, San Francisco, 1998.

[20] A. Darwiche and J. Pearl, Symbolic causal networks. *Proceedings of the 12th National Conference on Artificial Intelligence (AAAI-94)*, AAAI Press/MIT Press, Cambridge, 238–244, 1994.

[21] D. Dennett, *The Intentional Stance*. MIT Press, Cambridge, 1987.

[22] D. Dennett, Precis of *The Intentional Stance*. *Brain and Behavioral Sciences*, **16**(2), 289–391, 1993.

[23] F. Donini, M. Lenzerini, D. Nardi, and A. Schaerf, Reasoning in description logics. G. Brewka (ed.), *Principles of Knowledge Representation and Reasoning*, Studies in Logic, Language and Information, CSLI Publications, Stanford University, Stanford, 193–238, 1996.

[24] F. Donini, M. Lenzerini, D. Nardi, W. Nutt, and A. Schaerf, An epistemic operator for description logics. *Artificial Intelligence*, **100**, 225-274, 1998.

[25] F. Donini, D. Nardi, and R. Rosati, Ground nonmonotonic modal logics. *J. of Logic and Computation* **7**(4), 523–548, 1997.

[26] H. Dreyfus, *What Computers Still Can't Do: A Critique of Artificial Reason*. MIT Press, Cambridge, 1992.

[27] J. M. Dunn, Intuitive semantics for first-degree entailments and coupled trees. *Philosophical Studies* **29**, 149–168, 1976.

[28] P. Edwards (ed.), *The Encyclopedia of Philosophy*. Macmillan Publishing Co., New York, 1967.
[29] J. J. Elgot-Drapkin, Step-logic and the three-wise-men problem. *Proceedings of the 9th National Conference on Artificial Intelligence (AAAI-91)*, AAAI-Press/MIT Press, Cambridge, 412–417, 1991.
[30] J. J. Elgot-Drapkin and D. Perlis, Reasoning situated in time I: basic concepts. *Journal of Experimental and Theoretical Artificial Intelligence*, **2**(1), 75–98, 1990.
[31] H. Enderton, *A Mathematical Introduction to Logic*. Academic Press, New York, 1972.
[32] R. Fagin, J. Halpern, Y. Moses and M. Vardi, *Reasoning about Knowledge*. MIT Press, Cambridge, 1995.
[33] R. Fagin, J. Y. Halpern, and M. Y. Vardi, A nonstandard approach to the logical omniscience problem. *Artificial Intelligence* **79**(2), 203-240, 1996.
[34] N. Findler (ed.), *Associative Networks: Representation and Use of Knowledge by Computers*. Academic Press, New York, 1979.
[35] A. Frisch, *Knowledge Retrieval as Specialized Inference*. Ph. D. Thesis, Department of Computer Science, University of Rochester.
[36] A. Frisch, Inference without chaining. *Proceedings of the 10th International Joint Conference on Artificial Intelligence (IJCAI-87)*, Morgan Kaufmann, San Francisco, 515–519, 1987.
[37] A. Frisch, The substitutional framework for sorted deduction: fundamental results on hybrid reasoning. *Artificial Intelligence* **49**, 161–198, 1991.
[38] P. Gärdenfors, *Knowledge in Flux: Modeling the Dynamics of Epistemic States*. MIT Press, Cambridge, 1988.
[39] M. Garey and D. Johnson, *Computers and Intractability: A Guide to the Theory of NP-Completeness*. W. H. Freeman, San Francisco, 1979.
[40] M. Gelfond, Strong introspection. *Proceedings of the Ninth National Conference on Artificial Intelligence (AAAI-91)*, AAAI Press/MIT Press, Cambridge, 386–391, 1991.
[41] E. Gettier, Is justified true belief knowledge. *Analysis*, **23**, 121–123, 1963. Reprinted in [45].
[42] G. Gottlob, Complexity results for nonmonotonic logics. *Journal of Logic and Computation*, **2**, 397–425, 1992.
[43] G. Gottlob, Translating default logic into standard autoepistemic logic. *Journal of the ACM*, **42**(4), 711–740, 1995.
[44] C. Green, *The Application of Theorem-Proving to Question Answering Systems*. Ph. D. thesis, Department of Electrical Engineering, Stanford University, Stanford, 1969.
[45] A. Griffiths (ed.), *Knowledge and Belief*. Oxford University Press, London, 1967.
[46] J. Y. Halpern, Reasoning about only knowing with many agents. *Proceedings of the National Conference on Artificial Intelligence (AAAI'93)*, AAAI-Press/MIT Press, Cambridge, 655–661, 1993.
[47] J. Y. Halpern and G. Lakemeyer, Levesque's axiomatization of only knowing is incomplete. *Artificial Intelligence* **74**(2), 381–387, 1995.
[48] J. Y. Halpern and G. Lakemeyer, Multi-agent only knowing. *Proceedings of the 6th Conference on Theoretical Aspects of Rationality and Knowledge (TARK-VI)*, Morgan Kaufmann, 251–265, 1996.
[49] J. Y. Halpern and Y. Moses, Towards a theory of knowledge and ignorance. *Proceedings of the AAAI Workshop on Non-monotonic Logic*, 125–143, 1984. Reprinted in K. R. Apt (ed.), *Logics and Models of Concurrent Systems*, Springer-Verlag, Berlin/New York, 459–476, 1985.
[50] J. Hintikka,, *Knowledge and Belief*. Cornell University Press, Ithaca, 1962.
[51] J. Hintikka, Impossible worlds vindicated. *Journal of Philosophical Logic* **4**, 475–484, 1975.
[52] G. Hirst, Existence assumptions in knowledge representation. *Artificial Intelligence*, **49**, 199–242, 1991.
[53] G. Hughes, and M. Cresswell, *An Introduction to Modal Logic*. Methuen and Co., London, 1968.
[54] I. L. Humberstone, I. L., A more discriminating approach to modal logic. *Journal of Symbolic Logic* **51**(2), 503–504, 1986. (Abstract only.) There is also an expanded, but unpublished, manuscript.
[55] D. Kaplan, Quantifying-in. [94], 112–144, 1971.

References

[56] K. Konolige, A computational theory of belief introspection. *Proceedings of the 9th International Joint Conference on Artificial Intelligence (IJCAI-85)*, Morgan Kaufmann, San Francisco, 502–508, 1985.

[57] K. Konolige, *A Deduction Model of Belief*. Research Notes in Artificial Intelligence, Pitman, London, 1986.

[58] K. Konolige, On the relation between default logic and autoepistemic theories. *Artificial Intelligence* **35**(3), 343–382, 1988. (Errata: *Artificial Intelligence* **41**, 115, 1989.)

[59] K. Konolige, On the relation between autoepistemic logic and circumscription (preliminary report). *Proceedings of the 11th International Joint Conference on Artificial Intelligence*, Morgan Kaufmann, San Francisco, 1213–1218, 1989.

[60] K. Konolige, Quantification in autoepistemic logic. *Fundamenta Informaticae* **15**(3-4), 1991.

[61] K. Katsuno and A. Mendelzon, Propositional knowledge base revision and minimal change. *Artificial Intelligence*, **52**, 263–294, 1991.

[62] S. Kripke, A completeness theorem in modal logic. *Journal of Symbolic Logic*, **24**, 1–14, 1959.

[63] S. Kripke, Is there a problem with substitutional quantification? G. Evans and J. McDowell (eds.), *Truth and Meaning*, Clarendon Press, Oxford, 325–419, 1976.

[64] S. Kripke, *Naming and Necessity*. Harvard University Press, Cambridge, 1980.

[65] G. Lakemeyer, *Models of Belief for Decidable Reasoning in Incomplete Knowledge Bases*. Ph. D. thesis, Dept. of Computer Science, University of Toronto, 1990. (A revised version appeared as: Technical Report KRR-TR-92-5, University of Toronto, 1992.)

[66] G. Lakemeyer, All you ever wanted to know about Tweety. *Proceedings of the 3rd International Conference on Principles of Knowledge Representation and Reasoning (KR'92)*, Morgan Kaufmann, San Mateo, CA, 639–648, 1992.

[67] G. Lakemeyer, On perfect introspection with quantifying-in. *Fundamenta Informaticae*, **17**(1,2), 75–98, 1992.

[68] G. Lakemeyer, All they know: a study in multi-agent autoepistemic reasoning. *Proceedings of the 13th International Joint Conference on Artificial Intelligence (IJCAI '93)*, 376–381, 1993.

[69] G. Lakemeyer, All they know about. *Proceedings of the 11th National Conference on Artificial Intelligence (AAAI-93)*, AAAI Press/MIT Press, Cambridge, 662–667, 1993.

[70] G. Lakemeyer, Limited reasoning in first-order knowledge bases. *Artificial Intelligence* **71**, 1–42, 1994.

[71] G. Lakemeyer, A logical account of relevance. *Proceedings of the 14th International Joint Conference on Artificial Intelligence (IJCAI-95)*, Morgan Kaufmann, 853–859, 1995.

[72] G. Lakemeyer, Limited reasoning in first-order knowledge bases with full introspection. *Artificial Intelligence* **84**, 209–255, 1996.

[73] G. Lakemeyer, Relevance from an epistemic perspective. *Artificial Intelligence* **97**(1–2), 137–167, 1997.

[74] G. Lakemeyer and H. J. Levesque, A tractable knowledge representation service with full introspection. *Proceedings of the 2nd Conference on Theoretical Aspects of Reasoning about Knowledge*, Morgan Kaufmann, San Francisco, 145–159, 1988.

[75] G. Lakemeyer and H. J. Levesque, $\mathcal{AOL}$: a logic of acting, sensing, knowing, and only knowing. *Proceedings of the 6th International Conference on Principles of Knowledge Representation and Reasoning (KR'98)*, Morgan Kaufmann, San Francisco, 316–327, 1998.

[76] G. Lakemeyer and H. J. Levesque, Query evaluation and progression in $\mathcal{AOL}$ knowledge bases. *Proceedings of the 16th International Joint Conference on Artificial Intelligence (IJCAI-99)*, Morgan Kaufmann, San Francisco, 124–131, 1999.

[77] G. Lakemeyer and S. Meyer, Enhancing the power of a decidable first-order reasoner. *Proceedings of the 4th International Conference on Principles of Knowledge Representation and Reasoning (KR'94)*, Morgan Kaufmann, San Francisco, 403–414, 1994.

[78] H. Leblanc, On dispensing with things and worlds. M. Munitz (ed.), *Logic and Ontology*, New York University Press, New York, 241–259, 1973.

[79] H. Leblanc, Alternatives to standard first-order semantics. D. Gabbay and F. Guenthner (eds.),

Handbook of Philosophical Logic, Volume 1, Kluwer Academic Press, 189–274, 1983.

[80] H. J. Levesque, *A Formal Treatment of Incomplete Knowledge Bases*. Ph. D. thesis, Dept. of Computer Science, University of Toronto, 1981.

[81] H. J. Levesque, A logic of implicit and explicit belief. *Proceedings of the 4th National Conference on Artificial Intelligence (AAAI-84)*, AAAI Press/MIT Press, Cambridge, 198–202, 1984.

[82] H. J. Levesque, Foundations of a functional approach to knowledge representation. *Artificial Intelligence*, **23**, 155–212, 1984.

[83] H. J. Levesque, Knowledge representation and reasoning. *Annual Review of Computer Science 1986*, Annual Reviews Inc., Palo Alto, 255–287, 1986.

[84] H. J. Levesque, Logic and the complexity of reasoning. *The Journal of Philosophical Logic*, **17**, 355–389, 1988.

[85] H. J. Levesque, A knowledge-level account of abduction. *Proceedings of the 11th International Joint Conference on Artificial Intelligence (IJCAI-89)*, Morgan Kaufmann, San Francisco, 1061–1067, 1989.

[86] H. J. Levesque, All I know: a study in autoepistemic logic. *Artificial Intelligence* **42**, 263–309, 1990.

[87] H. J. Levesque, F. Pirri, and R. Reiter, Foundations for the situation calculus. Linköping Electronic Articles in Computer and Information Science, **3**, 1998, available at http://www.ep.liu.se/ea/cis/1998/018/.

[88] H. J. Levesque, R. Reiter, Y. Lespérance, F. Lin, and R. B. Scherl., Golog: A logic programming language for dynamic domains. *Journal of Logic Programming*, **31**, 59–84, 1997.

[89] D. Lewis, *Counterfactuals*. Blackwell, Oxford, 1973.

[90] V. Lifschitz, Minimal belief and negation as failure. *Artificial Intelligence* **70**, 53–72, 1994.

[91] F. Lin and R. Reiter, Forget it!, *Proceedings of the AAAI Fall Symposium on Relevance*, AAAI Press, 154–159, 1994.

[92] F. Lin and R. Reiter, How to progress a database. *Artificial Intelligence*, **92**, 131-167, 1997.

[93] F. Lin and Y. Shoham, Epistemic semantics for fixed-point non-monotonic logics. *Proceedings of the 3rd Conference on Theoretical Aspects of Reasoning about Knowledge (TARK-III)*, Morgan Kaufmann, San Francisco, 111–120, 1990.

[94] L. Linsky (ed.), *Reference and Modality*. Oxford University Press, Oxford, 1971.

[95] L. Linsky, *Names and Descriptions*. University of Chicago Press, Chicago, 1977.

[96] B. Liskov and S. Zilles, Programming with abstract data types. *SIGPLAN Notices*, **9**(4), 1974.

[97] J. Lloyd, *Foundations of Logic Programming*, Second Edition. Springer Verlag, New York, 1987.

[98] W. Marek and M. Truszczynski, Relating autoepistemic and default logics. *Proceedings of the 1st International Conference on Principles of Knowledge Representation and Reasoning (KR'89)*, Morgan Kaufmann, San Francisco, 276–288, 1989.

[99] W. Marek and M. Truszczynski, Modal logic for default reasoning. *Annals of Mathematics and Artificial Intelligence* **1**, 275–302, 1990.

[100] W. Marek and M. Truszczynski, Autoepistemic logic. *Journal of the ACM*, **38**(3), 588–619, 1991.

[101] J. McCarthy, Programs with common sense. M. Minsky (ed.), *Semantic Information Processing*, MIT Press, Cambridge, 403–418, 1963. Reprinted in [9].

[102] J. McCarthy, *Situations, Actions and Causal Laws*. Technical Report, Stanford University, 1963. Also in M. Minsky (ed.), *Semantic Information Processing*, MIT Press, Cambridge, MA, 410–417, 1968.

[103] J. McCarthy, *Modality - Si! Modal Logic - No!*. Unpublished manuscript, available at http://www-formal.stanford.edu/jmc/modality.html, 1999.

[104] J. McCarthy and P. Hayes, Some philosophical problems from the standpoint of artificial intelligence. B. Meltzer and D. Michie (eds.), *Machine Intelligence 4*, Edinburgh Press, Edinburgh, Scotland, 463–502, 1969.

[105] D. McDermott and J. Doyle, Non-monotonic logic I. *Artificial Intelligence* **13**, 41–72, 1980.

[106] D. McDermott, Non-monotonic logic II. *Journal of the ACM*, **29**(1), 33–57, 1982.

References

[107] E. Mendelson, *Introduction to Mathematical Logic*. Van Nostrand Reinhold Company, New York, 1964.

[108] R. C. Moore,, The role of logic in knowledge representation and commonsense reasoning. *Proceedings of the National Conference of the American Association for Artificial Intelligence (AAAI-82)*, AAAI Press/MIT Press, Cambridge, 428–433, 1982.

[109] R. C. Moore, Possible world semantics for autoepistemic logic. *Proceedings of the 1st Non-Monotonic Reasoning Workshop*, New Paltz, NY, 344–354, 1984.

[110] R. C. Moore, Semantical considerations on nonmonotonic logic. *Artificial Intelligence* **25**, 75–94, 1985.

[111] R. C. Moore, A formal theory of knowledge and action. J. R. Hobbs and R. C. Moore (eds.), *Formal Theories of the Commonsense World*. Ablex, Norwood, NJ, 319–358, 1985.

[112] I. N. F. Niemelä, Constructive tightly grounded autoepistemic reasoning. *Proceedings of the 12th International Joint Conference on Artificial Intelligence (IJCAI-91)*, Morgan Kaufmann, San Francisco, 399–404, 1991.

[113] I. N. F. Niemelä, On the decidability and complexity of autoepistemic reasoning. *Fundamenta Informaticae*, **17**, 117–155, 1992.

[114] N. Nilsson, *Principles of Artificial Intelligence*. Tioga Publishing Company, Palo Alto, 1980.

[115] P. F. Patel-Schneider, *Decidable, Logic-Based Knowledge Representation*. Ph. D. thesis, Department of Computer Science, University of Toronto, 1987.

[116] I. Pratt-Hartmann, Total Knowledge *Proceedings of the National Conference of the American Association for Artificial Intelligence (AAAI-00)*, AAAI Press/MIT Press, Cambridge, 423–428, 2000.

[117] Z. Pylyshyn, *Computation and Cognition: Toward a Foundation for Cognitive Science*. MIT Press, Cambridge, 1984.

[118] W. Quine, Quantifiers and propositional attitudes. [94], 101–111, 1971.

[119] R. Reiter, On closed world data bases. H. Gallaire and J. Minker (eds.), *Logic and Data Bases*. Plenum Press, 55–76, 1978.

[120] R. Reiter, A logic for default reasoning. *Artificial Intelligence* **13**(1–2), 81–132, 1980.

[121] R. Reiter, What should a database know?. *Journal of Logic Programming*, **14**(1-2), 127–153, 1992.

[122] R. Reiter, The frame problem in the situation calculus: a simple solution (sometimes) and a completeness result for goal regression. V. Lifschitz (ed.), *Artificial Intelligence and Mathematical Theory of Computation*, Academic Press, 359–380, 1991.

[123] R. Reiter, *Knowledge in Action: Logical Foundations for Describing and Implementing Dynamical Systems*. MIT Press, forthcoming.

[124] R. Reiter and J. de Kleer, Foundations of assumption-based truth maintenance systems: preliminary report. *Proceedings of the 6th National Conference on Artificial Intelligence (AAAI-87)*, AAAI Press/MIT Press, Cambridge, 183–188, 1987.

[125] R. Rosati, Complexity of only knowing: the propositional case. *Proceedings of the 4th International Conference on Logic Programming and Nonmonotonic Reasoning (LPNMR-97)*, 76-91, LNAI 1265, Springer-Verlag, 1997.

[126] R. Rosati, On the decidability and complexity of reasoning about only knowing. *Artificial Intelligence*, **116**(1-2), 193–215, 2000.

[127] R. Routley and R. K. Meyer, The semantics of entailment, I. H. Leblanc (ed.), *Truth, Syntax, and Semantics*, North-Holland, 194–243, 1973.

[128] R. Routley and V. Routley, Semantics of first degree entailment. *Noûs* **6**, 335–359, 1972.

[129] B. Russell, On denoting. *Mind*, **14**, 479–493, 1905.

[130] S. Russell and P. Norvig, *Artificial Intelligence: A Modern Approach*. Prentice Hall, Englewood Cliffs, 1995.

[131] R. B. Scherl and H. J. Levesque, The frame problem and knowledge producing actions. *Proceedings of the National Conference on Artificial Intelligence (AAAI-93)*, AAAI Press/MIT Press, Cambridge, 689–695, 1993.

[132] K. Segerberg, Some modal reduction theorems in autoepistemic logic. *Uppsala Prints and Preprints in*

Philosophy, Uppsala University, 1995.

[133] Y. Shoham (ed.), *Proceedings of the Conference on Theoretical Aspects of Rationality and Knowledge (TARK-VI)*, Morgan Kaufmann, San Francisco, 1996.

[134] B. Smith, *Reflection and Semantics in a Procedural Language*. Ph. D. thesis and Technical Report MIT/LCS/TR-272, MIT, Cambridge, 1982.

[135] G. Shvarts, Autoepistemic modal logic. *Proceedings of the 3rd Conference on Theoretical Aspects of Reasoning about Knowledge (TARK-III)*, Morgan Kaufmann, San Francisco, 97–109, 1990.

[136] R. Smullyan, *First-Order Logic*. Springer-Verlag, New York, 1968.

[137] R. F. Stärk, On the existence of fixpoints in Moore's autoepistemic logic and the non-monotonic logic of McDermott and Doyle. *Proceedings of the 4th Workshop on Computer Science Logic*, LNCS 533, Springer Verlag, Berlin, 354–365, 1991.

[138] R. Stalnaker, *Inquiry*. MIT Press, Cambridge, 1987.

[139] R. Stalnaker, A note on non-monotonic modal logic. *Artificial Intelligence* **64**(2), 183–196, 1993.

[140] D. Subramaniam, R. Greiner, and J. Pearl (eds.), *Artificial Intelligence, Special Issue on Relevance*, **97**(1–2), 1997.

[141] M. Truszczynski, Modal interpretations of default logic. *Proceedings of the 12th International Joint Conference on Artificial Intelligence (IJCAI-91)*, Morgan Kaufmann, San Francisco, 393–398, 1991.

[142] A. Waaler, *Logical Studies in Complementary Weak S5*. Ph. D. Thesis, University of Oslo, 1995.

[143] T. Winograd, Frame representations and the declarative/procedural controversy. D. Bobrow and A. Collins (eds.), *Representation and Understanding: Studies in Cognitive Science*, Academic Press, New York, 185–210, 1975.

Index

action, 245
action model, 252
adjunct, 149
admissible, 207
ASK, 81
atoms, 24

basic action theory, 251
basic belief set, 130
belief implications, 197
belief set, 99
bound, 24

candidate, 92
clauses, 180
closed world assumption, 85

definitions, 92
derives, 36
determinate, 134

e_0, 81
E-form, 69
effect axioms, 248
e-minimal, 181
entailment, 10
epistemic state represented by KB, 100
e-p-minimal, 181
equivalent, 97
exhaustive pair, 168
existential generalization, 205
existential-free, 207
explainable, 96
explicit basic belief set, 228
explicit belief, 196
explicitly representable, 238

FALSE, 113
finitely representable, 101
first-order satisfies, 148
fluents, 246
frame axioms, 248
free, 24

ground atom, 24
ground term, 24

implicit belief, 196
inconsistent, 36
initial situation, 253
instance, 69

knowledge base, 7

knowledge level, 12
Knowledge Representation, 5
knowledge representation system, 12
knowledge-based system, 7
known instance, 44

literals, 180
logical omniscience, 195
logical symbols, 23
logically implied, 28

maximal, 98
maximally consistent, 68
meta-knowledge, 47

non-logical symbols, 23

objective, 46
objective level, 225
objective monotonicity, 82
only-knowing-about, 179
ordinary, 225

potential instance, 44
precondition axioms, 247
prime implicates, 185
primitive atom, 24
primitive term, 24
proposition, 3
propositional, 73

quasi-finitely representable, 126

reasoning, 6
reducing, 117
relevant, 189
representable, 101
representation, 5
RES, 115

satisfiable, 28
sensed-fluent axioms, 247
sentence, 25
situation, 197, 246
situation calculus, 245
sk-functions, 226
sk-term, 227
standard names, 22
strictly relevant, 189
subject matter, 179
subjective, 46
subsumed, 191
successor-state axiom, 249

symbol, 5
symbol level, 12

T-set, 69
tautological entailment, 199
TELL, 84
term map, 212
terms, 23
theorem, 36
theta-subsumption, 219
TRUE, 113

valid, 29
value, 27

wffs, 23
wh-questions, 93
world state, 26

x-satisfiable, 169
x-valid, 169

OHIO UNIVERSITY LIBRARY